BUCKINGHAMSHIRE COUNCIL

9510000372203

DAVID ATTENBOROUGH

LIFE ON AIR

Memoirs of a Broadcaster

BY THE SAME AUTHOR

Zoo Quest to Guiana (1956)
Zoo Quest for a Dragon (1957)
Zoo Quest to Paraguay (1959)
Quest in Paradise (1960)
Zoo Quest to Madagascar (1961)
Quest Under Capricorn (1963)
The Tribal Eye (1976)
The First Eden (1987)
Life on Earth (1979)
The Living Planet (1984)
The Trials of Life (1990)
The Private Life of Plants (1995)
The Life of Birds (1998)
The Life of Mammals (2002)
Life on Air (2002; revised edition *2009)*
Life in the Undergrowth (2005)
Life in Cold Blood (2008)
Life Stories (2009)
New Life Stories (2011)
*Drawn from Paradise (*with Errol Fuller, *2012)*

This book was first published in 2002 by BBC Books.
Paperback edition first published in 2003

Enlarged edition first published in 2014
by BBC Books, an imprint of Ebury Publishing,
a Random House Group Company

Paperback edition published in 2016

ISBN 978-1-849-90001-0

Printed by Clays Ltd., Bungay

Contents

Illustrations

All pictures property of David Attenborough,
except where otherwise indicated

1

Joining Auntie

The clock on the south-western tower of St Paul's cathedral appeared to have stopped. I could just see it if I craned my neck and peered out of the very side of my office window. Maybe it was suffering from a delayed shock after the Blitz. It was, after all, 1950 and the war had only been over for five years. But ten past two, which was what the clock's hands indicated, could hardly be the correct time. I had eaten my sandwich lunch in the garden on the bomb-site just over the road and had been back at my desk promptly at two o'clock, as befitted the most junior and recently recruited member of the London publishing house for which I was working. And that seemed at least three-quarters of an hour ago.

I turned my attention to the galley proofs in front of me. It was a text about tadpoles for primary schools. Interesting in its own way but hardly, I felt, the sort of work that taxed my hard-won degree in natural sciences. Fitting in the illustrations, using a technology that had not changed significantly since Gutenberg's times, required counting the words, sometimes even the individual letters. I had to tot them up, work out how many would have to be carried forward to the next page and then wrestle with the various possibilities for trimming the next illustration.

Another half hour of this and I took another glance through the window. The clock hadn't stopped after all. Its hands had certainly moved. They had advanced a few minutes. But it was still not half past. This dismal revelation depressed me so much that I decided to turn my desk around so that I wasn't hypnotised by the hands of a clock. Instead, I stared at a blank wall. And it was then that I decided that this was not the way I wanted to spend the rest of my life. Maybe my future was not, after all, to be a gentleman publisher.

But what was it? I was twenty-four. I had studied natural sciences at Cambridge, thinking that somehow research would ultimately take me to remote and exciting parts of the world. Then I had to do my national service. I went into the Navy hoping that I might be sent somewhere romantic. At Gosport, where we did our preliminary training, I met old naval hands who talked a lot about 'Trinco' – that is to say, Trincomalee in what was then Ceylon. The Far Eastern Fleet was based there. It sounded good to me and I told whoever I thought might have some influence in the matter that Trinco was just the sort of place I would like to be posted, at the end of my training. In the event, I was sent to join an aircraft carrier that was being mothballed as part of the Reserve Fleet in the Firth of Forth.

By the time the Navy had finished with me, I had decided that I did not want to go back to university to read for a doctorate. Zoological research in those days was largely laboratory-bound and that wasn't the way I wanted to study animals. In any case, I had now married Jane, whom I had met at university, and I didn't fancy going back to subsist on a student grant, even if I could have got one. So I had decided that publishing would suit me, and had got myself a job as a junior editorial assistant with an educational publisher. But now that didn't seem to be a thrilling proposition either.

I consulted the appointments columns in *The Times*, which, as an embryonic city gent, I felt I had to carry to the office daily and occasionally read. There I saw an advertisement by the BBC for a radio talks producer. Since I had failed to find a job that would take me to remote places, I thought I might enjoy that sensation, second-hand, by seeking out people who had managed to do so and getting them to talk about it. So I applied. A couple of weeks later I got a polite note telling me that the job had gone elsewhere. I returned to my galley proofs and my tadpoles.

Then, to my surprise and embarrassment, I got a telephone call at the office. Personal calls were not approved of and I felt slightly criminal as I listened to a lady from the BBC. She explained that her name was Mary Adams and that she was not from Radio but from the Corporation's Television Service. She had seen my application form and although I hadn't been summoned by Radio for an interview, she nonetheless felt that I might be the right sort of person for Television. Was I interested? I had to confess that I hadn't actually seen much television. I had once watched a television play in my wife's parents' house, but they were the only people I knew with a set and I certainly had not got one myself. Mrs Adams didn't think that was necessarily a disadvantage. Would I like to call upon her to discuss the possibility of joining a training course? So I applied to my employers for one day's leave of absence to be deducted from my annual fourteen-day holiday and went up to Alexandra Palace in North London to visit Mrs Adams.

Alexandra Palace was – and still is – a sprawling Edwardian building, built on the top of a hill surrounded by a park in the north-eastern suburbs of London. It has a huge hall that had been built as a concert hall and fitted with a colossal organ, but had since done service as a ballroom, an ice rink and a glorified village hall, and then become semi-derelict. All around it were blocks with rooms for offices. One of these had been modified to accommodate two small television studios. On the tower above them stood a tall pylon-like transmitter from which the world's first public television signals had been radiated in 1936. Immediately beneath it Mrs Adams had her office.

She was in her mid-fifties, grey-haired, affable, with a smoker's cough, a darting glance and a heartening guffaw. I subsequently discovered that she was the widow of a Member of Parliament who, had he lived, seemed destined for a glittering parliamentary career. She had been educated as a geneticist at Oxford University before the war but had left academia to join BBC Radio's Further Education Department. She had had a hand in some of the very first television programmes. After the war she had rejoined television as an administrator with the task of building from scratch a department called, perhaps oddly for television, 'Talks'.

She explained that she could not offer me a job, but I could, if I wished, apply to go on a three-month training course. Thereafter, I would have to take my chance. There could be no guarantee of subsequent employment. I explained that I was a married man, with a young son, and already had a job. Mrs Adams had to admit, she said, that the thousand pounds I would be paid while on the three-month course, was not much, but it was the most the Television Service could afford. I gulped and tried not to fall off my chair. It was twice what the publishers were paying me for a whole year. I stammered my thanks and left, already walking on air.

I have little memory of the selection board. I spared them my views about the social function of public service broadcasting in a democratic society, since I hadn't thought about it. Nor could I offer them a critique of any recent programmes since I had not seen any. Nonetheless, a week later, I got a letter telling me that I had been accepted. With some relief, I sent in my letter of resignation to my employers. I was grateful to them for having given me a job, I wrote, but I had made a mistake. Publishing, after all, was not for me.

* * *

While I was working out my time with galley proofs, Mrs Adams telephoned again. Was I interested, perhaps, while waiting to join the training course, in doing a little work in the studios as an interviewer? Why not? Once again I travelled up to Alexandra Palace. I was met by John

Read, the producer in whose programme I would be appearing. It was the first in a new series called *Joan Gilbert's Weekend Diary*. Miss Gilbert, I discovered, was well-established on television as the hostess of an afternoon programme for women. Her *Weekend Diary* was to be an attempt to move into evening programmes and the big time. What I didn't discover at this stage was that Miss Gilbert was a considerable prima donna who had decreed that she needed a co-interviewer – junior to her, naturally – and that this job should be given to an actor who was the current object of her affections. Mrs Adams had decreed otherwise. A coolness had developed and Mrs Adams felt the need to nominate an interviewer herself. Which is where I came in.

John Read showed me to the make-up room where a lady plastered my face with the heavy make-up that television's blazing lights made necessary. Then he took me to the studio and introduced me to Miss Gilbert. She was a large somewhat florid lady with an elaborate hair-do and a considerable bust, wearing what I believe was known as a matinee coat. She shook my hand distantly, turned on her heel and returned to her dressing room.

My interviewee was the long-distance Olympic runner Gordon Pirie. We were shown where we had to sit and we rehearsed a stilted conversation while the cameras moved around in front of us. This was somewhat alarming for the primitive cameras were so insensitive that not only was it necessary to flood the set with intense light, but the lenses had to be very wide-angled, to collect as much of it as possible. That, in turn, meant that to get a head and shoulders close-up, a camera had to be within a yard or so of its subject. If it peered over the interviewer's shoulder it was too far away to get a close-up of the interviewee. So when I asked a question I had to stare, not at Mr Pirie, but at the camera a foot or so away. And he had to do likewise in giving his answer. I think he found this just as off-putting as I did.

Transmission began. Miss Gilbert welcomed viewers to her new diary and interviewed someone herself. I was so preoccupied with my own problems that I did not really register what she was saying or who it was she was talking to, until I suddenly heard her say, 'And now I have great pleasure in introducing you to a new but very dear friend of mine. Over to you, David.' Her dear friend, I realised with some surprise, was me. The red light on top of the camera nearest to me lit. I forced a smile and we started.

Gordon was not, it must be said, very forthcoming. His answers to my questions were largely monosyllabic and I found myself falling into the elementary error of putting the information I was trying to extract into my own questions since the interviewee seemed disinclined to provide it himself. Our conversation ground painfully on.

'Do you,' I said, looking straight into the camera, 'have any special training techniques, Gordon?'

'Yeah,' he said staring at his.

'I'm told,' I ventured, recalling the research notes that I had been given, 'that you do most of your training runs in hob-nailed boots?'

'Yeah.'

'Why do you do your training runs in hob-nailed boots, Gordon?' I said despairingly.

'Coz when I take 'em off, I go faster.'

At that point, Miss Gilbert took over with an interview of her own. And that was that.

Jane said loyally I had been very good, but no one else I knew seemed to have seen the programme. John Read didn't invite me to take part in the next in the series the following week and I didn't hear anything from Mrs Adams. It dawned on me that my first appearance on the television screen had not been a success.

There is, however, a pendant to this story. Several decades later, when John Read came to retire from the Corporation, it fell to me to make a speech at his farewell party. As was customary on these occasions, his personal file was sent to me, as the speech-maker, so that I could give a summary of his career, picking out the high-spots. So I found myself looking through the memoranda that had passed during the early stages of his career. And there it was – a note from Mary Adams. 'David Attenborough is intelligent and promising and may well be producer material, but he is not to be used again as an interviewer. His teeth are too big.'

* * *

The Friday finally dawned in September 1952 when I abandoned my desk in the publishers beside St Paul's. On the following Monday I found my way to an office block off the Marylebone Road where the BBC ran its training courses. There were a dozen or so of us. Some had experience in the theatre and hoped to go to the Drama Department. Others were interested in music and design. I was not the only hopeful sponsored by Talks Department. There was also Michael Peacock, a brilliant recent graduate from the London School of Economics. He was three years younger than I and he told me that he hoped to become involved in the political side of television.

My recollections of the course itself are not detailed. One of the first lectures was given by a gentleman who came armed with a large box of coloured chalks. He drew a series of rectangles on the blackboard, wrote initials in them and connected them by lines, some continuous, some dotted. The finished diagram, we were told, represented the structure of the BBC. It came as no surprise to me later to discover that he was also the author of an authoritative book on witchcraft in medieval England.

At the end of two weeks of lectures, Michael and I were sent up to Alexandra Palace. Talks Department people, I discovered, were crucially different from other production staff. The drama producers had come from the theatre and were steeped in theatrical traditions and practices. Light entertainment producers had experience of revue and music halls, and knew about ventriloquists and comedians and how to rehearse lines of dancing girls. Music producers were all trained musicians – one of them, Philip Bate, was a world authority on the history of the clarinet. But Talks Department producers had no identifiable qualifications for their jobs. George Noordhof, a large and effusive Dutchman, had been a research physicist; Norman Swallow, a journalist on a northern newspaper; Andrew Miller Jones had made educational film-strips; Peter de Francia was a promising painter; Paul Johnstone had been an academic historian; and John Read, whom I now met again after my encounter under his guidance with Joan Gilbert, was the son of the distinguished art critic Herbert Read and an expert in the visual arts. The most formidable of them all, Grace Wyndham Goldie, was a small, vivacious bird-like lady – though perhaps with rather more of the eagle than the wren – who had been theatre critic for *The Listener*, the BBC's journal which reprinted radio talks. Trying to keep all these diverse personalities in some kind of administrative order was the Department's organiser, Cyril Jackson, whose primary qualification, apparently, was that he was an expert on the Icelandic sagas.

I served out my three-month trainee-ship assisting any of the established talks producers who might need help in any way with any thing. George Noordhof was producing a programme about race and needed an example of a Caucasian racial type. So he put me in front of a live camera and I made my second television appearance, non-speaking but once full-face and once in profile.

Television had used me so far as an interviewer and an anthropological specimen. Now came my first chance as a programme producer. News came of an exciting scientific discovery – a coelacanth fish had been dredged up from the deep sea off the Comoro Islands. It was a living fossil whose entrails were likely to reveal important new evidence about the evolutionary history of all land-living vertebrates. Mrs Adams wanted a ten-minute programme inserted into the schedules as soon as possible in which Sir Julian Huxley, one of the most distinguished biologists of the time, celebrated for his lucid expositions on Radio's Brains Trust, would explain its importance to viewers. I, as the only one around with biological qualifications, was given the job of enabling him to do so.

I spoke to Sir Julian on the telephone. He had not yet decided exactly what it was he wanted to say, so I got hold of a fossil coelacanth and a mixed collection of pickled sharks, codfish and salamanders which I put in a bath in the dressing room allocated to Sir Julian so that he could take his pick. When he arrived in the studio, we discussed the general line he would take. There were no such things as teleprompters so he would speak without a script. I took my seat in the control gallery, the engineer in charge faded up the camera showing a caption card bearing the title of the programme (rather unenterprisingly called *Coelacanth*, as far as I remember), and we were off. Sir Julian spoke very eloquently but I had little idea of what he was going to say next. I could only do my best to illustrate his line of thought by cutting between shots of pickled fish in enamel trays and blown-up photographs of the newly-caught coelacanth itself. To my relief, Mrs Adams seemed pleased with the result.

By the time my three-month term as a trainee had come to an end, I had managed to make myself sufficiently useful to be offered a further six-month employment, this time as a full-blown assistant producer. So my time as a member of the BBC's staff began.

* * *

Talks Department dealt with pretty well everything that could be described as non-fiction. Books, current affairs, science, arts, gardening, do-it-yourself, archaeology, knitting, quizzes, medicine, politics, travel – the list was as long as we liked to make it. No one had made television programmes on any of these subjects before, so there was no accepted way of doing so. We were in the process of instituting our own visual conventions and creating our own traditions. We liked to think, therefore, that Talks was the quintessentially televisual department, which owed nothing to any other medium. We felt it was up to us to invent a new kind of visual grammar – and we took the task very seriously. Our discussions over coffee in the canteen were vigorous and never-ending. What, visually, was the best way to change shots – by a cut or a dissolve, the only other two visual transitions then electronically available? Should we perhaps restrict the use of the dissolve and so establish a special meaning to it – to suggest, maybe, a change of time, subject or location. When could we justifiably use music? Was it dishonest to mix film sequences shot earlier with live action without making it explicit that we were doing so? How best could we cue a speaker to avoid him or her appearing frozen on the screen and suddenly jerking into life?

The engineers who ran the electronic side of the studios had, of course, high professional standards. They tended to regard Talks programmes as somewhat amateur affairs since their producers had no identifiable qualifications. We ourselves were certainly eager to improvise or experiment. In the hope of getting very high angle shots, we might suggest that a camera should look upwards into a mirror suspended from the ceiling. That caused hideous lighting problems. We might optimistically instruct our three or four studio cameras to take such a variety of shots at such speed that the engineers would have to point out, wearily, that if their cameras moved in such a way their trailing cables would tie themselves into knots.

There were only two studios in Alexandra Palace. Each had three or, if we were lucky, four cameras. They were the same kind as those that had broadcast the first television pictures back in 1936. Some people maintained that they were, in fact, the very same pieces of equipment. They were large metal boxes mounted on bicycle wheels and filled with glowing glass valves. Each had only one lens. It is true that you could ask for that lens to be changed to one with a narrower angle – a kind of portrait attachment – but that took about ten minutes to do and was not normally something that could be tackled during transmission. The picture the cameraman saw was not an electronic one from the camera's lens but came from a separate optical view-finder attached to the camera's side. The cameraman viewed this on a ground glass screen, where it appeared upside down and in colour. If the camera got too close to its subject, the electronic picture and the one in his viewfinder did not coincide. That could cause arguments between the cameraman and the producer who was viewing the camera's electronic output on a monitor up in the control gallery.

Adjustments to a camera's picture were made by a second operator who sat in a small room alongside the studio. Here the rest of each camera's bulky electronic guts were fastened on to vertical racks so that any components that failed could be quickly located and replaced. If Racks, as he was known, passed the picture as technically adequate it was fed up to the control gallery and displayed on a monitor. If you, as a producer, wanted that picture to be transmitted to the public, you had to nominate it out loud. The vision mixer then switched the picture to a preview screen, the senior engineer gave it his final technical approval and the vision mixer, usually a lady of iron nerve, impassive visage and total reliability, a breed of which the BBC seemed to have a limitless supply, would then press the right button and the viewing public would at last see it.

The process was cumbersome, to say the least. The days of the finger-snapping whiz-kid director with hair-trigger reactions was still some time off – but we did our best. Sometimes, while we were on the air, the picture we were hoping to use would disappear altogether from its monitor in the gallery and we would look frantically through the gallery window down to the studio floor to see brown-coated technicians surrounding one of the cameras, unclipping its covers and probing in its interior while the programme's contributors continued talking. Then we had to abandon our carefully planned camera moves and improvise as best we could. Every now and then, disasters would befall two or even three cameras simultaneously. At that stage, the senior engineer in the gallery would call a halt and viewers would be treated to a film of a kitten playing with a ball of wool, waves breaking against a rocky coast, a windmill slowly turning or – a particular favourite among viewers – a ball of clay rising and falling beneath someone's hands on a potter's wheel. Hardly surprisingly, there were one or two occasions – to be talked about in the canteen for days – when a director finally lost his nerve and ran from the gallery saying that he could not take the strain any longer, while the programme contributors, unaware of the crisis, continued their performances and the unflappable vision mixer stayed at her post, restoring some kind of coherence to the pictures for benefit of the public.

As yet, there was no way of recording these programmes. The engineer in charge of telecine, the apparatus which transmitted film, had improvised a system consisting of a 16mm camera lashed to a television monitor, but the bleary image this produced was not good enough to be transmitted. It could however give a producer some idea of what viewers of his programme had actually seen. Unfortunately the process cost cash – for the film stock – and cash was not what Broadcasting House gave Alexandra Palace much of. So only one or two programmes a week from Talks Department could be selected for such immortalisation. These were then shown at departmental meetings and earnestly discussed.

The press in those days – as it still does to some extent today – referred to the BBC as Auntie. It wasn't a term we used among ourselves. We spoke of our employer as the Corporation, the Corp or the Beeb. Nonetheless the name was not inappropriate. If Auntie did exist, she lived in

Broadcasting House. She tended to regard the fashions and moral attitudes of yesterday as being eternal. She was dignified in language and manners and certainly knew what was best for people – even if they didn't. She regarded her young offspring in Alexandra Palace as feckless, irresponsible, and occasionally inclined to naughtiness.

It was also she, of course, who gave us our spending money since we weren't yet old enough to earn our own. The number of viewers was still very small, so the income from the television licence, which was then separate, was tiny. The Television Service, therefore, had to be very largely financed from the radio licence. Auntie had been very generous in giving us amusing toys to play with – our studios. She even gave us spending money which we could use for such things as artists' fees, costumes, scenery and – occasionally – a little film. But we mustn't ask for too much.

Auntie also had a view on what and when viewers should view. News, it was felt, was not really suitable for television. Pictures were not to be trusted. A relatively trivial event that produced spectacular pictures – a building on fire, perhaps – might be included and divert attention from something – say, a change in the bank rate – that had no visual component but was of greater significance to the nation. Pictures would also, inevitably, focus viewers' attention on a particular happening when a better impression of national and world affairs would be gained from a generalisation made by a skilled reporter.

And then there was the knotty problem of what a news-reader was to do. If he (as yet there was no she) simply read from a piece of paper in front of him, it would be very boring. But on the other hand, if he tried to engage the viewer by looking at the camera, his words might lack authority as it might seem that he was merely chatting or making it up as he went along. So there were no news bulletins on television. Instead a sound-only summary was relayed at the end of television transmission at around half-past ten, and once a week there was a weekly digest of film stories culled from the newsreels that were then part of the staple diet of the cinemas.

The Government, in the shape of the Post Office, determined how many hours of television could be transmitted in a week but Auntie decided which those should be. There was, of course, no television at all in the mornings when God-fearing people were working. A couple of hours were provided in the afternoon for housewives who, Auntie knew, would like to hear about cooking, knitting, home decorating and similar domestic subjects. But then, of course, the children would be coming back from school. They would have homework to do and their mothers would have meals to cook. Television had such a hypnotic effect that it might dangerously deflect those with receivers from such duties. So television stopped. The press called this the Toddlers' Truce.

The evening's programmes started at about half-past seven (it varied from day to day) and were introduced by a male announcer wearing a dinner jacket and black tie (though also, beneath his desk and out of vision an old pair of slacks), or a dignified young lady in evening dress. What followed was arranged in very much the same way as a good hostess might arrange an evening meal. That is to say, it started with something light – an hors d'oeuvre, as it were. There would be a main course which would be of some substance, a play perhaps, which would be followed by something less challenging. The evening would then be rounded off some time before eleven o'clock with an epilogue, and a view of Big Ben to give a check on the time (which, it has to be confessed, was in fact a model in a box, suitably lit and adjusted to the correct time according to the studio manager's wrist watch). And finally the national anthem.

Each Sunday there was a production of a major play – something from the West End or a classic – but seldom a piece written specifically for television. This, in the absence of any recording system, was then performed live a second time in the studio on the following Thursday. But apart from this, there was little that was routine or predictable about the television menu on any particular night – except perhaps that it did not resemble what had been shown on that day the previous week. This, coupled with the frequent breakdowns and a very liberal view about how long a programme might last, meant that the starting times of programmes could vary quite considerably from those printed in *Radio Times*. Fifteen-minute over-runs were by no means uncommon. Viewing, therefore, was an unpredictable activity. Those who indulged in it tended to take their position in front of their receiver early in the evening and to watch whatever came. It was known for them to ring up the duty office, where an official with a soothing voice was stationed to speak to them, and suggest that the particular programme being shown, which they found boring, should be brought to a swift end so that they could move on to the next item in the menu.

* * *

The BBC had acquired some old film studios that stood in Lime Grove, a shabby suburban street close to Shepherd's Bush in west London. They had been fitted out with a very different kind of camera from those that we were struggling with in Alexandra Palace. They did not require so much light. They had electronic viewfinders so that the cameraman could see exactly what his camera was transmitting. Most excitingly, they had a revolving turret which enabled the cameraman to change lenses during a production. (Zooms were still far in the future.) These technological delights were given initially to Light Entertainment Department, but soon Talks Department began to get a share of them.

One of the first of our programmes to do so was a quiz called *Animal, Vegetable, Mineral?*. This was one of the Department's big successes. In it, a museum challenged a team of archaeologists, art historians and anthropologists to identify some of its objects. Points were awarded to each side by the chairman, Glyn Daniel, an amiable Welsh Cambridge don and bon viveur. Its creator and producer was Paul Johnstone and I was allocated to work as his assistant. Paul gave me the job of visiting the chosen museum and selecting the objects to be presented to the panel. We aimed to find exhibits with entertaining stories – a moustache-lifter used by the Hairy Ainu of Japan was one, the knucklebones of a horse used by Romans as dice was another – or ones which, whatever their identity, were beautiful and worth looking at in detail. Baffling the experts wasn't really our main aim. We hoped instead that the participants would give some idea of the sort of details they looked for and the sort of logics they followed in identifying objects.

The star of *Animal, Vegetable, Mineral?* was undoubtedly a moustachioed extravert archaeologist, Sir Mortimer Wheeler. Whatever archaeological object we chose, it seemed to turn out that Sir Mortimer had himself personally dug it up. He played outrageously to the gallery, twirling his moustaches, pretending initially to be baffled, then discovering a clue and finally bringing his identification to a triumphant conclusion.

On the grounds that our experts ought to be in a good humour and the right party spirit, we booked a private room in a Kensington restaurant and all dined together before each programme. Waiting taxis then whisked us back to Shepherd's Bush so that the experts were talking on the air, live, before their bonhomie evaporated. Sometimes this worked only too well. On one occasion, Glyn had perhaps had one glass too many. He started perspiring rather heavily. He became confused about how many points he was awarding to whom. 'But what does that matter?' he cried, waving his hand gaily. When the experts failed to identify a small sharp stick made of human bone, he reclaimed it to reveal what it was. 'This,' he said, in rather slurred tones pointing the bone at the camera 'is used by Australian Aborigines to point at their enemies and bring them death and disaster – just as I am now pointing it at the viewer who sent me a very silly letter.' We had to explain to the journalists who rang up after the programme, that the lights were very hot and Glyn was suffering from a sudden bout of 'flu.

After a number of successful editions, Mary Adams was keen that we should extend the programme's scope and include natural history objects. I was against this. I thought it would be difficult to find things about which you could spin a series of logical deductions in the way that Sir Mortimer did so expertly when faced by a chipped flint or a fragment of pottery. You either knew that a stuffed bird was a bar-tailed godwit or you didn't. However, Mary was insistent and ruled that her friend Sir Julian Huxley should be invited. He was delighted to accept.

Finding something about which he could expatiate was my job. I noticed in the displays of the challenging museum a small white egg with a hard shell about the size of a pigeon's. What made this one interesting was that it had been laid, not by a bird but a mollusc, the Giant West African Snail. I felt sure that Sir Julian would recognise it immediately, for such eggs were regularly presented to first year zoology students in their practical exams. Even so, I thought that he might be able to construct an interesting story about it.

Things didn't go so well over dinner. Sir Julian and Glyn didn't hit it off. Sir Julian decided that the programme was rather silly. Spot identifications were not what he called science. Glyn explained that simple identifications were the last thing we wanted. We were hoping for a little chat. Archaeologists, Sir Julian said, were able to *chat* about objects, because archaeology wasn't a precise science; it was, in itself, largely chat. Glyn didn't immediately respond, but I could see that he wasn't his normal amiable self when we got the party back to the studio.

The programme started. Up came the egg. The turntable revolved so that we could see it from all points of view. 'Now, Sir Julian,' said Glyn, 'I think this is an object for you.'

'There is not much to be said about that,' said Sir Julian dismissively. 'There are only two classes of animals in the world that lay hard-shelled eggs – reptiles and birds. You can see from its shape that this is not a bird's egg, so it must be a small reptile's, probably some kind of lizard. There is no good scientific characteristic to enable me to be more precise than that.'

'I don't think so,' Glyn said, silkily. 'I don't think the name I have on my card is the name of a lizard.'

'Well that,' said Sir Julian, 'is because you know nothing about zoology.'

'Even so, I am fairly sure it is not even a reptile.'

Sir Julian started to lose patience. 'I will wager five pounds that it is.'

'The name on my card,' said Glyn 'is *Achatina*, the Giant West African Snail.'

My job as camera director at this point was to decide whether to cut to a shot of Glyn grinning delightedly, or apoplectic Sir Julian grinding his teeth. I decided the diplomatic thing to do was to cut to a shot of the egg sitting on its turntable. By the time we got down to the hospitality room after the programme, press cameramen were waiting to take a photograph of Sir Julian handing over a five-pound note. I didn't think they were likely to get it – and they didn't.

After the series had been running for some time, we thought we might give Sir Mortimer some richer meat. Why not a fake for him to detect? We found a flint hand-axe made by a famous nineteent-century forger known as Flint Jack. Sir Mortimer, as we might have guessed, had no problems with that and pointed out what characteristics enabled the tutored eye to see that it was modern and not ancient.

A couple more objects were produced and identified and then came one that I had selected for Sir Julian. It was a stuffed specimen of a great auk. The museum was keen to have it included for, as the director had explained to me, it didn't really fit into his collections and there was a chance that someone seeing it, might offer his museum several thousand pounds for it. I was happy to accept it as I hoped that Sir Julian, having identified it, would talk about when and why the species became extinct. He obliged. He was getting the hang of things.

'Is there any more to be said?' asked Glyn.

'Let *me* see,' said Sir Mortimer. 'Ah yes, I thought so. A better description of this, I believe, would be that it is a penguin's beak and a bundle of suitably dyed chicken feathers, put together to produce a rather unconvincing simulacrum of a great auk.'

From my position in the control gallery I could see Sir Julian looking infuriated and the curator of the museum looking aghast as his vision of thousands of pounds evaporated. Few, in those days, would have questioned Sir Mortimer's opinion on anything.

* * *

Animal, Vegetable, Mineral? continued for several years, but eventually it was deemed to have run its course and I was given the job of devising a successor. I thought we might replace objects with photographs of strange places and ask a different panel of experts to identify them from either the architecture, the plants, the geological features, the clothing of the people or some other visible clue. For the first programme I invited Osbert Lancaster, designer and cartoonist, Peter Fleming, explorer and author of several entertaining travel books, and John Betjeman, poet and architectural critic.

The first picture was something really easy, just to get them started – a nineteenth-century photograph of the Burning Ghats at Benares. Sari-clad women were bathing in the holy waters of the Ganges. Behind stood the tall pagodas of the temples. Wisps of smoke rose from pyres on which human corpses were being cremated.

'Well now,' said the chairman, 'Osbert Lancaster, what do you think?' Lancaster looked at it blankly. There was a long pause.

'No idea,' he said.

'Peter Fleming?'

'Stumped,' said Fleming.

'John Betjeman?' the Chairman said despairingly.

Betjeman looked long and hard at the picture, the pagodas, the veiled women, the smoking pyres.

'Got it,' he said triumphantly. 'The Thames just above Maidenhead.' We went on to the next photograph but things did not improve.

A series was not commissioned.

* * *

The phone rang. It was Mary. 'David. Konrad's in town. With his new book. We must get him on.' Mary seldom bothered to explain or provide surnames. One was expected to know. And indeed, on this occasion I did. It was Konrad Lorenz, the extraordinary Austrian zoologist who would eventually receive a Nobel Prize for his work on animal behaviour. The book was *King Solomon's Ring* in which he explained how he had decoded the way animals communicate to such an extent that he could converse with some of them, as King Solomon was said to have been able to do. Professor Lorenz, encouraged by his publishers, was only too glad to appear on television. The producer was to be George Noordhof who dealt with science programmes and I, in view of my zoological qualifications, was to be the interviewer. Apparently Mary had decided that the dimensions of my teeth did not, after all, disqualify me from the job.

Like any other programme of the time it was, of course, live. Professor Lorenz proved to be amiable, bearded and white-haired and spoke with a thick but nonetheless comprehensible Teutonic accent.

'You explain in your book,' I began, 'that you are able to talk to animals and that you have a particularly close empathy with greylag geese.'

'Ja. Dat is zoh.'

'Well, we have a greylag goose here. Perhaps you would care to exchange a few words with it.'

With that, a keeper from the London Zoo walked on to the set carrying a goose which he put down on a low table that stood between the professor and myself. The goose, naturally enough, was somewhat perturbed at suddenly being thrust under the bright television lights and began to flap its wings.

'Komm, komm, mein Liebchen,' said Konrad, soothingly, putting his hands on either side of the goose's body so that its wings were held folded down. He was holding it so that its head was pointing away from him. This was sensible in that he was not then within range of the goose's beak which it showed every wish to use, if it got the chance. But that, of course, meant that its rear was pointing towards the professor and the goose, in the flurry, squirted a jet of liquid green dung straight at him.

'Oh dear dear,' said Konrad. 'All over der trouserz.' He released the goose, which flapped off the set and was neatly fielded by its keeper, took out his handkerchief and carefully wiped his trousers clean. Then, finding his handkerchief in his hand, in his embarrassment, he promptly blew his nose on it.

He completed the interview with a green smear down the side of his face which I hoped was not immediately visible to the watching millions but which, I have to admit, somewhat hindered my ability thereafter to frame the right questions.

* * *

Mary decided that the reason many of our programmes were stiff and unrelaxed was that our contributors were daunted by the blinding lights, by the tension of being seen live and – particularly – by the fact that we as producers were over-rehearsing them simply in order to perfect our camera moves. We should take more risks. 'What we need,' she said 'are raconteurs, natural gems who could come straight to the studio and, with virtually no rehearsal, sit and sparkle.' I was given the job of discovering the first.

I heard of a natural gem in the East End of London. He was a rat-catcher named Bill Dalton. I went to see him and he was all that everyone had told me. He had riveting stories about rats running up duchesses' skirts, swarming in the kitchens of London's smartest hotels, emerging from lavatory bowls and biting users in the most painful manner. There was only one problem with these anecdotes. They all ended with Bill despatching the rats in the most blood-curdling fashion, pulping them with a blow from the back-side of a shovel, bisecting them with one strike of a cavalry officer's sword, or throttling them with his bare hands. The British public, I explained to him, were sentimental about animals – even rats. They wouldn't like to hear the gruesome details of their demise. Did he think he could conclude each reminiscence by explaining that, in the end, the rats concerned gently and painlessly passed away? Bill understood completely. It wouldn't be a problem. I had my first natural gem.

In accordance with the new philosophy, I arranged that the car bringing him from his home to the studio would not arrive until a few minutes before transmission. We waited for him in the studio, with the cameras positioned, warmed up and ready to go. He arrived on time carrying, rather unexpectedly, two huge wire cages, each packed tight with rats. The smell was dreadful. The cameramen cringed. The sound operator climbed hastily on to his boom platform.

'I thought viewers would like to see what I'm talking about,' Bill explained.

'Okay,' I said. 'But just remember – keep the violence to a minimum.'

'Don't worry Dave,' and I ran up the steps that led to the control gallery.

On the monitors I saw the previous programme ending. Central control gave me the signal to start. I cued the title captions and Bill began.

'I am a rat-catcher,' he said confidently, 'and I'd like to show you what it is that I catch. This one,' he pointed to one of the cages 'is Rattus Norwegicus, the brown rat. And this is Rattus Rattus, the black or sewer rat, and 'e is the one I would like to show you first.'

Then, to everyone's horror, he lifted the wire lid of the sewer rats' cage and plunged his hand into what I can only describe as a maelstrom of rat. Quickly he grabbed an enormous rat by its tail, pulled it out, slammed the cage lid shut and started to whirl the rat around as violently as he could. Then something seemed to occur to him. 'Now I don't want you to think,' he said leaning confidentially towards the camera while the rat he was holding continued to whirl like a Catherine wheel, 'that I am in any way *maltreatin'* this rat, but unless I get 'im slightly dizzy, the bugger will bite me.'

It was, I think, the first time that word had been heard on British television. And Bill was both the first and the last of our natural gems.

* * *

Short stories were very popular as brief fifteen-minute items towards the end of an evening. Mary Adams selected them and they were handed out to producers to put on the air. They were read, usually from an armchair, by an actor – and occasionally by the author. The producer's contribution to such a programme was not large or demanding. First we had to select a set – usually a few flats with bogus bookshelves and a fireplace with a flickering gas fire. Then we listened to the actor read the piece and timed it. When transmission came, we usually put one camera some distance from him and asked the cameraman to advance slowly towards him so that when, say ten minutes later, the actor came to the story's denouement, his face was in close-up. As a camera instruction, it was known in the trade as a 'trickle'.

Cyril Jackson summoned me to his office. He had a problem. Among the stories Mary had acquired was one by William Sansom. It concerned a fishmonger who had found fulfilment in life by arranging fish on his marble slab – until he fell in love with a pretty girl who was one of his customers. He asked her to marry him but then, in order to increase his earnings to support him and his new wife, he had to take a job at the cash till. So he was deprived of the aesthetic satisfaction of arranging his scallops and turbot, his lobsters and his eels. He became unhappy and in consequence, his marriage collapsed. Sansom was a poet and his description of the patterns on the marble slab undoubtedly had literary elegance. On the other hand, his high-flown words, spoken by an actor sitting in a library set, might well sound unconvincing, not to say risible. Hardly surprisingly, every producer Cyril had allocated to short story duties, had picked some other example from Mary's collection. Only this one remained. But now something had to be done.

'I'm fed up of carrying forward the fifty pounds copyright fee from quarter to quarter,' Cyril said. 'It has got to be discharged. I don't care how you produce it. But produce it.'

I too felt that an actor talking to camera was not the way to present it. Nor did it seem to me that the right answer would be to put a narrator in a real fishmonger's shop with real fish. We would have to stylise it. I decided to do it as a ballet.

At that time, Philip Bate, the clarinet expert, was producing an excellent series called *Ballet for Beginners* in which two dancers, Michel de Lutry and his wife, explained and demonstrated the basic steps of classical ballet. They were very popular. I got in touch with de Lutry and asked him if he thought it possible to choreograph a ballet to words, providing they were rhythmically spoken. He was willing to try.

I recorded an actor reading Sansom's poetic words in as rhythmic a way as was tolerable. Design Department was commissioned to create a marble slab which would in reality be a back-projection screen on which stylised, changing and semi-abstract pictures of lobsters and codfish would be projected. So it went on air. De Lutry in a fishmonger's apron and straw boater whirled and pirouetted around his slab with its changing designs and his wife was coquettish and ultimately suitably miserable. And Cyril Jackson was freed from carrying forward that particular fifty pounds ever again.

Only the *Daily Mirror* noted the programme's passing. 'All right,' it said. 'So Auntie had to try the experiment of dancing to words. Now we know. It doesn't work. Don't do it again.'

* * *

I could hardly hope to become involved in producing music programmes for television. There were, after all, properly qualified music producers to do that. On the other hand, no one seemed to take any notice of folk music. I happened to hear a series of talks on the Third Programme by an American folk-song collector named Alan Lomax. His father, John, had put together one of the most important of the early collections of cowboy songs. Alan himself had been responsible for taking one of the pioneers of jazz, Jelly Roll Morton, to the Library of Congress and there making some now historic recordings of Morton talking at the piano about his early days in New Orleans. The Third Programme had also commissioned Alan to collect European folk music and Mary Adams agreed to my suggestion that we should give him programmes on television as well.

Alan was a large heavy man in his early thirties, with a relaxed easy-going manner, a wide smile and a Texan drawl. He was himself an engaging performer, singing in a high, slightly nasal voice and accompanying himself on a guitar. The format of the programme I devised for him could hardly have been simpler. He was to sit in a studio at Alexandra Palace and sing, and then introduce two or three guests from different parts of Britain.

Alan, accustomed to the somewhat more sophisticated television studios in the United States, looked at our lumbering primitive cameras with some surprise. However, no matter what our technical limitations were, he acknowledged that the American television of the time would never have contemplated devoting six half-hour programmes to the still recondite subject of folk music.

The British public were hardly more prepared for the subject than the

American. Victor Silvester and his Ballroom Orchestra and other bands led by such as Billy Cotton and Geraldo, with their ranks of saxophones and violins, were the dominant taste. Singers with guitars, whether performing singly or in groups, were virtually unheard. Nor did Alan feel that he should try to make any concessions to popular taste. I went to see him to discuss who we might invite to the first programme to be told that he had already arranged for a group of old ladies from the Outer Hebrides to be flown down to London to perform one of the traditional songs they sang when 'waulkin' newly woven tweed. 'Waulkin' consisted of sitting in a line with a dozen or so yards of tweed in front of them thumping it with their fists. I was more than a little put out. Their air fares alone would take up my entire budget for the first three programmes. Nonetheless, the arrangement had been made and I had to go to Cyril Jackson, the holder of the Department's purse, to get a special additional allowance. In due course, the ladies turned up, bringing a huge quantity of tweed with them and sang their songs in Gaelic to, I fear, a somewhat baffled audience in the first of the *Song Hunter* series.

I had to save money on the next programme. Alan had discovered that one of the best of the Irish traditional fiddlers, Michael O'Gorman, was at the time working as a porter at Paddington. Old and grizzled, he charmed viewers by explaining that he might not be able to perform his first few pieces as well as he would have wished because his hands were still a bit stiff from carrying baggage, but he would warm up as he went along – and then played a series of reels and jigs at dazzling speed.

Ewan McColl, who sang with one hand cupped to his ear, came to perform uncompromising industrial ballads about strike-breaking. That provoked serious questions from Broadcasting House about whether a politically subversive element was sneaking into the Television Service and escaping from Current Affairs Department where it could be kept under proper control and supervision. That enquiry was smartly dealt with by Mary Adams as Head of Talks, who in any case leaned quite far to the Left herself.

Alan was particularly keen to include an Irish tinker woman, Margaret Barry, who accompanied herself on a banjo and had in her repertoire a version of a famous ballad about a ghost lover – *As She Moved through the Fair*. Margaret, however, being a true tinker, had no fixed address. How could we find her? Folklorist connections of Alan's said that the only way was to send telegrams to a range of police stations along the west coast of Ireland asking that the next time Margaret was arrested on a charge of being drunk and disorderly, she should be told that a return air ticket and a fee awaited her if she would come over to London. She would certainly be brought in, they said, before the series was over. And indeed she was, and soon after she arrived in London, a tall striking figure with a lined tanned face, a beak of a nose and long black hair that went far down below her shoulders.

Alan brought her up to Alexandra Palace. After a brief run-through, she and Alan retired to their dressing rooms to rest for half an hour before the transmission, which was, of course, like everything else at that time, live. There, I suspect, Margaret found a little Irish whisky. I remained in the control gallery, sorting out the last technical arrangements and camera moves, while the engineers made final adjustments to the lighting. Alan's policy was not to return to the set with his guests until the very last minute so that they did not have to spend too much time sitting under the grilling lights. This time, he left it dangerously close. At the very last moment, he rushed in, took his seat and sang *Travelling Along*, the song with which he introduced each programme. Then he introduced Margaret Barry. Margaret took up her banjo which had been left lying on her chair since the run-through. She struck a chord. Unhappily, the heat of the lights had so affected it that not one string retained its original tuning. Margaret seemed not to notice and sang the opening lines with the strength she normally used to cut through the conversation in a busy Irish pub. As I instructed a camera to focus on her in close-up I saw that, while in her dressing room, she had removed her false teeth. The sight of this gaunt strange woman, with few visible teeth accompanying herself with a jangling out-of-tune banjo was not, to put it mildly, the highlight of that week's television.

Viewers started to ring up in considerable numbers to complain. One military gentleman said that he knew the BBC was short of money, but bringing in aged buskers from the streets was no substitute for real music. Where was Vera Lynn? he asked. A few viewers, however, recognised that Margaret was the custodian of a powerful and important tradition. One, an expert in the art of canal barge painting and other folk traditions, was so impressed that she made contact with Margaret and took her to Covent Garden to hear *Carmen* – a work that did not greatly impress Margaret. She said it lacked spirit.

All the contributors to *Song Hunter* were not so daunting and the series ended with an omnibus edition in which many of the most popular performers reappeared. Alan, in order to conclude the series in a properly exuberant way, suddenly decided during transmission that the staid and cautious movements of the camera that were imposed on me by their hopelessly unwieldy nature, were not to be tolerated and without notice he jumped to his feet and led the whole company in a mad procession around the studio, in and out between the cameras and the sound booms and the lighting stands. As a programme it was technically my most incompetent so far. But thinking of the contributors now – Bob Roberts, captain of a Norfolk sailing barge, Seamus Ennis, a virtuoso on the tin whistle, Jack Armstrong playing the sweet-toned Northumbrian pipes, Ewan McColl with his protest songs – they seem to be the star names in the British folksong revival that ultimately, some musicologists believe, contributed a great deal to the pop music to come.

2

Zoo Quest

The most successful animal programme on television in the early 1950s was a series in which George Cansdale, then Superintendent of the London Zoo, exhibited his charges on a large table covered by a door-mat. The fact that his programmes were live gave them much of their charm, for the animals concerned – chimpanzees, lion cubs, monkeys, pythons, parrots – did not necessarily do what Mr Cansdale wished them to do. They might insist on showing nothing but their rear end to the cameras. They wet the front of his shirt. Occasionally they would escape and have to be grabbed by Zoo keepers who stood either side of the table. And just every now and then they bit Mr Cansdale. All of which made excellent entertainment. Nonetheless, in spite of his efforts to demonstrate details of their anatomy and to explain how his exhibits were adapted for their way of life, the programmes were hardly illuminating natural history. The animals had been taken from their enclosures in the middle of the night, dropped in to a bag or shut into a small travelling cage and then thrust into the studio lights. How could one expect them to behave naturally?

It seemed to me that the BBC, with its mission to inform as well as to entertain, ought to try something a little more ambitious. I didn't dream that there could be enough money to enable me to make films. My job was to produce programmes from the studio. But perhaps we might use zoo animals to illustrate some basic zoological principles. Television was a visual medium so maybe we should start with the visual aspect of animals and examine why they are shaped and patterned as they are. Why not a three-part series – one programme about camouflage, one about warning coloration, and one about courtship displays? I did a bit of homework and eventually produced a fairly detailed outline of what such a series might show. I called it *The Pattern of Animals*.

Mary was enthusiastic. 'We must get Julian to do it,' she said, 'I'll talk to him.' The message came back from her office that Sir Julian Huxley would consider the idea and that I was to go to see him.

I arrived at his elegant Regency house in Hampstead and was shown upstairs to his book-lined study. Outlining the basis for a serious zoological programme was going to be more demanding than inviting him to take part in a parlour game and I was somewhat apprehensive. I told him what I had in mind. Fortunately, Sir Julian thought it a good idea. Would he like to think about it and perhaps suggest the examples he would wish me to find to illustrate the points he might make? 'No,' he said, 'that won't be necessary. You seem to have a clear idea of what you want. Just write it out in detail and let me have another look.'

Technically, the programmes were hardly adventurous. Sir Julian was to sit in front of a camera behind a desk, while the other cameras moved up and down a line of specially built cages, taking pictures of their inmates to illustrate his words. It was, nonetheless, a risky procedure. Since the programmes were live there could be no guarantee that the animals we wanted to look at would be properly visible at the time we needed to do so. The fronts of the cages were completely open as glass would cause troublesome reflections. Curators at the Zoo, however, who were to provide the animals, assured me that in most instances, this would be no problem, provided that the lights illuminating the cages were much brighter than those in the studios. Birds and reptiles would be reluctant to flee into the dark unknown.

There was one exception. In order to demonstrate an aggressive display we had an Egyptian cobra. It seemed to me unnecessarily risky to allow such an animal the opportunity to slither out into the studio. The Overseer of the Reptile House knew perfectly well how to restrain it. We wound a belt of transparent sticky tape round the cobra's middle, tied a string to the underside of the belt, threaded that through a hole in the plywood bottom of the cage and tied it to a studio weight. The cobra could then move both its front half and its rear half quite freely but would be quite unable to escape.

Sir Julian had familiarised himself sufficiently with the content of the first programme to be able to improvise some appropriate comments on the pictures of the animals that viewers were seeing and that he was watching on a monitor. When it came to the second programme, however, he hadn't taken as much trouble and he stolidly read through the prepared text without even lifting his head to look at the camera. The third programme was much the same. Nonetheless, the series was a reasonable success and I was quite pleased with myself when I returned to the Natural History Museum some specimens I had borrowed to illustrate insect mimicry. I asked to see the Keeper of Entomology so that I might thank him personally for his help. 'Please don't thank me,' he said grumpily. 'Television is a waste of time. I certainly wouldn't have lent you any of my specimens had it been left to me. I was instructed to do so.' As I left, I thought I should have reassured him by saying that maybe the previous night more people would have seen the insects he had lent me than would have entered his museum in a year. But I didn't think of it in time.

The series had an important consequence for me personally. I had been greatly helped with selecting live animals by the Zoo's Curator of Reptiles, Jack Lester. After *The Pattern of Animals* was over I sat with him in his super-heated office in the Reptile House, surrounded by tanks and cages in which he kept some of his favourite animals – bush babies, bird-eating spiders, chameleons, parrots, even sunbirds. Jack was an all-round naturalist with a wide-ranging enthusiasm for animals of all kinds.

We discussed the Huxley programmes. Surely we could do better than that. Viewers needed to be given some idea of the animals' natural environment. On the other hand, I felt sure that the BBC would not send me away to make complete films of animals in the wild. For one thing, I had no idea of how to do so. And anyway, my job was to make studio programmes not films. But even a few short film sequences illustrating an animal's habitat before seeing the animal live in the studio would improve things. How could that be done?

We hit on the idea of an animal-collecting expedition. At that time, the London Zoo still retained many of its nineteenth-century attitudes and one of its aims was to show as many different species as possible. Breeding its animals was not high among its priorities. Most zoos assumed that there was an unlimited supply of exhibits in the wild. No one seemed to suspect that a time might come when that supply might be in danger of exhaustion. So it was not uncommon for big zoos to send out expeditions to look for rare creatures that had seldom if ever been seen before in captivity.

We agreed that Jack should try to persuade his bosses to mount such a trip, saying that, as a bonus, he might be able to get the BBC to send along a film unit which would give the Zoo valuable publicity. At the same time, I would tell Mary Adams that I had discovered that the Zoo was mounting an expedition and suggest that I might manage to get permission to accompany it and so benefit from their expertise. When we got back, I would produce programmes which would include a number of film sequences showing Jack catching something or other. After each we would dissolve to the studio, live, with the same animal in close-up which Jack would show to the cameras in detail, pointing out particularly interesting details of its anatomy. There might then follow some of the unrehearsed behaviour by the animal that had made the Cansdale programmes so successful. A bite or two maybe, perhaps an escape.

Exactly where should we suggest the expedition might go? That was easy. Jack had worked in a bank in Sierra Leone. It was there that he had acquired his passion for tropical natural history. He knew the country well and he still had friends there who would help us.

'But we really need a particular objective,' I said. 'Some animal that is virtually unknown and has never before been seen in a zoo. A rare ape, perhaps; a mysterious much-feared reptile surrounded by bizarre legends. Then we could call it a quest for something or other.'

Jack thought for a bit. 'Well,' he said, 'I can't think of a particularly rare or unknown mammal, or even a reptile. But there is a bird in Sierra Leone that very few people have ever seen alive and no one has any idea about its nesting behaviour. It's called *Picathartes gymnocephalus*.'

I didn't think that *Quest for Picathartes gymnocephalus* was a title with the sort of crowd-pulling potential I had in mind.

'Doesn't it have another name?'

'Certainly,' said Jack. 'Its common name is Bald-headed Rock Crow.'

Even that didn't sound particularly compelling to me. So we agreed to call the project simply *Zoo Quest*. We both took the idea to our respective masters. They lunched together and the word came back that we could go ahead.

The next question to be settled was who the cameraman should be. I wanted to use 16mm film. At that time, the limited amount of film that BBC Television originated was all shot on 35mm, the same gauge as was and still is today used for feature films. Most of the Corporation's staff cameramen came from that industry or from newsreel companies. They would not dream of handling any other kind and certainly not 16mm. That was for amateurs. They referred to it derisively as 'bootlace'. On the other hand, 35mm cine cameras were large and heavy and the film to go in them was both those things – as well as being very expensive. Neither the kind of budgets we could hope for, nor the sort of places we wanted to take cameras, would allow us to work on anything else but 16mm. I said as much in a memo to Mary. She sent it on to the Head of Films.

Angry memos came back. The project could *only* be filmed on 35mm. BBC Television was a professional organisation. To lower standards was unthinkable. I stuck to my guns. Eventually a meeting was called. The Programme Director took our side of the argument and Head of Films had to back down. 'But,' he said, 'if 16mm film is introduced as standard practice to BBC Television, it will be over my dead body.'

Now I faced another problem. None of the BBC staff cameramen had any use for 16mm film either. If I wanted someone to handle that sort of stuff, I would have to find him myself. One name came immediately to mind. A superb documentary film about the first ascent of Mount Everest had been going the rounds of the cinemas. That, I knew, had been shot on 16mm by a cameraman called Tom Stobart. I traced him. He had just returned from another Himalayan expedition – this time in search of the Abominable Snowman. He himself had other things to do, but he had been assisted by a young cameraman called Charles Lagus. Maybe he might join us.

Charles and I met in a pub at the top of Lime Grove. He was small, dark-haired and soft-spoken. His parents had come to Britain in the 1930s from Czechoslovakia. At school he had thought that he might become a doctor, but he hadn't done particularly well in his first examinations so he had given up that idea. He had always been interested in photography but had never taken any formal training of any kind. He had, however, spent three months in the Himalayas with Tom Stobart and he thought he would be able to handle the sort of trip I described. I guessed that he knew as much, if not somewhat more, about making documentary films as I did. We both seemed to laugh at the same things. We finished our beer. We would do the trip together.

* * *

A few weeks later, at the beginning of September 1954, four of us assembled at a small airfield in Hampshire from which the cut-price airline we were using operated – Charles and myself, Jack and Alf Woods. Alf was Head Keeper of the Bird House in London Zoo and was one of those rare people who have the animal equivalent of green fingers – that is to say he had an inexplicable skill in soothing disturbed or sick animals, divining what their problems might be, persuading them to eat, and generally settling them down. He would be invaluable in caring for the animals that Jack caught.

The 16mm camera Charles had brought with him, the most advanced and robust available at the time, was driven by clockwork and could only run for 40 seconds before it had to be rewound. It used hundred-foot rolls, so after taking two minutes and forty seconds worth of shots, it had to be reloaded. Since colour television had not yet been invented, we were going to use black and white film which, although it was much more sensitive than the colour film of the time, was still so slow that it required full sunshine to get a reasonable result. I would be recording the sound using open reel magnetic tape on the only portable battery-driven tape recorder that was then available. It was the size of a large document case, and driven by ten big torch batteries. Rewinding was done by shutting its lid, holding down a button and cranking a small handle. There was no way by which tape and film could be linked, so filming someone talking and hearing what was being said was not possible. Maybe the Head of Films had some reason on his side when he said that 16mm film was for amateurs.

The small Dakota aircraft we boarded could not fly directly to Sierra Leone. There were no navigational systems to guide it through the night down the west coast of Africa and in any case, it would have to land several times to refuel. The first night we spent in Tangier. I had not been out of Europe before and I wandered wide-eyed with the other three through the casbah. The next day we set out for Dakar, and so arrived in tropical Africa in the dark. And then, on the afternoon of the third day, we landed at Freetown, capital of what was then still the British colony of Sierra Leone.

* * *

Stepping out into the heat and humidity of a West African afternoon was like entering a heated sauna. The hedge beside the ramshackle airport was bright with the scarlet trumpets of hibiscus. Sunbirds whizzed from one to another, hovering in front of each to sip nectar, their chests flashing iridescent colours, green, purple and red. Among them I suddenly spotted, clinging to the branch but rigidly immobile, a bright green chameleon. As I took a step towards it to get a closer look, my foot trod on the grassy verge and the leaves, to my astonishment, suddenly hinged back to lie alongside the main stems. It was sensitive mimosa. All in all, that little strip of ordinary hedge was a revelation of the glory and fecundity of tropical nature from which I have never recovered.

Jack was a splendid guide. His knowledge of every aspect of the local natural history astonished me. He could recognise all the birds. He knew how to make a special rubbery adhesive from plant sap that, smeared on to a branch, would catch a sunbird. And having caught one, he knew how to induce it to feed from a bottle of artificial nectar. He delighted in scorpions, particularly the big black ones, the Imperial scorpion, which he picked up by their venom-loaded tail using a pair of forceps. He showed me the tiny funnel-shaped pits in dust made by ant-lions and together we watched hapless ants stray into such a pit, slither down the slipping side straight into the jaws of the ant-lion, the larva of a large dragonfly-like insect, which was buried at the bottom. At night, drifting through the mangroves in a swamp, he could catch crocodiles. He shone a torch and spotted one by its eyes, a pair of two coals glowing red in the torch beam. Then silently drifting towards it, he would lean over the side of the canoe, and while it was still dazzled, grasp it by the neck.

Snakes were his particular passion and expertise. Gaboon vipers, three feet long, lay among the shrivelled grey leaves on the forest floor, virtually invisible to my eyes, immediately picked out by his. They are one of the most dangerous of West African snakes, easily trodden on and quick to strike with a lethal venom. They are not, however, fast moving and Jack showed me how to catch them. They are fat and heavy. Push a pole under one, about a third of its length back from the head, and it will be balanced so that you can gently lift it. Then you dump it in a box with its hinged lid ready open to receive it – and slam the lid down very quickly. There was also a large number of snakes, many quite small, categorised by Jack as 'back-fanged'. They were venomous but their poison came from fangs at the back of their mouth and their gape was too small to enable them to give you a proper bite. So they were not dangerous and Jack would pick them up by holding their heads down with a small supple stick while he grasped them around the neck with his thumb and forefinger.

We made our base at a station maintained by the Agricultural Department. There a large thatched hut surrounded by a neatly kept lawn was put at our disposal and in it Alf Woods slowly accumulated a miniature zoo. As well as snakes, there were bush babies and bird-eating spiders, tortoises and chameleons, sunbirds, owl chicks and parrots. Jack was particularly pleased with a dozen or so emerald starlings, spectacularly beautiful glossy green birds, that are only found in a small area of West Africa, for they would be a species new to the Zoo's collections.

Charles and I did our best to film Jack and Alf as they worked assembling this collection. But we had considerable problems. The very first time we followed Jack in to the rain forest and prepared to film, Charles held out his light meter.

'The only way we can get enough light to film by here,' he said bitterly, 'is to chop down the trees.'

Nor was there any chance of our getting shots of birds or monkeys high above in the canopy of branches. Powerful long-focus lenses for 16mm cameras did not then exist. I now understood why it was that virtually all films of African wildlife at that time were shot out in the open grassy savannahs of east Africa where huge animals like elephant, rhinoceros and giraffe wandered about in the open under a blazing sun. But my programme plan had always been that we would use film, not to show the animals but Jack catching them – and there was just about sufficient light to do that. For close-ups we would rely on getting Jack's captures back to London and then transmitting detailed pictures of them live in the studio.

Even so, I hankered after some genuine wildlife filming. We would have to concentrate on those species which spent most of their time outside the forest, such as weaverbirds and crocodiles, or creatures like scorpions and chameleons that are so small that we could take them to places where, one way or another, there was enough light for filming. Such a plan at least had the advantage that most of these creatures had been ignored by previous wildlife film-makers.

Driver ants were an example. Columns of them appeared, marching across grassy clearings on their way to hunt. There it was possible for Charles to sit within a foot or so focusing on the big black soldiers that lined the sides of the columns with their powerful jaws agape. If we waited long enough, we could also watch the dismembered parts of the victims that had been caught by the ants at the head of the column and were being taken back to the bivouac in the forest where the queen and her entourage were camped. As we filmed them in detail and at length, we became aware of miniature dramas that until then I knew nothing of. Robber flies the size of bluebottles swooped down on the columns and pounced on a worker transporting one of the colony's pupae. They stabbed the cocoon, sucked out its contents and then dropped it back in the column with its porter still clutching it.

Although these ants provided us with good programme material, they were a real hazard for Jack and Alf. If a column got into the hut housing the animal collection, they could easily invade the cages and kill the inmates since they were unable to get out of the ants' way. Accordingly Jack stationed watchmen to sit by the hut with a can of kerosene, day and night. If ants appeared from the forest and headed across the lawn towards the hut, the watchman had to deflect the column by pouring kerosene across its path and setting light to it. Nothing else would turn them back.

One evening, we returned from the forest to find the watchman fast asleep. A column of driver ants was already inside the hut. As quickly as we could, we took out each animal, picked off the ants that were biting it and put it in a new cage of which Alf still had a few in readiness for new captures. This was easy enough to do with chameleons and millipedes. It was a different matter with the snakes. Jack held one by the head, Alf held the tail, and Charles and I worked our way along its length, removing every ant we could find. Some had their jaws so firmly fixed between the snake's scales that, as we pulled, their heads separated from their thoraxes. Gaboon vipers, spitting cobras, black and white cobras, green mambas, black mambas, all had to be treated in this way. We worked late into the night and dealt with everyone. Even so, a third of our snakes did not survive the attack.

* * *

We also tried our hand at filming bats. Jack had told us before we left that there were reports of an unknown species of bat that lived in a cave in the forest. He wanted to collect specimens for the Natural History Museum. I thought that might make a good sequence. But how did one film in a cave? There were then no portable batteries to power hand-lamps. The answer, I decided, was magnesium flares. Getting permission to transport such things by air was not easy. They had to be packed in metal boxes and sealed with solder. However, we managed to have that done and I was confident of getting a dramatic and unexpected sequence.

The entrance to the bat cave was a long horizontal cleft in the ground about five feet across at its highest. Inside, the rocky floor shelved quite steeply downwards and then, after about fifty yards, dropped away very sharply so that it became nearly vertical and widened out into an abyss. We shone our torches down this but the bottom was far beyond the range of their beam. That didn't matter, for the bats were hanging in huge numbers from the roof above. I imagined that they were well within range of the light from a flare, so once Charles was ready, I opened the sealed metal case and took one out. It looked like the head of a very large firework, a fat cylinder about twelve inches long. Charles focused on the bats by the light of the torch beam and then I lit the flare and held it up.

It fizzed for a few seconds. Then suddenly the whole cave was filled with dazzling bright light that illuminated the bats perfectly. But only for a few seconds. Choking white smoke of magnesium oxide from the burning flare blanketed us. 'Throw it away,' yelled Charles, by now invisible in the smoke. I hurled the flare down into the abyss but it must have landed on a ledge a few feet further down that I hadn't seen. At any rate, the illuminated smoke continued to billow all around us. Thousands of bats flew in panic through it on their way out of the cave. They could escape by using their echo-location, but we could not move. The smoke was so thick that we couldn't see well enough either to climb out or to descend to wherever it was the flare had lodged in order to throw it still farther away. There was nothing to do but to sit it out. The air in the cave must have been almost static for the smoke got thicker and thicker as the flare continued to burn a few yards below us. Both of us were coughing violently. I thought I was close to choking. The flares, I remembered, were guaranteed to last for two minutes. When I had bought them in London, I thought that an absurdly short time for filming. Now it seemed that I had never known a longer two minutes in all my life.

At last the brilliant light of the burning magnesium flickered and died. Coughing and spluttering, Charles and I managed to pick our way up the steep cave floor and into fresh air. A bat sequence was one element in the programmes we would have to do without.

* * *

Wherever we went, we asked the local people for information about the animals that lived in their neighbourhood. Jack carried specially commissioned drawings of the creatures he was particularly keen to collect, including *Picathartes*, and showed them to local hunters to see if they recognised the animals represented and knew where we might find them. As the weeks went by, so we got to know some villages particularly well and became friends with their chiefs. One of them was exceptionally helpful and offered to stage a performance by his musicians and dancers in our honour. We were delighted and decided to try to film and record it.

The first item was a solo by the chief's *balange* player. The *balange* is a xylophone with a gourd suspended beneath each wooden key to act as a resonator. As the musician played, I recorded him. The villagers had not seen a recorder before, so when the piece was over, I wound the handle in the lid and replayed the sound of the *balange* through the small tinny speaker. Everyone listened, entranced and amazed – except perhaps the *balange*-player who suddenly found that he was no longer the star of the show. He was not surprised, he said dismissively, that my box had learned that piece of his. It was, as a matter of fact, a very *easy* piece. He would play something altogether more difficult that would give it something to think about – and with that he started on another performance of impressive virtuosity, his two sticks, tipped with local gum, flashing at great speed across the wooden keys. I replayed that too – and everyone except him seemed extremely impressed that my box could be such an adept pupil.

Next came the dancers, two wearing spectacular helmet masks and raffia costumes. By now it was evening and the dancing arena was lit by kerosene lamps. These attracted clouds of insects. A particularly large cricket landed on Jack's shirt. His reaction was immediate – it would make excellent food for the bush babies that he had collected the previous evening. With a swipe of his hand, he caught the cricket and, not wanting to disturb the performance, put it in his breast pocket and buttoned it down. A villager standing beside him noticed what had happened and furrowed his brow. A few minutes later the dance came to an end. Jack opened the same pocket and this time took out a small currency note to give to one of the dancers. The villager applauded wildly. Clearly if Jack could turn a cricket into money, we really *were* magicians.

After the performance was over, the Paramount Chief presented me with a splendid helmet mask that had been worn in the dance. It was the first piece of genuine tribal sculpture I had ever owned, and the foundation of a collection that was to grow only too large over the next fifty years.

* * *

Jack wanted to go to a patch of primary rain forest away to the east of the country where there were said to be black and white colobus monkeys. That suited us. The monkeys were just about big enough for Charles to get some pictures, limited though his lenses were, and we thought we might be able to find them in well-lit trees beside a river. Getting there was not easy. It would take a couple of days driving along rough dirt tracks and there had, in any case, been a lot of trouble with rebellious local people. However, all now seemed quiet and leaving Alf to look after the collection, we set off. Eventually we reached the small Government station, built in a clearing in the forest and went immediately to pay our respects to the District Commissioner. He was exactly as I had imagined an empire-builder to be. Tall, barrel-chested, bronzed, wearing khaki drill with very short shorts.

'I've given you an empty bungalow,' he said gruffly, 'with a couple of servants to look after you. My ADC has the bungalow next door. He'll keep an eye on you.'

While we were still unpacking empty cages, nets, camera gear and stores, the ADC appeared. He was the exact opposite of the DC, small and pale, one of those people that the sun does not love. He hospitably invited us to join him for dinner that evening. His bungalow was a typical West African Government building, with huge rooms surrounded by enormous verandas. We dined in some state, waited on by dignified bare-footed African servants wearing smart scarlet cummerbunds.

I started dinner time conversation by asking the ADC about the colobus monkeys. Had he seen them? No, he hadn't. What about the birds? Well, there were some around but he didn't know what they were. Were the local people of particular interest? Not as far as he knew.

The conversation faltered. I was wondering what interest he had that had induced him to start on a career as a colonial officer when he said abruptly, 'Are you interested in model railways at all?'

'I'm sure they are fascinating,' I said, cautiously and somewhat baffled, 'but I'm afraid I don't know much about them.'

'Perhaps you would like to see my set-up.'

He got up from the table and I followed him into the next room which was even bigger than the dining room. And there, running on a specially built ledge all around it, was a model railway track. There were points, signals and several stations. One section of the track ran through a little landscape with model cows standing beside trees made from bits of sponge painted green. Close to the door there was a control panel, complete with rows of switches and levers for the points.

I searched desperately for something to say and then I noticed that on the far side of this immense room, there were loose wires and a soldering iron.

'Are you building another control panel over there?' I asked.

He blushed slightly.

'Yes,' he said, 'I got engaged the last time I took home leave and my fiancée will be coming out in a month or so to join me. We are hoping that at weekends we will have enough spare time to run a complete timetable together.'

* * *

Everywhere we went, we asked about *Picathartes*. Eventually, in the east of the country we found someone who recognised Jack's drawing of the bird. It lived in the deep forest, he said, and built its nest on the sides of immense rocks. But it was a dangerous creature and not to be interfered with, for it was the servant of a one-eyed, one-legged devil that lived inside the rocks. Whether this was a story aimed at increasing the reward Jack would pay to anyone who showed us the bird it was difficult to tell, but after some negotiation, he agreed to take us to the site.

We walked through the forest for an hour or so and came at last to a great overhanging rock face. On the side of it, eight feet above the ground, were three nests, made of mud, like that of a swallow but a foot across and eight inches deep. There was no sign of the birds themselves, but one of the nests contained a couple of eggs and they were still warm. The site was certainly occupied. Filming was not going to be easy for the forest was dark, but there seemed a good chance that the rays of the sun might just shine through one gap between the branches of the tree and the rock face in the mid-morning. That afternoon, we built a hide using sprays of leaves draped around a nearby bush.

Early the following morning Charles and I returned. As we got close, we caught a glimpse of a couple of the birds as they flew away. We sat down in the hide and waited. After only a few minutes, one of the birds reappeared. It was, as its common name suggested, about the size of a crow with a blackish back and white underparts. But its most striking feature was its extraordinary bare head. This was bright yellow except for a black patch at the back which made it look as though it had a small skull-cap fitting over its ears. It perched high in the trees. Within seconds another joined it. One of them flew down to the ground not far from us. Then it sprang up and landed on the edge of one of the mud nests. Charles' camera whirred and made a noise that I thought was alarmingly loud, but *Picathartes* seemed to take no notice and eventually settled down on its eggs.

We stayed for two and a half hours before the sun moved behind the rock so that the nests were in shade and too dark for our film. We went back to rejoin Jack. That night our guide returned to the rock and netted one of the birds. So Jack at last got a living specimen of his mysterious bird.

Keeping *Picathartes* alive was the next problem. It was very fussy about its food. Alf tried every thing he could think of, but the only things that interested *Picathartes* were small frogs the size of a thumb nail. It needed at least a dozen a day and Alf organised local children to keep him supplied until gradually he managed to wean the bird on to a more easily obtainable diet.

We had now been away for two months. The animal collection was about as big as Alf and Jack, with Charles and me acting as unskilled labour, could properly care for. The whole collection was loaded into a huge rickety lorry and we drove back down to Freetown. There Jack booked space in a freight plane that would take us and it back to London.

We had three days before the plane left and I decided on one last jaunt. I thought we had enough film sequences for the six programmes I was committed to produce but we were decidedly short of big animals. No lions or leopards, giraffes or rhinoceros. Such creatures simply didn't exist in the Sierra Leone forest. But there was one possibility – the pygmy hippopotamus. It is only found in one or two West African rivers and looks significantly different from the well-known species found elsewhere in Africa. I couldn't hope to catch one, but I thought we might get a few shots which would come in useful. So leaving Jack and Alf to make final preparations for the journey back, Charles and I set off for the swamps near the Liberian border.

We managed to hire a small launch with an outboard engine and for two days travelled up the river asking if anyone had seen hippos. No one had.

On the second night, the last we would have out in the African bush, we drifted slowly down the river through mangroves towards the village where we had left our truck. Charles and I lay on the roof of the launch looking at the glory of the stars in the cloudless black-velvet sky. On either side of us, and even brighter, lines of fireflies in the mangroves switched the lights in their abdomens on and off. I went through in my mind all the sequences we had shot.

'You know,' I said to Charles. 'I reckon we are going to be able to get away with it. And if we play our cards right, we might even bamboozle them into letting us do another of these trips.'

'That's okay by me,' said Charles and he went to sleep.

* * *

The first *Zoo Quest* programme was transmitted, live, six weeks after our return. It featured chameleons, emerald starlings, weaverbirds and pythons. Jack did the talking in the studio and handled the animals. I directed the cameras from the control gallery. But Jack was not well. He had been feeling ill and feverish in Africa and had put it down to an attack of malaria which he believed he could throw off. Now the stresses of the studio together with his illness, whatever it was, made him unsure and hesitant in presenting the programme. There was a real question as to whether he should appear again the following week. But a decision on that score was not needed. He worsened seriously. He was taken to the Hospital for Tropical Medicine and doctors there said he would have to stay for some time while they took tests. The second programme was already advertised in *Radio Times*. Someone else had to take over the job of talking in the studio.

By this time Mary Adams had moved on to other things and the new Head of Talks Department was a gentle but shrewd ex-newsman from the BBC's office in New York, Leonard Miall. He decided that for the second programme, I would have to leave the control gallery and take Jack's place on the studio floor and that Nancy Thomas who had just joined the Talks Department from Planning, would take over the camera direction. Cyril

Jackson, in his capacity as keeper of departmental finances, summoned me.

'Just one thing to make clear about your studio appearance,' he said. 'It comes in a special category in my books called Staff – No Fee.'

* * *

A week or so earlier, another series about African wildlife had also just started on television. It was called, accurately enough, *Filming Wild Animals* and featured a Belgian film-maker Armand Denis and his blonde and glamorous wife Michaela, who had both lived in Kenya for many years and had an extensive library of wildlife footage they had filmed over that time. One programme in their series had already been shown. Another was scheduled to be transmitted three days after the second of the *Zoo Quest* programmes. This singularly inept piece of scheduling was caused by the fact that my series was studio-based and came from Talks Department whereas *Filming Wild Animals* was entirely on film and purchased from an outside contributor. It was therefore handled by a different Department. I feared that our film would suffer badly by comparison with the Denises' shots of big dramatic animals, but I did not expect the trouble this juxtaposition caused.

The day after the second *Zoo Quest* episode was transmitted the Denises, who had come to London especially to launch their new series by talking to the press about it, rang the Controller of Programmes, Cecil McGivern, and demanded that *Zoo Quest* be immediately cancelled. There must not be another programme about African natural history to compete with theirs. McGivern decided that the best way to solve the problem was to have a conciliatory dinner to which the Denises and I would both be invited.

It was held in a rather shabby garage-like room next to the Lime Grove canteen. I waited there with McGivern, Leonard Miall, a couple of planners and various other Television Service dignitaries. Armand and Michaela arrived full of affability to show what reasonable people they really were at heart – until, that is, they reached me. I held out my hand and Michaela spun on her heel and walked away. It didn't bode well. We sat down to dinner.

I was placed opposite Michaela who pointedly talked to those on either side of her, and indeed on either side of me, but continued to ignore me until suddenly, half way through the first course, she leaned across to me, eyes blazing, and said with all the venom she could muster 'Who is the spy who sold you my script?'

It seemed that the sequence about weaverbirds that we had included in the first *Zoo Quest* was similar to one that they were going to show in their next programme. That was certainly unfortunate. But it was hardly surprising that both commentaries sounded similar. There is only so much one can say about weaverbirds weaving. I did my best to apologise and explain, but Michaela was not to be placated. Eventually, Armand joined in and said in his rather engaging Belgian accent 'Zat is enough, Michaela.'

'But that man has stolen our script,' shrieked Michaela.

'Enough!' and Armand took her by the elbow and led her out of the room.

He came back a few minutes later, and sat down and we resumed our meal. Shortly after that, Michaela too returned and sat down as though nothing had happened. She had made her point and in the years that followed our two series were scheduled more sensibly.

Zoo Quest continued with its run and began to get good notices in the papers. In the third programme, I introduced a young chimpanzee who had been presented to us as a gift for the Zoo from one of the District Officers we had met. She and I had become particular friends and she behaved so engagingly during the programme that viewers were almost as smitten by her as I was. At the end of each programme, I explained that we had still not managed to find the main objective of our search, the little-known *Picathartes* bird. But, I said, if viewers watched again the following week they would find out what happened next. The size of the programme's audience grew. Just before the fifth programme, I was travelling with Charles in his open red two-seater car down London's Regent Street. A big London bus drew up alongside us at the traffic lights. The driver wound down his window and leaned out.

''ere, Dave,' he said, 'are you going to catch that *Picafartees gymno-bloody-cephalus* or aren't you?'

I reckoned that now was the time to suggest to Leonard Miall that we should try to make a second series.

3

Guiana

Having been to Africa, I had no doubt as to where we should go for a second *Zoo Quest* trip – to South America, the other great jungle-covered continent. The last programme in our Sierra Leone series was transmitted on January 25th 1955. Exactly two months later, we set out again for Guyana, then still part of the British Commonwealth and known as British Guiana. Jack seemed to have recovered from his mystery illness and he had been given permission to take part by the Zoo's doctor. With him came Tim Vinall, one of the Zoo's most experienced Overseers, who was to set up a mini-zoo in the capital, Georgetown, and look after the animals as Jack collected them in the same way that Alf Woods had done in Sierra Leone. Charles and I as before would film the whole enterprise.

Today, anyone starting on such a trip would, of course, have carefully researched scripts, long lists of contacts and pages of detailed itineraries. There would have been lengthy telephone conversations by satellite half way round the world and exchanges of e-mails with scientists defining exactly the optimum times to film particular species. Chartered aircraft and ferries would have been booked, accommodation reserved, detailed costings approved, film editing suites and recording theatres allocated and completion dates agreed. In March 1955, it seemed sufficient for me to say airily that we would be back some time in June with, I hoped, enough material for six half-hour programmes; and for Cyril Jackson, on behalf of the Talks Department, to tell me not to spend more than a thousand pounds while I was away.

The limping nature of air travel at the time can be judged from the fact that to get to Georgetown, we had to travel first to Amsterdam, where we spent the night, and continue the following day by way of Nice, Madrid, Lisbon and the Azores. Then, as I noted excitedly in my journal, 'we come to the big hop – the Azores to Surinam. Into the tropics at one go and overnight.'

From Surinam, we took another plane to Georgetown. The town stands on the coast close to the mouth of the Demerara River. There was little here that was authentically South American. The countryside was dominated by vast fields of a giant grass that came originally from India – sugar cane. The canals that ran through the plantations and the defensive walls that prevented the sea from flooding them were built by the Dutch, for they had been the first to colonize the area, in the nineteenth century. And the people we met were British, Africans from the West Indies, or Indians from India. Of the original inhabitants of the area who, due to misapprehensions on Columbus' part as to where he had landed, are of course also known as Indians, we saw no sign. Nonetheless, there was enough of the continent's native wildlife even here to provide us with exciting sequences.

Many of the creeks that wind idly through the plantations towards the sea provide a home for a particularly interesting bird, the hoatzin. Birds, as fossils make very clear, evolved from reptiles. Over millennia their forelegs sprouted feathers, lost several of their digits and changed into wings. The hoatzin shows how such ancient birds might have appeared at an intermediate stage in the process for they still have claws on the fore-edge of their wings. You can't see these easily in the adult bird for they are buried deep in the feathers. But they are very conspicuous in the unfeathered chick which uses them for clambering around in trees in just the sort of way one imagines a creature that was half-reptile half-bird might do.

The adults were easy to find. They squatted in some numbers among the dense beds of a kind of giant arum that fringed the creeks, plucking at the leaves with their large bills. They were big birds, the size of chickens, rich chestnut in colour with glittering eyes rimmed with a bright blue skin and permanently erect bristly crests. As they live entirely on leaves, they have to eat a great deal to get enough nourishment so they have very large stomachs. In consequence, they are rather clumsy in the air and Charles had little difficulty in getting all the film of them that we wanted as we chugged slowly up the creeks. But what we really needed to film were the chicks.

I spotted a likely nest through my binoculars and we rowed over in a dinghy to see how we might best film it. As we neared, two little chicks craned down to look at us from the untidy raft of twigs that constituted their nest. At the first touch of our bows on their bush, they dived straight over the edge of their nest and into the water beneath. Valiantly flapping their little semi-naked wings, they floundered towards one of the branches of their bush that dangled in the water and, to our delight, clambered on to it, visibly gripping it with the little claws on the front edge of their wings. It seemed clear that we would have no trouble in filming this fascinating behaviour. We've only been here a few days, I thought complacently, and we are going to get another television first. The nest was not ideally placed for filming, but I was sure we would find another.

But we didn't. We searched hard, but we never saw any more chicks and when we returned to the first nest a few days later it was empty. It would be twenty years before I got another chance to film that remarkable performance.

Jack's main objective in this cultivated coastal country was to catch a manatee. These creatures, also known as sea-cows, drift along the bottom of the creeks, ripping up the vegetation with their cleft muscular upper lips. We had little hope of filming them for we did not have any underwater gear and the creeks where they were said to live were hopelessly cloudy. The most we saw of them was a pair of nostrils that very occasionally and totally unpredictably, broke the surface of the water, snatched a breath and disappeared. Jack however let it be known among the fishermen that he was in need of a manatee. Before long a man arrived to tell us that he had accidentally captured one in his nets and had dumped it in a nearby pond in case we wanted it.

We went along with Jack to watch its collection and filmed a great deal of splashing and shouting as men waded around the pond groping in the café-au-lait water for the invisible monster that was cruising around their legs. Eventually one of them managed to tie a rope around the base of the manatee's tail. Soon it had been hauled out and was lying motionless on the bank, like a large sack of wet sand, with a spade-like tail at one end and an immense bristly moustache at the other. The municipal water lorry was summoned. The manatee was heaved into it and driven off to the Georgetown Zoo for temporary lodging. There it stayed, munching its way through endless supplies of lettuces until the end of our trip when it was transferred to a canvas swimming-pool specially erected for it on the deck of a merchant ship and taken back with the rest of the collection to London.

* * *

Cultivated land fringed the Guianan coast but nowhere was the strip more than eight miles wide. Inland from it lay rain forest that stretched for a thousand miles into the heart of Brazil. That green blanket, however, is not entirely unbroken. Two hundred miles south, it is replaced by a patch of open grassy plains, the Rupununi savannahs. They are the home of animals that can be found nowhere else in Guiana. We clearly had to go down there and see what we could discover.

A plane flew to Lethem, the main settlement in the Rupununi, once a week. It was piloted by an elderly American, Colonel Art Williams. No one was quite sure how old he was but it was rumoured, darkly, that he only managed to keep his pilot's licence by having it renewed in some small Central American republic where officials were not too particular about such pedantic details as date of birth. He was a tall, grey-haired, leathery-skinned man wearing a khaki eye-shade and deep green sun-glasses. His plane was an antiquated twin-engined Dakota, the same kind as he was said to have flown throughout the Second World War. He invited me to sit alongside him in the co-pilot's seat.

With Charles and Jack and all our baggage and stores strapped down in the back, we thundered along Georgetown's bumpy strip. The Dakota seemed to have more parts capable of independent movement than I had seen on any other plane. Separate panels on the fuselage and wings shuddered and trembled. Even the dials in front of me seemed to be wobbling. As the engines roared at full throttle, I watched a line of palm trees beyond the far end of the strip approaching us with alarming speed. The Colonel was leaning forward with his jaws clenched tight. Suddenly when it seemed that we must have run out of runway and were about to crash into the trees, he pulled the joystick towards him, and leaned backwards. The plane surged steeply into the air. At this critical moment the Colonel started to snatch frantically at his breast pocket.

'Can't find my gol'darn bifocals,' he shouted.

To my relief he managed to extract them from his pocket and while we were still climbing steeply upwards, changed them for the pair he had preferred for the take-off. Perhaps I looked a little shaken. At any rate he thought it necessary to add an explanation.

'In bush flying,' he yelled, 'I always like to get the old girl moving as fast as possible along the runway, so that if she loses one engine, I can still get airborne.'

The Rupununi was the land of the Melvilles. Back in the 1890s, H.P.C. Melville, a young prospector born in Guiana of Scottish parents, was travelling up one of the colony's main rivers, the Essequibo, looking for gold and diamonds. When he was many days' journey away from the coast, he fell ill with malaria. A travelling party of Wapisiana Indians on their way to the Rupununi, found him collapsed in his camp on a sand bank. They took care of him, dosed him with their own herbal medicines and took him with them to the Rupununi. He liked it so much he settled there. He bought hammocks, pottery and other things that the Wapisiana people made and took them back down-river to the coast where he sold them. In return he brought back steel fish-hooks, axe-heads, bush knives and other things the Indians needed. He married a Wapisiana girl. He bought cattle from Brazilian territory on the other side of the Takutu River which forms the frontier between the two countries and he started a ranch. His first wife bore him five children. When she could no longer do so, he took another who bore five more. And there, in his own private empire, he lived in virtual isolation until 1919. Then a long track was cut through the forest that connected the savannahs to the coast two hundred miles away to the north. That allowed Rupununi cattle to be driven up to Georgetown for sale and the ranches began to flourish.

In the late 1920s, with his ranches well established, Melville left the savannahs, and sailed off to Scotland, the land of his fathers. The story goes that when he first arrived, he was baffled by the absence of hammock hooks in his hotel bedroom and for the first few nights slept on the floor. However that may be, he soon acclimatised and before long he had acquired a Scottish bride. When the news reached the two old ladies back on the savannahs, they are said to have smiled and said they were not in the least upset. After all, they had had him while he was young. Thirty years later, at the time of our visit, his sons and daughters still occupied all but two of the ranches on the Rupununi.

Teddy Melville, who ran the hotel at Lethem, took us under his wing. He asked Jack to list the sort of animals he was looking for. Giant anteaters were at the top of it. That, Teddy said, was easy. And indeed it was. The next day he sent out a couple of the Wapisiana cow-hands on horses. We followed in his jeep. Within an hour, they had found one. It had been sleeping within a patch of tall grass, sheltering from the sun beneath its long banner-like tail of coarse bristly hair that it had laid over itself like a blanket. It shambled off across the savannah. We started to run after it. It looked harmless enough. After all, you can't be bitten by anteaters because they have no teeth. Their jaws are no more than a long tube housing the black cord-like tongue with which they flick up insects. But Teddy told us to be careful. A giant anteater may not be able to bite but it can hug. Its forelegs are bowed and very muscular as they need to be in order to rip apart termite hills. And they have huge curved claws. Better to leave it to him. Whirling his lasso around his head, he neatly roped the anteater round its neck and soon it was tethered in the back of the jeep and on its way back to Lethem. There Jack decided that raw minced meat mixed with eggs would be a suitable replacement for the termites the anteater normally ate and it soon settled down.

Next – an anaconda. A suitable one was located in a swamp and duly caught by a couple of men who between them managed to bundle it into a sack. Then word came that Indians, fishing in a nearby lake, had found a huge caiman, the South American equivalent of the African crocodile. It was lying in a hole at the lake's edge. Away we went once more. All we could see of it was the tip of its jaws in the entrance to the hole. A noose was carefully negotiated over them and drawn tight, so preventing it from biting anyone. Then fishermen pushed a long pole into the hole along the animal's back and slowly pulled it out, tying the reptile to it every yard or so. It proved to be ten feet long.

After a week or so at Lethem, we drove fifty miles north across the savannahs to Karanambo, to stay with one of the few Rupununi pioneers who was neither descended from, nor married to, a Melville. Tiny McTurk, born and bred in Guiana, had come down here from the coast thirty years earlier. He had built a temporary house of thatch and mud brick. Just as soon as it was finished, his newly-wed wife, Connie, came down to join him. They still lived in it. He had added a few spacious verandas in which to store guns, outboard engines, books, felling axes, Indian feather headdresses, grindstones, jars of water, saddles, fishing lines, paddles, and other such things that a man needs to have at his elbow in these parts. There were orange-boxes to sit on and he apologised for them. They weren't as good as the old ones before the war, he said, neither so steady nor the right height. But that didn't really matter, because most people, he reckoned, would prefer to lie in one of the several hammocks that hung permanently between the house-posts.

Tiny explained that he had chosen the place because the fishing in the Rupununi River close by was so good. There you could catch arapaima, one of the largest freshwater fish in the world, growing to over six feet long. He had also brought in a few cattle, though to begin with, keeping them was not easy. For the first few years, he said, he had to shoot a jaguar every couple of weeks or so. Otherwise he would have lost the lot.

He took me down to a swamp to look at some of the birds that lived there. To reach it, we had to walk through a patch of forest. Walking with him was an education.

'Know what this is?' he asked, pointing to a small stick with a hole in it. 'The work of a carpenter bee.'

'Smell anything? Howler monkeys are sleeping close by.'

'See this track? – savannah deer.'

'Know what ripped the bark from this tree? – tapir.'

'And this smell? – something dead; not very big; probably a lizard.'

'Now *that* smell is the smell of a dead arapaima. Indians have been here. Yes, here's their encampment – and there's the arapaima skeleton. They were here yesterday.'

The lake, when we reached it, was thick with egrets, jabiru storks, ducks, lily-trotters and – loveliest of all – roseate spoonbills, feeding by swishing their bills from side to side in the shallows. The two of us crouched in the bush peering through a concealing curtain of leaves. I had never seen such a spectacular assemblage of waterfowl. Tiny must have seen it a thousand times. I think the sight still thrilled him just as much as it did me.

That evening after a meal of fresh fish and delicately roasted duck cooked for us by Connie, we lounged in his hammocks listening to his stories. A year or so ago, he said, a Wapisiana man had come up to him as he sat fishing and asked him if the three small crystals he had picked up were diamonds. Tiny looked at them, told him that they were nothing but worthless quartz and threw them into the river. 'They *were* diamonds, of course,' he said. 'But I wasn't going to have all the riff-raff from the coast down here searching for more.'

I recorded a number of his stories on my clumsy tape machine, night after night. Maybe sound radio would be able to broadcast them. Sadly, however, I could not show him telling them on television. We still, in the 1950s, had no way of synchronously linking sound tape with 16mm film.

* * *

Our main work, however, was in the rain forest. For that we flew up to Kamarang in the north-west of the country. There a District Officer, Bill Seggar and his wife Daphne had just established a station beside a stretch of the Mazaruni River that was long and straight enough to allow a float plane to land on it.

As far as I knew, no films about South American natural history had ever been shown on television, in Britain at any rate. So even the commonest creatures would be new to viewers and therefore interesting. Leaf-cutter ants were everywhere in the forest and the bane of Daphne Seggar's life since she was trying to establish a small orchard and a vegetable garden. Day and night, they marched along their trackways, holding above their heads segments of leaves an inch across that they had just scissored from a tree. These they took down into their underground nest where they chewed them up and used them as compost on which to grow the fungus they feed on. Charles filmed them as he had the driver ants in Africa, lying on his stomach alongside a column. As long as his tripod legs didn't cross the margin of their trackway they took no notice of him.

Sloths were not as easy to find, but once located, were equally tolerant of Charles and his camera. Two-toed sloths, then as now, are commonly kept in zoos even though they make somewhat unexciting exhibits since they hang upside down asleep for most of the daylight hours. When they do wake they lack charm, having wet snuffly noses and shaggy uniformly brown fur. The ones that were brought to us, however, were *three*-toed – and they are very different. Their fur is a delicate grey, mottled with patches of cream. The males have a brilliant yellow star-burst in the middle of their back. Some even have a vague greenish tint. That comes from microscopic algae that grow in grooves along their hairs. And they have most engaging furry faces, cream-coloured with small bright eyes and a pert black nose, all framed by a circle of orange-yellow fur. Speed is not one of their talents. They move one limb at a time and so slowly that you have no difficulty in catching them if they are within reach. As you unlatch them they blink at you with a distant but permanent smile, and an air of having just been woken up. But none of these charms had ever been seen in London Zoo – nor were they likely to be, for whereas the two-toed sloth will eat fruit and a wide variety of vegetables, the three-toed is more fastidious. It will only take leaves – and even then only a very few kinds that are virtually unobtainable outside the rain forest. So Jack did not collect three-toed sloths. But we could film them.

It is not easy for an unpractised eye to spot sloths in the forest, but the Akawaio Indians at Kamarang found them without difficulty, and could quickly clamber up a tree and bring them down. The first one they brought to us we re-hung in an isolated tree on the station. It was unlikely to risk descending and struggling across the ground back to the forest, so we were in no hurry to film it. We went off in a leisurely way to collect the camera gear and find ladders. When we returned, we noticed a small wet patch on its stomach. It had just given birth. An adult three-toed sloth may be charming, but a baby one is bewitching and we spent considerably more time filming and photographing mother and child than we really needed to have done.

* * *

The Seggars suggested that one of the first things Jack should do would be to tour the local villages, for the Akawaio people are enthusiastic pet-keepers. That suited Charles and me as well and day after day we travelled up the Mazaruni and several of its tributaries in canoes fitted with outboard motors.

Sometimes the going was easy. As we cruised, cormorants took off from their fishing perches along the banks and flew ahead of us. So reluctant were they to leave the river and so unwilling to allow us to get near them that twenty or thirty would accumulate in a flock flying ahead of us, until one summoned enough courage to plunge into the bush on the bank and the rest of them followed. As the rivers narrowed, so we would make our way through drifts of yellow blossom that had fallen on to the clear brown water from flowering trees arching above. Sometimes we came to rapids and had to get out and wade up to our waists to haul up the canoes.

The villagers we met were initially impassive and not given to extravagant gestures but they always received us most hospitably. Many women had tame parrots. Their husbands, returning from hunting in the forest, sometimes brought them back young chicks and their wives had reared the little birds, feeding them with chewed-up cassava. Curassows, like slim black turkeys, strolled around the village. There were tame capybara too. These are the biggest of all rodents, the size of pigs, and spend much of their time in the water. Most villages had young ones as pets with which the children would go swimming. Jack, as Curator of the Zoo's Reptile House, kept a specially sharp eye open for possible occupants of his own enclosures. He caught water turtles, Surinam toads that look as though they have been run over and squashed flat, and snakes of many kinds – lethally poisonous fer-de-lance as well as more amiable boa constrictors and young anacondas. Before long, he had almost as many animals of one kind and another as he could properly care for.

Charles and I after a week or so found a village where the people were particularly welcoming. We decided to stay with them for a few days so that we might get to know the people a little and earn their confidence. We filmed the children splashing and diving in the river with their capybara playmates. We recorded in detail all the stages of making a woodskin, a bark canoe, starting with the removal of the bark, dripping with sap, from a forest tree, and its curing over a fire, and ending with the shaping of its ends, the insertion of thwarts and its final launching. We watched the women weaving their bead aprons, which was all that many of them wore, and documented the way in which they squeezed the poisonous prussic acid from their grated cassava using long woven tubes before they baked it into cakes.

One day the men took us on a long march through the forest to see a cliff covered with ochre paintings, showing sloths, monkeys and other animals, geometric signs and a wide scatter of hand prints. We took detailed photographs, which later the British Museum took into their archives since the paintings had not been documented before.

The villagers patiently helped us find the nests of parrots and the sleeping places of howler monkeys. As always I was astonished at the sharpness of their eyes which enabled them to spot animals that were invisible to me until they pointed them out. Their hearing was also extraordinarily acute. They were able to detect and identify the faintest whistle which I could only just distinguish. But one of their talents I found completely inexplicable. They were able to predict the arrival of a canoe from down-river long before it arrived. 'Canoe come,' they would say in a conversational way, and invariably, exactly an hour and a half later, a canoe driven by an outboard engine would appear around the bend in the river immediately downstream. How they managed this puzzled me for a long time but eventually I discovered the answer. The river beside which we were staying meandered a great deal. One of its bends quite far downstream came so close to the village that for a minute or two an outboard engine coming round it could just be heard. As the canoe travelled on it became inaudible again until, an hour and a half later, it swept round the bend immediately below the village.

The night before we left, our friends gave us a farewell feast. We all sat in a circle in one of the communal huts. We ate fish that had been caught that afternoon in the river. To drink we were given cassiri ladled out from an enormous calabash. I had heard about this drink. Its main constituent is cassava. To make this ferment, women chew cassava bread and spit it into the calabash. It is then left for several days bubbling gently away. I was handed a gourd-full. It looked like a lumpy white liquid porridge. As I lifted it to my lips, its smell reached my nostrils and I felt my stomach heave. I knew that politeness would demand that I drank it all. I equally knew that I might not succeed in getting it to my mouth a second time without my losing control of my stomach. So I drank it down in one long gulp and smiled wanly. The man who had given it me smiled back. Clearly I liked it – and he handed me another. That I felt able, politely, to decline.

* * *

Our filming went so well that after three weeks based at Kamarang, I considered that we had sufficient footage to sustain the animal-collecting theme to which we were committed by our title of *Zoo Quest*. And we still had a week to spare. There might just be time for a trip that had nothing to do with catching animals but might nonetheless be smuggled into one of the programmes.

We were within fifty miles of one of the most romantic mountains in the world – Roraima. It is the biggest of a group of table-shaped mountains, called locally *tepui*. Its sides are vertical rock precipices two thousand feet high. When Europeans first saw it in the nineteenth century, it was considered to be unclimbable. Scientists speculated that all kinds of animals from earlier geological epochs, that had died out elsewhere, might still survive on its summit plateau. Conan Doyle, it is said, was inspired by these thoughts to write his novel *The Lost World,* in which explorers found dinosaurs and pterodactyls roaming around on the top of just such an isolated table mountain in South America. In fact, when Roraima was climbed in 1884 there were no such creatures there, but even that knowledge cannot rob the mountain of its magic.

The standard route up the mountain is along a ledge that slopes up its south-western side. From all accounts it is an exhausting climb though not a technically difficult one. But to reach that ledge from where we were at Kamarang would take far more time than we had. Even so, we thought it would be wonderful simply to see the mountain.

Charles and I pored over Bill Seggar's maps. There had been no aerial survey of the country between us and Roraima, so their accuracy was not necessarily great. As far as we could see, the Kako, a river that flowed into the Mazaruni a few miles upstream from Kamarang, might get us close enough to see the mountain. Bill could not tell us whether it would or not. He knew of no one who had ever been up the Kako for any distance. That only made it more exciting. So Charles and I loaded a canoe with all the food we had left and set off.

The first day we made good progress. The second brought us to the edge of a savannah. We looked across and there was the monumental rectangular outline of the mountain, hazy blue in the distance. We were still twenty miles or so from it but it still looked immense. Maybe a few more hours of travel up the river would bring us even closer. But the Kako, against all indication from our maps, turned north away from the mountain. If we went on for much longer, we risked missing the float-plane that had been chartered to take Jack, all his animals – and us – to Georgetown. We turned back. I was not going to set foot in the Lost World – on this trip at least.

* * *

A few days later, Charles and I were on a plane flying back to London. Jack and Tim would follow by sea in due course with the animal collection. Jack, however, had not been feeling well for some time. Soon after we left, it became apparent that his old illness had returned. He became so sick that he had to fly back to London urgently to see specialists once more and someone else from the Zoo came out to help Tim care for the collection on the return voyage.

Back in London, Jack went straight into hospital. Once again he was unable to introduce the television programmes of which he should have been the star and I had to present them while he watched from his hospital bed. He died, his illness still not identified, a few months later. He was only 47.

4

Infant Empires

Television was now growing fast. As broadcasting hours and the number of programmes needed to fill them increased, so producers began to specialise. Empires started to grow in the Talks Department. Each was centred around a programme and its producer. *Panorama* had started out as a general interest programme but it was taken over by Michael Peacock, my fellow student on the training course and he turned it into a weekly examination of current affairs. A young, rebellious and brilliant Welshman, Donald Baverstock, recruited from the Overseas Service, had been producing a short ten-minute programme every weekday called *Highlight*. He had done so with such verve that he was given the job of creating a nightly half-hour programme called *Tonight*. That enabled him to assemble a team of people as iconoclastic as himself, among them a future Director General, Alasdair Milne, and a brilliant writer, Tony Jay, who invented one of the wittiest of television comedies, *Yes, Minister*.

Huw Wheldon arrived in the Department from the Corporation's Public Relations Department. A tall voluble Welshman, he was older than most of us and an ex-serviceman with an impressive war record. On D Day he had parachuted into France with the Airborne Division and been awarded the Military Cross for bravery under fire. He had already established a reputation in Children's Programmes, presenting a series called *All Your Own* in which he interviewed children about their hobbies and achievements. His manner was peremptory and no-nonsense to which his young interviewees, for the most part, responded with spirit. With his military connections he managed to persuade General Sir Brian Horrocks to make programmes about the important battles of the war, including the action at Arnhem in which the General himself had played a crucial part.

The Department wanted to start an arts programme, the first in British television. It was to be fortnightly and called *Monitor*. Several pilots were made which were generally considered to be catastrophic and those responsible were moved to other things. How justified this was, most of us could not tell, for pilots, like the programmes, were live and there was still no way to record them electronically, so most of us, including me, did not see them. Then, surprisingly to those who had so far only known Huw for *All Your Own* and his military programmes, the job was given to him. He too recruited a substantial team which included many who subsequently became key figures in the British arts world, among them Melvyn Bragg, John Schlesinger, Ken Russell, John Drummond and Humphrey Burton.

I felt it was time I started to follow the same empire-building trend. Since my interests and qualifications were in natural history, I suggested to Leonard Miall that the Department should have a regular programme, perhaps topical and monthly, about wildlife in the British countryside, while at the same time I could continue with *Zoo Quest* expeditions every year or so. This idea got nowhere. Later I discovered the reason. Down in West Region, Desmond Hawkins, a radio producer who had more or less invented natural history programmes on radio, was busy manoeuvring to extend his output into television. He was hampered by the fact that BBC Bristol had not yet got a television studio. Indeed, even the transmitter network had only just reached the west country. Desmond, however, was such a wily operator, so skilled in BBC politics, that this was no real obstacle to him. Bristol did have a large sound studio for concerts and recitals. He devised a way by which an Outside Broadcast Unit and its cameras could be driven in alongside it and so effectively turn it into a television studio. So the *Look* series started, introduced and narrated by Peter Scott, already famous as a naturalist from Desmond's radio programmes.

A clash seemed to be on its way. Desmond thought of a flattering solution to the problem. He suggested that I should go down to Bristol and take charge of a new group he would call the Natural History Unit. It wasn't an idea that appealed to me. Jane and I were well settled in London and had started paying for a house in Richmond. We now had a second child, Susan. Robert, our first, was starting school. We both had ageing parents and brothers or sisters living in either Richmond or nearby Kew. And anyway, London was where exciting things were happening in television, even in spite of Desmond's enterprise in Bristol. I wanted to stay put.

A meeting was arranged between Leonard and me and the Head of Programmes from Bristol who brought Desmond Hawkins with him. We agreed that British natural history should become the preserve of a Natural History Unit in West Region. In return, Talks Department in London could have a group to be called the Travel and Exploration Unit. Its brief would be to make programmes about exploration overseas. It might even include exotic natural history in the shape of further *Zoo Quest* series. It would consist of me and a secretary and perhaps – in due course – if all went well – an assistant producer. As empires go it was not much, but it was enough for me.

* * *

I felt reasonably confident about handling the exploration side of my brief. If I had had a passion as a boy that rivalled that for natural history, it was for mountaineering. My favourite aunt, my mother's sister Margaret, had married a mountaineer of some distinction, a burly black-haired Yorkshireman called Gilbert Peaker. He had made some spectacular ascents in the Alps before the war. He was a champion marathon runner. By profession he was a mathematician and on the outbreak of war he was drafted into the Treasury. Since he could no longer get to the really high mountains he so adored, he spent all his spare time in British hills, rock-climbing. He asked my mother whether I would like to go with him. I was fourteen. I thought it was a wonderful idea.

We climbed mostly in Wales. Usually we went there by the erratic war-time trains, taking bicycles with us and pedalling up the Nant Ffrancon Pass from Bangor to stay at Helyg, the Climbers Club Hut in the Ogwen Valley. Gilbert taught me how to walk properly over rough country, slowly but rhythmically. He persuaded me that drinking at every mountain stream one crossed on a hot day was a waste of time since as fast as one took in water, one sweated it out again – not a view shared by medical fashion today. He demonstrated, to my initial dismay, that if one was caught in a short shower without adequate rain gear it was better, even in chilly Easter weather, to take off one's shirt and put it in a rucksack so that one had dry clothing when the shower was over. And it was in his company that I first experienced the deep pleasure that follows in the wake of great physical exhaustion.

At Helyg I met and even climbed with some famous mountaineers. Most celebrated of them was Eric Shipton, who had been on Everest in the 1930s and had explored both the Himalayas and the Karakorum. He had made his reputation climbing immense peaks of snow and ice and would not normally have bothered to tackle the gymnastic problems that we sought out on the rock faces of the Welsh hills. Using this as an excuse he charitably allowed me to lead him up some quite difficult climbs. I nearly burst with pride. I even helped Major J.O.M. Roberts of the Gurkha Regiment, who after the war was to become one of the leading organisers of Himalayan expeditions, to train British paratroopers in the basic essentials of ropework in climbing. At the end, he gave me his camouflaged parachuting smock which I wore on every outing to the hills for years thereafter.

I continued to climb with Gilbert and others throughout my years at university and while on leave from the Navy, but when Jane and I were married in 1950 I decided to give it up. It seemed scarcely fair to her, who did not climb, to spend my annual two-week holiday on cliffs while she waited at the bottom with a small child. Nonetheless, I still had mountaineering friends, and enough contacts to find my way around in that community and assess what opportunities it might offer for programmes. So I took on the responsibility of producing programmes about major mountaineering expeditions when they returned to Britain – Charles Evans' team that climbed Kanchenjunga in the Himalayas, Vivian Fuchs and Ed Hillary's expedition across Antarctica.

* * *

I started a series called *Travellers' Tales* in which people who had returned from exotic parts of the world came to talk about their experiences and showed 16mm film that they had shot during their adventures. This required a sharp editorial eye. A self-proclaimed explorer of the deepest jungles of South America arrived with a can of film under his arm to tell me that he had reached such a remote part of the Brazilian rain forest that the animals there had never seen human beings of any kind and therefore were totally tame. I was intrigued and we went down to look at his film together in a small viewing theatre. After some protracted but disconnected shots of tramping through forest, we saw a grassy clearing.

'Watch this,' he said.

A savannah deer stepped out of the trees and walked towards the camera.

'You see! Totally fearless but totally wild.'

'Yes, but why has it got that mark of a halter around its neck?'

'Has it?' he said incredulously, peering more closely at the screen. 'So it has! You know, I never noticed that before.'

I believed him.

At around this time, an enterprising tourist operator in what was then still the Belgian Congo had cut a road into the rain forest and built a large hotel at the end of it. He had recruited a group of local pygmies who, on a rota system, provided special facilities for his guests. For a small fee, visitors could have their photograph taken with a group of them. For a rather larger one, they could go into the forest and visit a genuine pygmy encampment and watch a ritual dance.

I came to know one of these pygmies rather well. He kept appearing on the screen of our viewing theatre. He had a particularly gnarled face and was the smallest of his group. He was therefore always the one who was selected to stand next to any explorer wishing to record how small pygmies really were by sticking out his arm and tucking a pygmy beneath it. Little did these travellers realise that I regarded his face, grinning from within an armpit, as an automatic disqualification for showing the film of their unique encounter on television.

The age of cheap, worldwide air travel had still not dawned. Even the most unremarkable shots of places with romantic names – Constantinople and Lhasa, Bangkok and Macassar, Samarkand and Timbuktu – made beguiling viewing. Among the best of the Exploration Unit's programmes was a short series made by half a dozen students from Oxford and Cambridge who became the first group to drive overland from London to Singapore. In the 1950s the world had not yet shrunk into a global village.

5

Dragons

I was anxious to keep the *Zoo Quest* series as a regular event. The trips were far too enjoyable to let slip. Pondering on the location for a third series my mind went back to a book that had made a great impression on me as a boy – *The Malay Archipelago* by Alfred Russell Wallace.

Wallace was one of the greatest of nineteenth-century naturalist travellers. As a young man he developed a passion for entomology and yearned to go to the tropics. Without either the private money or the social connections of his great contemporary, Charles Darwin, he decided that he might be able to finance his journeys by collecting butterflies, beetles, bird skins and other natural history specimens to sell to collectors with more money – but less enterprise – than he. His first trip was a joint venture to Brazil with another self-taught naturalist with a similar idea, Henry Bates. After four years of gruelling work in the Amazon rain forests, he set off on the journey home. But after three days at sea, a fire broke out on board his ship and all his collections were destroyed. Wallace was lucky to escape with his life. Nothing daunted, after a short time back in England, he set out again, this time for the islands of South-east Asia.

I knew his book well enough to be sure that there was more than enough material for an exciting *Zoo Quest* series in the lands through which he travelled – today's Indonesia. The wood engravings with which it was illustrated were by themselves enough to convince anyone. There was a picture of a giant orangutan attacking one of his carriers; another of a huge python entwined round a post of his hut with one of his men pulling on its tail, trying to dislodge it. The frontispiece to the second volume was for me the most thrilling of all. It showed a group of birds of paradise displaying in a tree with, in the foreground, native hunters crouched on a platform, armed with bows and arrows.

Such animals were excitement enough, but there were also other dramatic creatures to be found there that even Wallace never saw. In 1910 a colonial officer, stationed in Java, then part of the Dutch East Indies, heard stories of a gigantic lizard that inhabited a tiny island called

Komodo in the middle of the island chain that stretches eastwards from Java towards New Guinea. He sailed there to investigate and brought back the first specimens known to western science of a gigantic monitor lizard. It was over ten feet long, the largest lizard in the world, and it quickly became known as the Komodo Dragon. With birds of paradise and a dragon to quest for, how could a series fail?

Charles Lagus agreed to take on another trip. The London Zoo, however, was not prepared to send anyone in Jack Lester's place, though they said they would gladly accept any creatures we might bring back. The collecting ingredient was a very valuable element in the *Zoo Quest* programmes. Not only did it provide a strong narrative line, but having animals live in the studio gave us those close-ups which we were still unable to get with the lenses and the film stocks that were then available. But who would catch the animals? Charles would be busy behind his camera. The only person available to do that was me.

I had learned something about the business from Jack. But not much. Then I heard that one of the great, almost legendary animal collectors, Cecil Webb, who had in the past brought back many birds of paradise, had just arrived at the London Zoo with another consignment of animals. I hurried off to see him. We lunched together at the Zoo. If, as some say, people tend to look like the animals they own, then Webbie, as everyone called him, was more like an adjutant stork than a bird of paradise. He was tall and dignified and the last thing he would indulge in was self-important display. He was full of advice and information – how to bottle-feed a cheetah cub, what to do when a duck-billed platypus went off its diet of worms, and how to pacify a boa-constrictor. He had been to the very places where Wallace himself had seen birds of paradise. He drew me a little sketch map of the village in the Aru Islands where he had captured specimens of the Greater Bird of Paradise and even told me the name of the headman. I went away my head full of adventures in tropical jungles, but no very clear idea of how I might wrestle with an orangutan or what to do when faced by a Malay bear.

Nonetheless, with what now seems to me to be absurd optimism, Charles and I set off on our third expedition together in September 1957. Neither of us spoke any Indonesian. Apart from a formal letter of accreditation from the London Zoo addressed to the Director of the Surabaya Zoo in Java, we had no personal contacts or introductions to anyone in Indonesia. We had been given visas by the Indonesian Embassy in London and they had promised to write to the Minister of Information in Jakarta to ask people to assist us, but that was about all the preparation I had been able to make. I imagined that, somehow or other, after landing at Jakarta in Java we would go north into Borneo and then make our way eastwards, filming and collecting animals as we travelled along the island chain, through Bali, Sumba and Sumbawa until we got to Komodo. Then we would carry on until we reached the Aru Islands and the birds of paradise. I reckoned we would be away for about three months.

* * *

The first hitch came even before we left London. I called at the Indonesian Embassy to ask what response they had had from Jakarta. At that time, telephoning there was impossibly difficult and somehow the letter they meant to write had not been written. But they would type something right away. Since the post took so long, it would be quicker and safer, they said, if we took it with us and delivered it by hand.

When we landed in Jakarta, our letter did us no good whatsoever. As the immigration official rightly pointed out, it was simply a request. No one in Indonesia had given us permission to do anything. Our filming gear was accordingly impounded. No one seemed able to explain what we had to do to get it back. I went up to see people in the Ministry of the Interior to get permission to travel from island to island. Could they have an exact itinerary of where we wanted to go and when? Well, I said vaguely though perfectly truthfully, we had not got a precise schedule, but we would like to make our way eastwards to the Aru Islands. Aru! That was certainly out of the question. The islands were close to the western end of New Guinea which then was still a Dutch colony. Indonesia was claiming it. Foreigners were not allowed to go so close to the frontier. There were plenty of other places we could visit. I repeated over and over again that we wanted to go to Aru to film birds of paradise which occurred nowhere else in Indonesia, but plainly no one in the Ministry believed me. Maybe, I said to Charles, they thought that no one in his right mind would want to travel so far just to film a particular kind of bird.

It was not until twenty years later, when a book about international espionage was published, that I discovered that Webbie, who had been so helpful, had himself during his travels been working for British Intelligence. Perhaps the officials of newly independent Indonesia had good reason to be suspicious of our motives. At any rate, we were expressly forbidden to go to Aru. So *Paradisea major*, the bird that Wallace had illustrated in his dramatic frontispiece, could not become our new *Picathartes gymnocephalus*. It was a major blow to our plans.

What could take the birds of paradise's place? There was still the

Komodo Dragon. So could we go to Komodo? Officialdom seemed baffled by this. Where was it? We pointed to a little dot between Sumbawa to the west and Flores to the east. Well, yes, there was nothing of any consequence there, so if we wanted to go there, we could.

Next, the Ministry of Natural Resources. They too were unimpressed by our letter from the London Zoo. Would I list the animals I proposed to catch? I groped around in my mind. Bears, orangutan, pythons maybe, I said airily. And a Komodo Dragon. The official did not know what that was, but the more I described it, the more he was impressed. But I overdid it. By the time I had finished, he was *so* impressed that his mind was made up. We would not be allowed to export such a creature. Nothing I could say would change his mind. It looked as though our expedition was a disaster before we had even shot a foot of film. I decided that we would have to go to Komodo and at least film its dragon – even if we were not in the end allowed to take one back to London.

It was becoming increasingly clear that applying for permits was a self-generating process. The more permits we asked for, the more permits were discovered that we needed. There was only one way out of this. Stop asking. So having at last extracted our equipment from Customs, we got on a train and started off eastwards with tickets to Surabaya.

* * *

Surabaya is the second biggest city in Java. Being near the eastern end of Java it was that much closer to Komodo and the wilder less populated islands of Indonesia. We took our letter of introduction from London Zoo to the Director of the Surabaya Zoo. He was welcoming and kind and showed us around his institution, but was unable to help us directly. He did, however, throw a little party for us in a Chinese restaurant and there we had a great stroke of good fortune. We met Daan and Peggy Hubrecht.

He was Dutch, born in Indonesia, educated in England and the owner of two large sugar-cane plantations just outside the city; she, a large imposing English lady with a society accent and a booming laugh. Over a meal together in a restaurant, they decided that we should be taken under their wing. First, we must stay with them. Second, we should need a jeep and there must be a spare one lying around the sugar factory. And third, could Daan come with us, at least for a short time? We said yes, with profound gratitude, to all three propositions.

The following day we started making plans. We wanted to go to Borneo, away to the north. Daan would like to come with us for a couple of weeks but couldn't leave immediately. In any case, it would take time to book berths on one of the cargo ships that sailed from Surabaya up to the east coast of the island. Meanwhile, we could hire one of his factory's jeeps. In that we could travel down to the eastern tip of Java where there was still some relatively wild country. Then we could cross the narrow strait to Bali where Daan had friends who could put us up. By the time we got back from that, he would have freed himself from his business commitments and would be able to come with us to Borneo.

The jeep turned up two days later. It was an elderly machine and had had an eventful life. The voltmeter on its dashboard, according to the name on its dial had started its life as part of an air-conditioning plant. There were strange improvised wires snaking around its cylinder head. The horn was sounded by touching a place on the shaft of the steering wheel, specially scraped clean of paint, with the bare end of a piece of flex. The process not only startled those ahead but also shocked the driver – literally though mildly.

The following morning, we thought we might accustom ourselves to its idiosyncrasies by taking it out for a spin. This was something of an anticlimax. Charles put it into gear, the engine roared but the machine itself remained totally stationary. Inspection proved that the half-shafts – rods that fitted through the hub down the centre of the axle and linked the wheels to the engine – were missing. Daan was not in the least put out. It was foolish of him to have left the jeep outside in the drive. Someone had stolen them. We would be able to find them down at the thieves' market, where the stall-holder who had them would keep them for a day or so. It was accepted that it was only fair to give original owners a chance to be the first to buy their own property. And there they were.

The day after that, Charles and I set off from Surabaya, driving eastwards. For the next three weeks we filmed Javanese wild life – pythons and pangolins, green tree ants and cockatoos. We watched and filmed turtles laying their eggs on remote beaches, and wild peacocks sailing across the sky trailing their huge trains on their way to roost in the tops of trees, a most surprising sight for those more accustomed to see them strutting around on the well-clipped lawns of English country houses. Eventually we arrived in Banjuwangi, on the far eastern tip of Java, and caught a rickety ferry that took us across the strait, less than two miles wide, to Bali.

Bali, at the time, was the very epitome of a remote oriental paradise. Its only airfield was so small that it was incapable of taking international traffic and there was only one hotel in its main town, Denpasar. It was famous for the splendour of its music, its dances, its painting and its temple ceremonials. Once, a thousand years ago, all of Indonesia had been ruled by Hindu kings who had their capitals in Java. In the fifteenth century, however, Islam began to invade Java from the west. There were violent battles between followers of the two religions. The Hindu sultans retreated eastwards, taking their courts with them and eventually crossed into Bali. The supporters of Islam pursued them no further. So five centuries ago, the island received a rich cultural transfusion that generated the love of the arts that still obsesses the Balinese today. Almost every village has its own orchestra, its gamelan. The instruments are mostly metallophones – like xylophones but with bronze keys – but a full-scale Balinese gamelan includes also bamboo flutes, huge gongs and drums. The music it creates is full of plangent ripples, subtle rhythms and crashing chords.

The gamelan plays an essential part in temple rituals and accompanies thrilling dances and theatrical portrayals of episodes from Hindu legends, so there was plenty of visual excitement for our camera to record. Even though we could not synchronise sound recording with film, I could still record the music 'wild' and fit it with the picture in a roughly correct position. I wondered whether I could really justify including such sequences in a television series that, according to its title at least, was concerned with questing for animals for a zoo. I decided to risk it. After a blissful two weeks, we drove back to Surabaya.

* * *

We stayed once more with the Hubrechts. Everything was arranged for the Borneo trip. Our ship was leaving in three days' time, bound for the port of Samarinda on the south-east coast of the island, and Daan had discovered the name of a man there who had collected animals for Surabaya Zoo. If we could recruit him, he would provide the expertise that we so clearly lacked.

Both Charles and I were anxious for news from home. There were several letters from both our wives, though the most recent was over two weeks old. We tried to telephone, but discovered that calls to Europe had to be booked at least three days in advance and take their turn on the land lines. Even then, many never got through.

Charles was particularly concerned. His wife, Biddy, was expecting their first child. The night before we were to set off for Borneo, however, a telegram arrived for him. Biddy had produced a daughter. We all went out for a celebratory dinner and set about composing a congratulatory message to send to Biddy in return.

'What is the baby to be called?' asked Peggy.

Charles had no idea.

We drank a little more rice wine as we considered a number of possibilities. None seemed to suit Charles.

'Well,' I said, 'we are sailing for Samarinda in the morning and I think that is a lovely name.' So we sent a telegram to Biddy welcoming the arrival of Samarinda and that indeed became one of the names of Charles' first-born.

We landed there four days later. With Daan to help us, we bought stores – sacks of rice, tins of meat, palm sugar wrapped in segments of banana leaf, strange fruits that I did not recognise, and sticks of tobacco and cakes of salt that Daan assured us we would need for trade with people in the interior. He also managed to find a launch that would take us up-river. It was called the *Kruwing* and came complete with a five-man crew.

The first night we moored beside a jetty outside a small village where we had been told we might find Sabran, the helper recommended by Surabaya Zoo. It was intolerably hot and stuffy on board the *Kruwing* and Charles and I put our camp beds on the jetty. The mosquitoes were abundant and ferocious but we had rectangular nets supported by steel rods that fitted on each corner of the bed. With them tucked in around us, we were reasonably immune from insects.

I woke in the middle of the night and saw on the roof of my mosquito net and silhouetted against the moon, a huge rat. The spring in the metal rods supporting the net was such that the rat was bouncing up and down to within a few inches of my face. The image was so surreal that for a moment I thought it was a nightmare and lashed out with the back of my hand. I caught the rat squarely on its underside and it sailed in a great parabola across the night sky and landed with a plop in the black river beyond the *Kruwing*.

The next morning I was woken more pleasantly. A young round-faced Indonesian man was peering down at me. 'Sabran' he said, pointing to his chest. Daan emerged from the *Kruwing* and with his help we discovered that Sabran had heard that people down at the jetty had been asking for him and he had bicycled through the night to find us. We quickly came to an agreement about wages and Sabran joined us.

Talking to him was not easy for I still had not acquired much conversational Indonesian, but he and I soon evolved a pidgin of our own which served our particular requirements. One morning as we were steaming slowly up-river, he called me and pointed excitedly to a tree on the bank.

'*Burung ada,*' he said. That I understood. There had been a bird there.

'*Apa?*' I said, meaning – what kind of bird?

His answer to that, however, defeated me. He repeated it several times but I was baffled.

Then he said '*Irena puella puella.*' And that I understood immediately. It had been a fairy bluebird. Sabran knew the scientific name from his time as a zoo collector and I from my study of the field guide to the birds of South-east Asia. I was delighted by the thought that he, who had never left his native Borneo, and I, a stranger from England, should have been able to make ourselves mutually understood by speaking the dog-Latin that European scholars had used during the eighteenth century.

I was particularly keen that we should film proboscis monkeys for they are extremely spectacular creatures and as far as we knew they had never been filmed. The male has a long pendulous nose, rather like a squashed banana, that hangs down well below his mouth. It must have some sexual significance, for the nose of the female is only short and upturned, but exactly what it is no one has explained. The Indonesian name for the species is *orang belanda,* which means 'white man', an understandable nick-name when one considers the delicate features of most Indonesian people. Proboscis monkeys feed only on a very restricted range of leaves, those of mangroves being among their favourites, and as a consequence they are very rarely seen in zoos. That, together with their extraordinary appearance, meant that film of them would be a great prize for us.

The monkeys fed early in the morning in the mangroves close to the river's edge but filming them was not easy. They took fright as the *Kruwing* motored up towards them. We tried continuing past them, anchoring and then drifting quietly back downstream towards them in a dinghy close to the shore. But even when they stayed, it was virtually impossible for Charles to film from a moving boat. We tried landing and picking our way through the mud of the mosquito-ridden swamp to get close to them from the landward side. That was only marginally more productive and very much more uncomfortable. We spent several days at it, and eventually got some shots. Today they would be regarded as fifth rate. Nonetheless, the sequence when it was shown created a considerable sensation.

Wading through the swamps, we encountered crocodiles. But there were no giant potential man-eaters to play the villain in a dramatic sequence. All we could find were mere babies. That being so, I thought we might construct a mild filmic joke. Charles photographed the little creatures resting among the floating water plants in extreme close-up so that they looked immense and fearsome. He then filmed me as I slowly took off my shirt and holding it out in front of me like a matador, waded onwards, apparently preparing to throw it over a giant crocodile's head. The two scenes would be cross-cut as I approached closer and closer. All over Britain, we hoped, viewers would be holding their breath at my bogus bravery. Then when I pounced and my capture was seen to be less than two feet long, there would be a big laugh. In the event, it seems that most viewers thought I was seriously trying to persuade them that I was dicing with death. I got one or two letters accusing me of over-sensationalising. I learned that it is not easy to make any except the most heavily signalled jokes in a natural history film.

After a couple of weeks, Daan had to return to his sugar factory and hitched a lift on one of the boats that were sailing down to Samarinda with cargoes of rattan canes and wooden roof shingles. We continued upstream in the *Kruwing*. Slowly the people we met changed in character. There were fewer settlers from the coast and more Bahau, forest-living Dyak people. Eventually we reached a fully traditional settlement where the people were so welcoming and kind that they suggested we might stay.

They lived in a single longhouse, a huge construction built on stilts. A veranda ran across the entire front, facing the river. Behind this lay an open communal living space. Here, around the line of gigantic wooden pillars which supported the roof, most of the community spent most of the time. At the back, a line of doors opened into large rooms, each of which accommodated a family. Many of the men still wore long loin-cloths. Their straight glossy black hair was cut in traditional fashion – shoulder-length with a fringe across the forehead – and many had spidery blue-black tattoos on their arms, legs and necks. Most of the older women had pierced earlobes carrying such heavy brass rings that the lobes themselves reached down to their shoulders.

We slept on board the *Kruwing*, that lay moored to the longhouse's landing stage of logs. Our days were spent in the forest with some of the hunters as guides. In the evenings we sat with them and their families on the longhouse veranda, drinking our canned beer or their rice wine and struggling, with Sabran's help, to improve our Indonesian. Gradually, we began to assemble a collection of animals – squirrels, lizards, civet cats, crested quails and, most engaging of all, small emerald green parakeets, scarcely bigger than finches, with scarlet bibs and rumps and blue stars on their foreheads. Many families kept them as pets in wickerwork cages. When evening came, the little birds would climb to the top of their cage and go to sleep, hanging downwards, suspended by their toes – which is why they are called in English 'hanging parakeets' and in Indonesian *burung kalong*, bat-birds.

I was hoping that we might film orangutan, the great red ape of Borneo. The hunters said they knew where to find them. Eventually with their help we discovered one, high in the branches of the canopy. We followed and filmed it for several hours until Charles felt he had got as much material as he was likely to and we started on the way back to the longhouse. As we left, one of the hunters raised a gun and fired at the animal. He did not hit it, but I was outraged. 'Why did you shoot?' 'He no good,' the hunter said. 'He eat my banana. He take my rice.' There was little I could say. It was not I who had to feed my family in the forest.

One evening, sitting on the longhouse veranda after a hard day walking and filming in the forest, relishing the cool of the evening with a mug of rice beer in my hand, my eye was caught by a speck in the far distance down-river. Through my binoculars I saw that this was a small canoe. Whoever was in it was in a hurry for it was being paddled with great vigour. It got closer. Now I could see that the paddler was naked, except for a small head cloth and a sarong. Closer still – and I noticed what looked like a tiny white flag attached to a little mast, stuck in the bows. The paddler headed for our landing. He tied up his canoe, reached forward and picked up the white flag from the bows. Holding it upright, he ran up the notched logs that linked the landing place to the longhouse veranda. Only then did I realise what it was he was carrying. It was one of those famous elements in nineteenth-century chronicles of empire – a message in a cleft stick. I had often wondered why anyone should send a message in such a way. Now I understood. With messengers clad only in loin-cloths there was no other way of carrying a letter that would leave it in a readable condition.

To my alarm, he came up to the veranda and gave it to me. I opened it apprehensively. It was from Television Film Department. 'Latest tests show essential use reflectors with Kodachrome,' it read. 'Contact Sarawak Museum soonest for advice.' We had already shot an experimental roll of colour within a few days of starting filming as we had been asked to do. Since then, of course, we had been filming entirely in black and white. Contacting Sarawak Museum was certainly possible. I had heard of an expedition that had travelled from roughly where we were across the mountains of central Borneo and down the Rajang River. As I remembered it had taken them six months. I had another sip of rice wine. The office seemed a long way away.

Eventually the time came for us to leave. We had heard of a hunter down-river who some days earlier had killed a female orangutan that had been raiding his crops. She had a baby which the hunter had taken and was keeping in a wooden crate. We located him on our way back and he agreed to exchange the orphan for all our remaining salt and tobacco. Charlie, as we called the baby ape, had cuts and grazes on his limbs which we coated with antiseptic cream. We ingratiated ourselves by giving him spoonfuls of sweetened condensed milk which he plainly adored. Soon he became so tame that we allowed him the freedom of the boat and he spent much of his time sitting in the wheelhouse alongside whoever was at the helm.

One evening I told Sabran of our plans to go to Komodo and asked him whether he wanted to stay in Samarinda or to come on with us. He didn't hesitate. He would come with us. He too wanted to see a dragon.

* * *

Back in Surabaya, Sabran went to stay with relatives and we rejoined the Hubrecht household. Peggy claimed to be charmed to have the hanging parrots on her veranda. Charlie and the bear cub and the rest of our collection were given their own quarters in the grounds and the gardeners took on the job of caring for them. They said they were glad to have this variation to their regular work.

According to the lady in the airline office in Surabaya, the nearest we could get to Komodo by air during the next week or so, was Maumere. It was the main town on the island of Flores, close to its eastern end. Inconveniently, Komodo lay off its western end, but the map showed a road running along this long banana-shaped island, and I felt sure that we would find some way of getting along it. So we booked tickets.

The plane was a small one. Charles, Sabran and I were the only passengers. As its wheels touched Maumere's grassy strip, water spurted from the turf on either side. It was drizzling. We squelched to a halt. The Dutch pilot helped us off with our baggage.

'Good luck to you,' he said. 'Personally, I wouldn't want to spend three months in a dump like this.'

'Oh, we will only be here a couple of weeks,' I said.

'In that case,' he replied, 'you won't be leaving by air. This strip is already too wet for safety. The rains have started early. As soon as I get back in the air, I shall be radioing Jakarta to say that no more planes should land here until the rainy season is over. If you want to leave here by air, you had better come now while you have got the chance.'

Charles and I had a quick consultation. Going back now, after all the trouble we had taken to get here, was out of the question. All the same, I had a slight sinking feeling as we watched the plane trundle down the grass strip in a cloud of spray, climb laboriously into the sky and disappear into the low grey clouds.

Daan had given us the name of a Chinese trader who owned the main store in Maumere. From him we discovered that the road to the west end of Flores was not quite as continuous as the line on the map suggested. Land-slips had broken it in several places. In any case there was only one lorry on the island that was capable of making the trip and it, at the moment, was not roadworthy. The only alternative was to go there by sea. We went down to the harbour, but it was largely empty. Most of the praus were out fishing. Only one was moored by the jetty. I suppose I should have questioned why this was the case, but I was so glad to see a sea-going vessel of any kind that I did not.

The captain was a surly looking man. With Sabran interpreting, I explained what we wanted. He said he would take us to Komodo and we negotiated a price. We loaded our gear on board and bought supplies of food from our Chinese trader. The captain found two young boys to act as crew. Neither could have been more than twelve years old. That afternoon, we set sail.

The prau was single-masted and only some twenty feet long. A small thatched roof amidships provided a little shade and covered the otherwise open hold where our gear was stored and the captain and the two boys slept. Charles and I would have to sleep on the open deck. That suited us. The hold was filled with the stench of rotting fish and copra and it was nicer in the fresh air. We went to sleep with the pleasant sound of water slapping the bows and the sail creaking in a gentle wind.

We were woken in the middle of the night by an alarming grinding noise. A violent shudder ran through the ship. I looked over the side and saw, clearly in the moonlight, that we had struck a coral reef. Charles, Sabran and I grabbed bamboo poles and helped the captain push ourselves off it. Fortunately there was little wind so we had been moving quite slowly and the contact had been a relatively gentle one. It could have been much worse. Eventually, we went back to sleep, but my confidence in our captain's seamanship was somewhat shaken.

The second day, what wind there had been dropped entirely. Our little ship wallowed gently in the swell. The northern coast of Flores lay along the horizon several miles to the south. Charles and I sat in the stern, dangling our bare feet over the side, smoking (people did in those days) and throwing the stubs overboard. Depressingly, several still floated on the surface of the sea, swilling about the stern. Not even the water was moving.

I tried to ask the captain about the weather. Would this calm last long?

'Tidak tau.' He didn't know.

How far was it to Komodo?

'Tidak tau.'

Were there many more coral reefs on the way?

'Tidak tau.'

He had been to Komodo, hadn't he?

'Belum.'

This was a new word for me. I clambered down into the hold, found my bag and excavated a dictionary.

'Belum: not yet,' it said.

I decided that I had better take a closer interest in navigation than I had so far.

The only map I could find was in an airline brochure. Komodo was the size of a full stop. Flores was less than half an inch long. How far had we progressed along its northern coast? Once more the captain didn't know. Clearly it would be up to Charles and to me to decide when to turn south.

The following morning, Flores was still there, a thin streak along the horizon. Our progress continued to be painfully slow until the afternoon when the wind picked up. By early evening, I thought I saw a gap in the coast to the south. It could, of course, be the entrance to a bay, or it could be the opening of the strait between the end of Flores and Komodo. I decided we should investigate. If it was a bay, then we would be sheltered from the freshening wind and we could continue east in the morning. Better that than unwittingly overshoot the strait leading to Komodo. So we turned south.

It should have been obvious to me, had I thought about it, that the few narrow gaps in the long chain of the Indonesian islands would inevitably be torn by tidal races. As it was, I was quite unprepared for what I saw ahead. The surface of the water was ripped and blistered by fearsome rapids. The wind was near gale-strength and right behind us. There was no going back. We skimmed past a huge whirlpool. The sea was now roaring. We were driven towards the side of another giant eddy in the centre of which I could see sharp fangs of coral. One of the boys was on the tiller. The Captain sat beside him clutching his head with a shocked look on his face. Charles, Sabran and I and the other boy grabbed bamboo poles and pushed ourselves away from the edge of the reef. The boat was spun sideways by the eddy and careered into a patch of boiling water, heading straight for another reef, white with breaking waves.

Had it not been for the wind blowing hard behind us we could not have made any headway and doubtless would have been captured by one of the whirlpools. As it was we were blown from one to another and all our energies were given to trying to fend ourselves off from the coral at their centre. It was already getting dark when at last the strait ahead began to widen. To the right, in the failing light, I could see the entrance to a bay. We turned into it, dropped anchor in blessedly calm water and lay down on the deck, totally exhausted. All three of us fell asleep immediately. We had no idea whether we had reached Komodo or not.

* * *

We woke at dawn. We had anchored just inside the entrance of a wide bay. Beyond a white beach, there were hills covered with parched brown grass with a few scattered palm trees, standing upright like hat pins. At the far end, I could just see a small village of thatched huts. The tide had gone out since the previous night and the water was now shallow enough for us to wade across the coral to the village.

The headman received us in his thatched house. Yes, it was Komodo; and yes, there were *buaja darat*, land crocodiles, roaming around in the hills. One had visited the village only a few nights before and had killed some of his chickens. Were they dangerous? Well, the headman said in a matter-of-fact way, one of them had killed an old man some years ago, but he didn't really count for he was very feeble. He might even have died before the dragon found him. There would be no problem in our seeing them. All we needed was the carcass of a goat, preferably one that was beginning to smell a little, which he could provide.

The next day, with men carrying a dead goat, we walked up a small bush-filled valley. We suspended the goat carcass from a tree that hung over the dry bed of a stream so that its smell, which was now satisfactorily powerful, would spread as far as possible. Up on the bank, we built a small hide of branches behind which Charles set up his camera.

We waited, scanning the opposite bank through binoculars, anxious for the first glimpse of the extraordinary reptile for which we had come so far.

There was a rustle immediately behind us. Thinking it was the carriers who had returned for some reason, I turned round with my finger to my lips to ask them to keep quiet. It was not the carriers. It was the dragon. It was no more than ten yards away. I nudged Charles who was still looking at the river-bed. He turned and we both stared at the dragon. The camera had a long lens on it so Charles could not even film, for the dragon was too close to focus. There seemed nothing to do but sit there.

The dragon remained quite motionless, with the total statue-like immobility so typical of reptiles. Its head was raised. The scales had been worn away at the edge of its jaws, so that in places the flesh beneath showed pink. Otherwise its armoured skin was a gun-metal grey. The line of its mouth sloped up slightly at the angle, giving the animal the hint of a grim smile. A long yellow tongue, deeply forked, slid out from the front of its mouth and slid back again. It was tasting the air, savouring our smell, no doubt.

A butterfly flapped gently down and settled on its nose. Still the dragon did not move. Then, very slowly, it turned away. With great deliberation, it walked in a semi-circle around us, down the low bank, out on to the dry river-bed towards the bait.

We had miscalculated. The dragon was so big that as it reared up it could just reach the lower edge of the goat carcass. Now it was far enough away to be within Charles' focal range and he filmed it as it tried to claw at the bait. It got a hold and started to pull, jerking backwards with all its strength. If it pulled the carcass down, it might make off with it. BBC Accounts Department would have difficulty in approving the purchase of *two* rotting goats. I jumped up from behind the hide and waved my arms. The dragon let go and charged off into the bush.

How big was it? Nine feet? Ten feet? It certainly seemed much bigger than any water monitor I had ever seen. They were its nearest relations and certainly grow very large. But whereas a water monitor's tail is so long that its body is only a third of its length, the dragon's body amounts to about half. So in terms of weight the dragon is a more massive animal and much more powerful than any water monitor. Sabran was keen to catch one. He said he could make a trap using only material that he could get from the bush, apart from a length of rope. I explained that we had no permit to take it away. On the other hand, if we did trap one, we would be able to measure it. And anyway, the process of dragon trapping might in itself make interesting film.

Under Sabran's instruction, we drove stakes into the sandy floor of the river bed, lashed the stems of saplings around them, using rope made from twisted lianas. We suspended a trap door at the front and attached it to a simple trigger connected to a tilted platform. Inside we put a hunk of the now powerfully stinking carcass.

The dragon behaved perfectly. It entered the trap, and the gate fell. We filmed it at close quarters as it hissed and peered through the bars of our trap. We measured it – just over nine feet. Not a world record for the species but satisfactorily large. And then we released it.

That evening we had a celebratory meal with the headman and his family. Sabran, as usual, helped me understand what was being said. The captain of our prau, the headman told us, was no good, '*Tidak baik.*' That, I said, we knew. But in what way was he no good? Well, he wasn't a fisherman. He wasn't even a Flores man. He was a gun-runner from Java who bought armaments in Singapore and smuggled them to rebels in the island of Sulawesi, away to the north. The army was chasing him, which was why he had taken refuge in Maumere harbour. That explained a lot – his ignorance of these waters and his poor seamanship. But there was worse. He had told the villagers that we had lots of money and valuables and had asked if a couple of them would like to join him on board when we left Komodo. Together, they would be able to overpower us.

'Is anyone going to come?' I asked.

'No,' the headman said and smiled.

The following day we filmed the dragons again, tearing away at the goat carcass. Then, well pleased with ourselves, we sailed away. We felt that the captain would be unlikely to try to cause trouble with no one to help him except the two lads. Even so, we decided that we would not all go to sleep at the same time on the rest of the voyage.

There was no point in going all the way back to Maumere. The captain could not claim that it was his home port and we knew that no plane was likely to pick us up from there for several months, so we continued our journey westwards to the little port of Bima on Sumbawa, the next island in the chain. Our airline map marked it so boldly that there seemed a fair chance that there might be an all-weather strip there. Twenty-four hours later we reached it and, with some relief, paid off the captain.

The airstrip was indeed concrete – a step up from the sodden grass of Maumere – but no one could tell us when a plane was due to land or in which direction it might be going. We decided to camp in the concrete box with a desk and a set of scales that served as the airport office. We didn't want to miss a flight, wherever it was going. Sabran went into town to buy some food and we made ourselves comfortable on the floor. Soon we were on joking terms with the one or two people who worked for the airline and occasionally turned up for a couple of hours. One or two flights were expected daily. Sometimes they were cancelled. Sometimes they didn't go to where the schedule said they were going to go. None came that was going to Java. Soon we became something of an institution. People came in from the nearby villages to sit down alongside us and watch, at close range and without inhibition, the odd way in which we ate our food.

On the third day, a plane landed that was heading for Sulawesi away to the north. We debated whether or not to take it, in case there might be more flights from there to Java, but it was full anyway. The pilot turned out to be the one who had deposited us in Maumere. When he saw how we were living he returned to his plane and came back with six packed lunches for us. They were the best meals we had had for weeks. The following day a small plane arrived bound for Surabaya and that we did get on. An hour or so after that we were back in the luxury of the Hubrechts' house. Daan and Peggy had been so worried not to have had news of us for so long that they were on the verge of organising an Air-Sea Rescue search for us from a nearby US Air Force base.

Sabran helped us gather together our collection of animals from their various lodgings in the Hubrechts' garden and the Surabaya Zoo. Together we loaded them on to a goods train for Jakarta and then he left us to sail back to Samarinda. My Indonesian was still very poor but even if it had been better I would have found it hard to find the right words to thank him for all he had done for us. But I hope he understood what was behind my threadbare phrases.

In Jakarta we transferred the animals to a cargo plane and travelled back with them to London. Three days after that I started editing our film.

* * *

Charlie the orangutan proved to be the star of the programmes. He also acquired considerable distinction in the history of the London Zoo. He became the father of the first orangutan to be born there and thus the founder of the zoo's flourishing breeding colony. As for the dragon, no one seemed to mind that it was not in fact in the London Zoo but still where we had first seen it – in its own island. I certainly did not.

6

A Plunge into Politics

Politicians in the 'fifties did not pay much attention to television. Its audiences were still small and in any case its programmes did not aspire to covering politics in any depth. But as our audiences grew and the programmes broadened their scope, both Government and Opposition became more watchful and soon the arguments began. The BBC's commentators and interviewers were said to be too hostile. It was irresponsible of the Corporation to question politicians so closely that they were unable to make a coherent presentation of their policies. Each political party claimed that its opponents were either let off too lightly or given more air time than they were.

It was to silence this kind of criticism that the Corporation – and I suspect, in this case, that this meant Grace Wyndham Goldie – proposed that there should be an annual ration of short programmes in which politicians would be able to speak directly to the audience without any kind of interrogation. They were called Party Political Broadcasts and they became one of Talks Department's responsibilities.

None of us were enthusiastic about taking them on. There were no ha'pence to be earned and plenty of kicks to be endured. So we took it in turn. We did our best. The politicians, however, were convinced that the BBC, in some unexplained way, was failing to provide the best expertise. Maybe they suspected that some producers, if they did not agree with the message, might deliberately undermine their efforts. Soon people from within their ranks were appointed as mass communications experts and started to come along to advise and check on the BBC's efforts. This led inevitably to confusions.

The one I recall most vividly concerned housing. The broadcast, by the Opposition, was going to explain that they would build five times (I forget the exact figure) more houses in a year than the Government had succeeded in doing. I was told that their television consultant had devised a special 'visual' to hammer home this point, but exactly what it was must remain secret from me until just before the broadcast.

On the day of the programme, the lady consultant, the speaker and the 'visual' all arrived together. The 'visual' proved to be in two parts. One was a small doll's house and the other another of exactly the same design, but gigantic. The lady consultant explained to me that she wished to have a shot in which the small one was placed alongside the big one so that the huge disparity in size would suddenly be dramatically evident to all.

'You see,' she said 'we are going to build five times as many houses – so this model is five times bigger.'

'But it's five times longer, five times wider and five times higher,' I said.

'Of course,' she replied, unperturbed.

'But that,' I said, 'is a hundred and twenty-five times bigger.'

There was an appalled pause.

'What are we going to do?' she said so quietly that it was almost a whisper.

'I think the best thing is for me to make sure that the two are never seen together in the same shot.'

'What a good idea,' she murmured.

* * *

Sir Anthony Eden, when he became Prime Minister in 1955, was quick to take up the opportunities offered by the new medium and asked for a Prime Ministerial Broadcast. It was my turn to supervise such programmes. I waited just inside the entrance at Lime Grove for him to arrive. Mr Stacey, the Senior Commissionaire and an ex-sergeant-major from the British Army, was on hand in his best uniform, white gloves under one epaulette. Quite a large crowd had gathered around the steps outside and in due course, the Prime Ministerial Rolls purred up.

The Prime Minister alighted and smiled to his fans. Some waved pieces of paper at him asking for autographs. I thought it rather demeaning that the Prime Minister of Great Britain, one of the world's most powerful statesmen, should be treated like a mere film-star. But Eden knew how to deal with the situation.

'Collect them,' he said, 'and I will sign them before I leave.' Mr Stacey did accordingly.

The broadcast was unexceptional and straightforward, delivered by Eden sitting behind a desk. When it was over we went back to the hospitality room. I gave the Prime Minister a whisky and told him how good he had been. He had not forgotten about his fans and asked for their papers and autograph books. I handed them to him, one by one and he signed. Eventually we came to a crumpled receipt of some kind. I decided that it should be beneath a Prime Minister's dignity to be asked to sign such a thing, and dropped it surreptitiously behind the sideboard.

Eden finished his drink and waving at the throng still waiting for him outside, departed. The rest of us trailed back to the hospitality room to finish up what little was left of the drinks provided for these occasions by BBC hospitality. Producer's perks. While we were dissecting the broadcast, the door opened and Mr Stacey came in.

'Excuse me, sir,' he said. 'We've got a bit of bother outside. A gentleman says that he sent in a paper to be signed and it hasn't come out. And he's now in trouble because it has the address of his hotel on it and he doesn't know what it is.'

I grovelled behind the sideboard and handed it to Mr Stacey.

'But it's not signed,' he said.

'Of course not, I wasn't going to give the Prime Minister a grubby piece of paper like that.'

'Well, never mind,' said Mr Stacey. He leant against the wall, took a pencil out of his top pocket and carefully wrote 'Sir Anthony Eden,' grinned – and left.

Sir Anthony, it transpired, was quite pleased with the way his broadcast had gone. So much so that when the next one was scheduled, he asked that I should once again oversee it. That too was regarded as a success by No.10. Soon afterwards I was rung up by Philip de Zulueta, Sir Anthony's Parliamentary Private Secretary.

'The Prime Minister,' he said, 'wonders if you would like to come down to Chequers at the weekend for a spot of tennis.'

I hadn't played for years but clearly this was not an invitation to be refused. I dug out my ancient racket. Half a dozen of its strings were broken. I should have to buy a new one. Jane said I should buy new tennis clothes as well, but by the time I had paid for a racket I decided I couldn't afford new kit. My old shorts and shirt would have to do. Jane was horrified and insisted that at least they should be freshly washed and ironed.

Early that afternoon, a Government car arrived and collected me. As we drew up in front of Chequers, Philip de Zulueta came out to greet me. A footman opened the boot of the car, took out my battered case and bore it away. There was tea and scones with Sir Anthony and Lady Eden and then the Prime Minister was keen for tennis. We would change and meet on the court in half an hour. I was shown up to my room. My case had been unpacked. My toothbrush and razor had been placed in the bathroom and my few extra clothes laid out on the bed. I washed and only then noticed that my shirt and shorts were missing. Wearing only underpants, I looked in my case and in the wardrobe – but no tennis clothes.

The sound of racket striking ball came from outside. I looked through the window and saw Sir Anthony practising his serve. I was keeping the Prime Minister waiting.

I peered out of my door into the empty corridor.

'Is anyone about?' I called softly.

Philip de Zulueta stuck his head out of a door farther up the corridor.

'Trouble?'

'My tennis clothes have disappeared.'

'Don't worry. I'll lend you mine.'

This was extremely kind of him and I accepted gratefully. Unfortunately, Philip was about twice my weight but I had no alternative. When at last I appeared on court I was, to say the least, somewhat unstylish, with a balloon-like shirt and shorts that reached below my knees.

The Prime Minister won fairly easily and generously said it had been a good game. Only later did I discover that the Chequers footman who unpacked my case had decided that my clothes were still too damp from Jane's washing to be worn and had taken them away to iron them.

That evening, Philip and I dined with the Prime Minister. Lady Eden had a headache and wouldn't be joining us.

'What do you think of the wine, David?' Eden asked.

I dealt with the question cautiously. I knew nothing about wine.

'Excellent, Prime Minister. What exactly is it?'

'Guy Mollet, the French Prime Minister gave me a few cases of it. It's Amboise.'

Now I knew where Amboise was. Jane and I, only the previous year, had driven down the Loire valley on holiday and we had visited the Chateau of Amboise. So I now had one conversational card in my hand. Just one. The question was – how to play it.

'Really!' I said. 'It's not a wholly typical *Loire* wine, is it?'

'Isn't it?' said the Prime Minister, in shocked tones.

I was aghast. 'Well, I just thought...'

The Prime Minister called for the bottle. It proved to be a very odd one indeed, with a waist to it.

He took out his spectacles and looked at the label.

'You are quite right,' he said. 'How foolish of me. It's a double fermentation as one can see from the shape of the bottle. It's Arbois.'

I smiled modestly. Sometimes one does not deserve one's good fortune.

* * *

Several months later, in 1956, I was summoned again to handle a Prime

Ministerial broadcast. This time, it was for more serious and urgent reasons. An Outside Broadcast unit was on its way to Downing Street and I had to join it there as soon as possible. I knew why. President Nasser of Egypt had taken over the Suez Canal and expelled the British management. The Suez crisis had started.

Jane drove me up. Football crowds blocked the road. Chelsea were playing at home and we couldn't get through them. For a mile or so we had to drive at a walking pace. When at last we got to Downing Street I went straight in to No.10 and asked for William Clark. He had been a distinguished journalist, writing for the *Observer*, and was now the Prime Minister's public relations man. I was shown down to his office in the basement.

'What's going on, Willy?'

'No idea. I think the old man has gone mad. He won't even see me. You had better go up to the flat.'

In the flat on the top floor, Eden was in bed, looking desperately pale and drawn, surrounded by Rab Butler, Selwyn Lloyd, Harold Macmillan and other members of the Cabinet sitting on chairs. Two pretty but solemn girls from the Prime Minister's office were sitting with open notepads on their knees. Behind Eden's head on a shelf stood an alarming number of bottles of tablets and medicines.

'Ah, David, come in. We are just going through my television message. See what you think.' And he continued reading aloud from the paper in front of him. It had been typed in the large print that, in those days, seemed to be used only by the most senior levels of Government.

'There are times for courage,' he intoned, 'times for action. And this, my friends – ' He paused.

'Do you think I should say "my friends" at this stage, David?'

'I am not certain, Prime Minister, perhaps I might hear a little more.'

And so it went on. As he came to the end of a page on which he had marked changes, he handed it to the secretary sitting on one side of his bed who took it away to retype, and the girl on the other side handed him the next. I gathered that this process had been going on for some time. The retyped pages were brought in and when Eden came to the end, he started again from the beginning. After I had heard a complete cycle, and he was launching on to yet another hearing, I nervously suggested that it was time the Prime Minister had a rest, if he was to have the energy needed for the broadcast, and we all filed out.

The cameras had been set up in what I assumed to be No.10's main living room, turning it as they always do, into a shambles, except for the small pool of light around the desk behind which the Prime Minister was to sit. The primitive teleprompter of the time, which needed most delicate adjustments if it was to work properly, was already in place, with the operator fussing about its exact positioning. Lights on tall stands were heating the place up. Video monitors stood here and there. And among it all wandered Lady Eden looking understandably very cross at this despoliation of her home.

I went outside into the control van to check on pictures and came back ready to receive the Prime Minister and get him settled and relaxed as far as I could. In due course, he appeared looking pitifully frail but still treating the cameramen and electricians to his Hollywood smile. Threading his way through the forest of light stands and tangle of cables, he took his seat behind the desk. The make-up girl powdered his face to remove any trace of perspiration and I leaned over to give him the few trite words of assurance and encouragement that directors routinely give on these occasions. Suddenly, Lady Eden rushed across.

'Is this a conspiracy?' she cried. 'I've just seen a monitor. The broadcast can't go ahead. The Prime Minister's moustache is invisible.'

And with seconds to go before the world was told that Britain was about to risk starting World War Three, she opened her handbag, took out her mascara, and started to brush it on to the hairs of his upper lip.

7

New Guinea

I still yearned to film birds of paradise. The Indonesians had refused us permission to go to the Aru Islands, where Wallace had first watched them display. But Aru is only on the fringes of bird of paradise territory. The headquarters of the family lies farther east in New Guinea itself. That was the next place to try.

In the 1950s, the western half of New Guinea was ruled by Holland and the eastern by Australia. It made obvious sense for us to go to the eastern half. At least the officials there would speak English. But there was a snag. Animals, shown live in the studio between film sequences were still an important element in *Zoo Quest* programmes. I did not aim to capture birds of paradise, only to film their courtship displays. But there were many other exciting things that could be brought back from New Guinea – small mammals, unusual parrots, colourful snakes and lizards, extraordinary stick insects, brilliantly coloured frogs. No animals of any kind could be taken back to London by air, for the only flights between Australian New Guinea and the rest of the world ran through Australia, and Australia had, as it still has, strict and unbendable quarantine regulations. Regardless of the fact that wild birds migrated back and forth between New Guinea and northern Australia, any *captive* birds that entered Australia had to spend several months in quarantine.

They could, however, leave New Guinea by ship and be taken across the South China Sea to Hong Kong. From there they could be flown to London. But they could not travel by themselves. Someone would have to accompany them and keep them properly fed and cleaned. I put the problem to the London Zoo. They said that they would, of course, be overjoyed to have a collection of New Guinea animals, but they could not spare anyone to help with its transport. It looked as though I would have to deal with that myself.

To do so would add several weeks to the trip and I was not sure how the BBC would react to the idea that one of its staff producers should spend a month of his time acting as an animal attendant. When I put in a proposal for a *Zoo Quest* to New Guinea, I mentioned this arrangement somewhere in the detail, but no one raised any objection. Perhaps no one noticed.

The Zoo, however, did produce a trump card for me – Sir Edward Hallstrom. He was an Australian multi-millionaire who had made a fortune from the manufacture of refrigerators. He was also an enthusiastic aviculturalist and a major benefactor of Sydney Zoo. And, very importantly from my point of view, he had a passion for birds of paradise. He had set up an experimental farm at Nondugl in the central highlands of New Guinea which was attempting to establish sheep in those high pastures. At least, that was its ostensible purpose. But Sir Edward had also installed there a famous Australian animal collector, Fred Shaw Mayer, and given him the job of catching birds of paradise and sending them down to Sydney Zoo. There they could be kept in quarantine for the necessary months until it was legal to put them on display – or add them to Sir Edward's own private collection. London Zoo wrote to Sir Edward telling him of our hopes and he agreed that we might use his New Guinea establishment as our base. In June 1957, Charles Lagus and I arrived there.

* * *

The Wahgi Valley, in which Nondugl lies, was completely unknown to the outside world until 1933. Up to that time, the unexplored interior of this part of New Guinea was thought to be filled entirely by a tangle of mountains. But in that year, two Australian brothers, Mick and Dan Leahy, panning for gold and travelling up the Purari, one of the great rivers of eastern New Guinea, came across human corpses floating down one of the tributaries that flowed into it from the west. From the wounds on the bodies, it seemed that these men had been killed in a tribal war. They were tall and heavily bearded. The Leahys had never seen anyone like them before. Who were they and where had they come from?

Three years later, Mick Leahy, back once more on the Purari, noticed distant cloud formations farther to the west over the centre of the island. He thought they were more typical of grasslands than mountains. That, and his memory of those strange corpses, led him to believe that somewhere in the interior there must be an unknown valley with a substantial population. In 1937, he and Dan accompanied by a Government officer, set out to try and find it.

They travelled into unexplored country in the headwaters of the Purari, climbed a steep mountain barrier and then walked into a wide upland valley, the Wahgi. They found a people who had never heard of the sea. They had never seen metal tools of any kind. They carried stone axes and wore spectacular headdresses of bird of paradise feathers. Mick Leahy's account of their discovery, of his reactions to them and theirs to him, is one of the great classics of twentieth-century exploration.

Two years after this, the Second World War broke out and the valley remained virtually untouched by outsiders until 1945. Even at the time of our visit, twelve years later, it was still little changed. The men who worked on the farm at Nondugl and the hunters who helped Fred Shaw Mayer catch his birds still wore traditional costume – a broad cummerbund with, at the front, a long apron woven from animal hair that hung down to the knees and at the back, like a bustle, a spray of leaves. And almost everyone wore headdresses of bird of paradise plumes.

At first I thought this was a good sign. It must surely mean that there were lots of the birds in the area. Fred Shaw Mayer put me right. The true situation was quite the reverse. Paradise plumes were not just splendid ornaments. They were money, essential currency for many tribal transactions. In consequence, the birds were intensively hunted. A decade earlier, men had seldom wandered any distance from their village to look for birds, for inter-tribal warfare was rife and the danger of ambush a very real one. So then the birds in the wilder parts of the forest flourished in considerable numbers. But within the last few years, the Australian administration had brought law and order to the Wahgi. Now a hunter could go more or less where he liked. As a consequence, the plumed birds had almost disappeared from the Wahgi – and the headdresses worn by the men were more spectacular than they had ever been.

We soon had dramatic evidence of the intensity of their hunting. Minj, a village on the other side of the valley from Nondugl, was due to receive a ceremonial visit from people who lived up in the mountains to the south. There would be trading between the two communities. Pigs and pearl shells would be exchanged. Old debts would be paid off and new ones incurred. And, as always on important occasions in New Guinea, there would be a dance – a sing-sing. We went to see it.

People started to gather on an open patch outside the village in mid-morning. We heard distant drumming and chanting and through binoculars I saw a line of tiny figures descending a steep grass-covered ridge of the southern mountain wall. Normally it would have taken a man no more than half an hour to walk down to the valley floor, but the visitors were stopping so frequently to dance and sing that it was two or three hours before they had all come down and everyone had assembled outside Minj.

The din of drumming and chanting, that had started with the arrival of the first contingent, did not stop. It only grew. By midday, it was deafening. There were now hundreds of visitors. They formed up in platoons, five in a rank and ten ranks deep, and stamped ferociously back and forth. Their bodies were smeared with red clay, soot, and pig fat. They wore crescent pendants of pearl shell around their necks. Shell discs hung from their pierced noses. And every man had a feather head-dress of indescribable splendour. I recognised the yellow gauzy plumes of the Lesser Bird of Paradise and reddish ones that came from a similar, closely related species, Count Raggi's Bird. But there were others that were quite new to me, belonging to species that had never been seen alive in Europe – the long black tail feathers of Princess Stephanie's Bird, the iridescent triangular green bib of the Magnificent Bird, and the blue flank plumes from Prince Rudolph's Bird. Most extraordinary of all were long quills with small blue triangles like chips of shell along one side. A pair of these sprout from the head of the King of Saxony's Bird. The dancers had inserted these through their noses and looped them up to fasten on their head-dress above the centre of their forehead, so that they formed a frame to their face.

It was simple enough to calculate how many birds these represented. A King of Saxony male only produces two head plumes. Only two bunches spring from the flanks of Count Raggi's Bird. A Superb Bird has only one green bib. So I could easily see that each man wore the plumes of between twenty and thirty birds. And there were at least five hundred dancers. At that one performance, the men of Minj had decorated themselves with the feathers of twenty thousand slaughtered birds of paradise.

* * *

Fred Shaw Mayer lived in a little shack close beside his aviaries. Whereas Cecil Webb, the collector I had met who had told me all about Aru, was tall and almost stately, Fred was small, stooped and apparently timorous. He wore a strange deerstalker hat with its ear-flaps permanently lowered, and several long woolly cardigans, for it was cold at night up in the Wahgi. Yet this was the man who earlier in his life had travelled alone through the wildest jungles of the Pacific and South-east Asia. He himself had discovered several new species of birds of paradise. The most splendid of them he had first identified from two white pennant-like tail feathers nearly three feet long being worn by a Wahgi man in his headdress during a sing-sing like the one we had seen at Minj. Eventually when the bird itself was discovered and scientifically described, it was named in Fred's honour, *Astrapia mayeri*.

Fred's entire life was devoted to his birds. Every morning, he rose well before dawn to mix their food. He said he liked to put it in the aviaries before it was light so that the birds might eat at the same time as they do in the wild. Each species had its own particular tastes. Fred knew them all and provided each with the meal it enjoyed most, mixing local fruits with specially gathered spiders, wasp grubs or tadpoles.

I asked him what chance we had of seeing the display of these birds in the bush around Nondugl. He was gloomy. Not so long ago Count Raggi's Bird had been quite common in the Wahgi. This species was of great interest to me because it is similar to the one which Wallace had seen in Aru. The main difference is that the species Wallace had described has golden plumes whereas those of Count Raggi's Bird are a dusky-red. The habits of both birds are similar. The males assemble in groups of half a dozen or so in particular trees in order to dance and attract females. Immature males, which have not yet developed plumes, go there too. That was the scene pictured in the frontispiece of Wallace's book that had fired my boyhood interest in the birds.

Fred explained to me that men who owned these display trees guarded them with great care. They would wait until a male grew plumes and then shoot it before its daily dances tattered its feathers. Most display trees in the valley – and even farther away in the forest – had already been shot out. Even if a hunter knew of a more distant display tree, he would be unlikely to show us in case others discovered it and killed the birds before he could.

Our best chance of seeing displaying birds would be to go to the valley of the Jimi River which ran more or less parallel to the Wahgi to the north beyond a range of mountains. But the Jimi Valley was still uncontrolled territory. To enter it, we would have to get the permission of the District Commissioner who had his office in the settlement of Mount Hagen farther up the Wahgi Valley.

Hagen, today, is a large and thriving town. Then, in 1957, it was no more than half a dozen corrugated-iron buildings. There was a medical station, an administrative office, and the DC's home. You could get to it by road, along a twisting trail that only too frequently degenerated into deep mud – but that could take a long time and there was a sizeable risk of getting stuck. Most Europeans travelled from one settlement to another by hitching a lift in one of the planes that every few days made the journey along the valley calling at all the stations with mail and supplies. In one of those, you could make the trip in a few minutes. We took one.

The District Commissioner, a bluff burly Australian, sat behind his desk in Hagen and looked at me hard. He was immaculately dressed in neatly pressed khaki but it was clear to me that the office was not the setting in which he was most at home. I had little doubt that he knew as much as any European about the wild country he ruled. He had probably planted the Australian flag in about half of it himself. He was a daunting figure – and I felt thoroughly daunted. I explained as best I could that we wanted to go into the Jimi Valley to film birds of paradise. His expression did not change as I talked. He seemed neither surprised, nor impressed.

My speech trailed rather lamely to an end. Then he spoke. The Jimi was a rough place, he said. A few years earlier, pilots flying across the valley on their way from the Wahgi to Madang on the north coast, had reported seeing villages in flames. He had despatched a young patrol officer with half a dozen locally recruited armed policemen to find out what was happening. The patrol walked into an ambush. Several of the police were wounded by arrows and they all came back in a hurry. So he had gone in himself with another patrol officer called Barry Griffin. Together they had found a site for a permanent station at a place called Tabibuga and he had left Griffin there to set it up. That was over a year ago. Although Griffin had returned briefly to Hagen a couple of times, no one else had been into the Jimi since. If we started to wander around there, we would have to have an armed escort. The only person who could provide it would be Griffin and he probably had his hands quite full enough without having to put up with stray Poms who were interested in taking snaps of dickie-birds. Nonetheless, he would send a message by radio asking Griffin if he could cope with us. He would not instruct him to do so. But if Griffin agreed, we could go in with the next carrier line taking him supplies. It was due to leave in a week or so.

Four days later, we heard, over Nondugl's radio, that Patrol Officer Griffin had agreed to accept us. He had suggested that after we had walked in to Tabibuga, we should march down the Jimi valley and then over the Bismarcks, the next range of mountains to the north, and finally go down to Aiome, a station on the Ramu River, which had an airstrip from which a charter plane could pick us up.

Suddenly there were dozens of things to be done. We took the next plane down to the coast and bought a month's supply of food. We went to Government stores and bought trek gear – tarpaulins which could be rigged as tents, and canvas tubes which, with poles through them, would serve as beds. We bought goods to use as barter – bags of salt, gold-lipped pearl shells, mirrors, mouth-organs and combs. And we arranged for a small plane to come and collect us from Aiome in five weeks' time.

Eight days after my interview with the DC, with all our purchases in a trailer towed by Nondugl's truck, we drove to a small settlement at the foot of the mountains that formed the northern wall of the Wahgi Valley to start our march. After the truck had unloaded everything and was driving away down the red muddy track, I realised that from now on, and for the next five weeks, the only way we would get anywhere would be to walk.

A tall and very impressive policeman, bare-footed and bare-chested, wearing a neat khaki wrap around his waist, stepped up and handed me a note. It was from Barry Griffin. The bearer, Wawawi, it said tersely, was one of Barry's most trusted policemen. He would recruit porters and guide us to Tabibuga.

We set off the following morning at dawn. Wawawi had assembled forty Wahgi men to carry our gear. He allocated each load to two of them. They tied it to a pole which they cut from the bush. When all our loads had been dealt with, Wawawi, with a rifle over his shoulder, gave the order to start. He led the way. The men shouldered the poles, another more junior policeman took up the rear, and we were off.

To begin with, the muddy track wound up alongside a small stream but soon the going became steeper. And harder. And colder. The track – it could hardly be called a path – threaded its way between bushes that brushed our shoulders with rain-laden leaves and zig-zagged its way up steep rock faces, running with water. Every hour or so, we stopped for a rest. By about midday we were at an altitude of around eight thousand feet, so high that mist was drifting through the forest.

The gradient slackened a little and we emerged on to a bare col. I had been near the back of the column and as I walked out on to the col, I found that the carriers ahead of me had put down their loads and were clustered around Wawawi, arguing vociferously.

By now I understood enough New Guinea pidgin for me and Wawawi to communicate. The carriers, he told me, were refusing to continue. They wanted to go back down the mountain to the Wahgi. I said that I realised that they had been carrying heavy loads and were tired, so I suggested that from now on we should all have longer and more frequent rests than the ones we had been taking. But that, said Wawawi, was not the problem. The men would not go further because this was the edge of their tribal territory. The people ahead were bad people. 'Ol 'e kai-kai man,' as Wawawi put it. Charles was standing close by filming the discussion. He was quicker to understand the phrase than I was. 'He means they are cannibals,' he muttered.

At that moment I saw a movement behind a pile of rocks a couple of hundred yards ahead. I looked again. It was the flicker of a feather head-dress. Then I saw another. Suddenly fifty or sixty men burst from hiding and were charging along the track towards us, yelling at the tops of their voices and brandishing bush knives and axes. The only thing I could think of doing was to walk towards them with my hand stretched out, trying to look as peaceable and welcoming as I could manage. Within seconds, they were upon me. And to my astonishment, they seized my proffered hand and started pumping it up and down and clapping me on the shoulder. They were, apparently, simply *delighted* to see me. Why they should have put on such a ferocious display I could not at first imagine. And then it dawned on me that since this was the tribal frontier, it was sensible when they met people from the other side to appear as strong and as warlike as possible. If they seemed to be gentle and peaceable, the Wahgi people might think them weak and ripe for raiding.

Whatever the Wahgi men's opinion on this point, it was clear that they wanted to get back down to their warm home valley as quickly as possible. With men from the country ahead apparently positively eager to carry our loads, there was no reason to keep the Wahgi porters any longer. I took a bag of coins from my patrol box and Wawawi gave each carrier the standard Government wage for a day's porterage. It was the last time for a month that we were able to use coinage. From now on, payment would have to be in salt or glass beads, calibrated by the tablespoonful.

Our new friends seized the loads and set off with great enthusiasm, singing at the top of their voices. The path ran down the crest of a grassy ridge. I thought to begin with that we were about to descend into the Jimi Valley itself and that the hardest part of the walk was over, but soon we were climbing up the flank of another ridge. We were in the headwaters of the Jimi and having to travel across the grain of the country. Up and down we went. In the late afternoon, we walked wearily into a small settlement – two lines of huts strung out along the crest of a ridge. There we would spend the night. We were only half way.

The march the next day was a kind of crescendo. Wawawi had sent out word that we wanted carriers and far more than we needed came in from surrounding settlements to see us. Those we didn't engage decided to come along anyway. The porters sang, a happy never-ending yodel, and everyone joined in. As we approached each village on the trail, people came running out to meet us, forewarned by the singing, and then cavorted alongside about us. Around midday, we crested yet another ridge and started swinging downhill once more when I saw through a gap in the trees, far below us, a tiny speck of ochre red in the middle of the dark green of the forest. It was a clearing. It must be Tabibuga.

By now we had an escort of hundreds. A platoon of several dozen warriors were running ahead of us in short spurts, waving spears, and then stopping to dance, stamping with their right foot and shouting triumphantly until we almost caught up with them, whereupon they made another run ahead and repeated the performance. Even our porters, laden though they were, did their best to break into a trot. Wawawi had shouldered his rifle in correct military fashion and was marching and swinging his free arm with formal parade-ground stiffness.

The path ahead broadened and opened out into the red-earthed open arena I had seen from the forest above. At least a thousand people were there awaiting us. As we emerged from the forest, they all screamed and shouted excitedly. Beyond them, on the far side of the arena, stood a thatched open-sided building and on its veranda, in a chair, sat a European in neat white clothes, reading a book on his knee, apparently impervious to the bedlam that surrounded him. The crowd parted as I walked towards him. When I was no more than twenty yards away, he with some deliberation closed his book, put it on his chair and walked slowly towards us. It was difficult to avoid thinking of that famous meeting at Ujiji and I only just stopped myself from saying that I presumed he was Dr Griffin. Instead he spoke first.

'Griffin,' he said, shaking my hand.

'Attenborough,' I duly replied.

'Sorry about all the noise,' he continued. 'My chaps are a bit excited because you are the first Europeans who've come in since I've been here. Guess they thought I was the only one in existence and they are probably pretty shattered to find there are a couple more.'

* * *

Barry lived in a house he had built up on a ridge overlooking the main station. It had just one room which he kept meticulously tidy. His footwear – boots, shoes, slippers – were lined up along one wall; two piles of Australian magazines along another, one consisting of those that he had read and another of those that he had not. A camp bed, with a neat counterpane stood in a corner, a small trestle table in the middle. That was it. There were two small out-houses – one a kitchen, the other a wash-house with a suspended canvas bucket that served as a shower.

I suggested that we should put up our own camp and sleep there, but Barry would not hear of it. We should stay with him. Dumping our film gear and trek boxes in the middle of such controlled neatness seemed a kind of desecration. That evening we ate tinned lambs' tongues and drank bottled beer. I appreciated only too well what it had cost in human energy to put such things on a table in front of us in that remote place, and I knew that I had seldom eaten a more expensive meal.

Barry had already sent out messages to the surrounding villages, telling people of our arrival and asking them to bring in any animals that they might be able to find for us. Two days later a hundred or so Milmas arrived. Their country was a day's walk away. They were the traditional enemies of the local Tabibuga people, the Marakas, and it was the flaring war between the two groups that had led to the establishment of Tabibuga in the first place. Barry had sited the station in the middle of Maraka country since they seemed if anything to be the more aggressive and war-like of the two and he dealt severely with anyone who started fighting again. To minimise the possibility of further violence, he only allowed the Milmas to come in to the station once a week on a specified day. This time he had asked them to bring in animals.

They looked a tough lot. Their dress was basically the same as the Wahgi people's, woven aprons at the front and leaf bustles at the back, but they had a much wilder appearance. Their feather headdresses were smaller and more business-like and did not have the dandyish extravagance of those worn by the swanky Wahgi people. Many also sported a kind of outsize furry cravat made from the back half of a tree kangaroo skin, the hind legs fastened around the neck and the chocolate-brown tail dangling down their chests. They had collected all kinds of things for us – beetles, spiders, snakes, and cassowary eggs. Some we wanted, most we didn't, but we took all and paid for all. The Milmas could hardly be expected to know what was likely to make a good exhibit in the London Zoo.

Best of all, one man produced three tiny nestling parrots, the size of baby sparrows. They were only just acquiring feathers. Rearing them would not be easy. I should have to chew up banana or some other vegetable and then feed it to them from my lips just as their mother would have done from her beak. It would take a lot of time and labour, but I could hardly refuse to accept them, and anyway I suspected from their size alone that they might be dwarf fig parrots, intriguing little birds that occur only in New Guinea and had seldom been seen outside it. So I committed myself to a daily duty that would last for the next four weeks – but one that ended successfully with a small group of hand-tame miniature parrots that delighted all who saw them.

Two days after that, it was the Marakas' turn to come in from outlying villages to trade. They too had got the message and brought us animals – huge prickly stick insects a foot long, tree frogs and green tree pythons with a line of white scales down their spine. One man also took me to a bird of paradise display tree outside his village but all the plumed birds that had once visited it had been shot, just as they had been in the Wahgi. We were still not far enough from human settlement, apparently, to find an un-plundered display tree.

I explained to Barry that filming displaying birds of paradise was our priority. He agreed that we might stand a better chance of finding them in the less populated forest still farther down the Jimi Valley. We hatched a plan. He did not feel he could allow us to go further into the Jimi without some protection. Though he had not heard of trouble recently, various groups were still at loggerheads and there could be no guarantee that they might not ambush strangers passing through their territory. He had things to do on the station and could not leave for over a week, but if we wanted we could leave sooner than that with Wawawi escorting us. Barry would set out later with more armed police and would catch up with us before we got too far away. We could travel together to Aiome and he could then take advantage of our chartered plane and come back with us to the Wahgi for a few days' holiday.

* * *

After six days at Tabibuga, Charles and I set off again. Our first target was the village of Menjim on the banks of the Ganz, one of the Jimi's tributaries, two days' walk away. There at least we might be reasonably certain of finding something to film for there men still made stone axes.

We found a group of them at work sitting beside a small stream. The stone they needed occurred as boulders in its bed and they broke them by hurling one on to another. The larger flakes produced in this way they then chipped into a rough axe shape. Then they finished them by grinding them with small slabs of rough sandstone. The blades were so thin, with such a wide extravagant flare that they looked very fragile. They would certainly have splintered had anyone tried to use them to cut down a tree, and they seemed too large and unwieldy to be used in battle. These were objects used for display in sing-sings or as essential currency in certain transactions. If we were looking at a scene from the Stone Age, then we were seeing it in its last days.

One of the Menjim men told us he knew a tree where the birds of paradise danced and the next day, full of hope, we went to see it. It was a giant.

Its lowest branch was about a hundred feet up. We had no ropes. I could see no way of climbing it and there was no chance of our getting even the slightest glimpse from the ground of any bird that might be displaying in its crown. Perhaps we might construct some kind of ladder up the trunk to get into the branches a hundred and fifty feet above and there build a filming platform. All that would take time, but even that possibility disappeared in the evening when Barry caught up with us.

Our plans, he told us, would have to change. The DC in Hagen had contacted him by radio just before he left Tabibuga. A gold prospector named Jim McKinnon, working on the other side of the Jimi in the Bismarck Mountains, had been labouring away for months building himself an airstrip. He had chartered a plane to bring in machinery with which to work his claim but the plane would not land unless there was a signal laid out on the ground by a Government officer to say that the strip was usable. Barry was the only one around who could do that so he had to get there as soon as possible. He didn't know the country or the people between and would have to take Wawawi and the rest of the police with him. So we would have to go too.

The next day, we continued down the narrow Ganz valley towards the Jimi River itself. By now we were getting quite fit. Marching was no longer the exhausting trial it had been when we had started. We were, it is true, travelling downhill but the rain held off, and my legs seemed to swing forward effortlessly. We strode through groves of magnificent tropical conifers, araucarias, with upright cylindrical trunks the size of factory chimneys. The forest rang with the metallic calls of birds. Occasionally, we heard the slow rhythmic swishing noise made by hornbills flying invisibly above the canopy. Three times I saw a flash of golden feathers in the forest ahead – there were Lesser Birds of Paradise here. It was frustrating that we could not stop and spend time to see where they might be displaying; but the forest was so beautiful and pristine, the air so fresh and clean, the Ganz River, so cheerful and beautiful, cascading and purling beside us, that it was one of the most enjoyable days marching in a forest that I have ever had.

Now that Barry had joined us we had nearly a hundred porters who were carrying between them a formidable amount of baggage. We had not been able to rely on finding either food or shelter in the forested country on our journey so we were carrying everything we might need – tents, cooking pots and pans, empty cages for any animals we might find, filming gear and, above all, food – food for us, food for those who carried the gear, food for those who carried food.

We were headed for a hamlet called Tumbungi on the banks of the Jimi. Barry had been there once before. Just once. There people had built a suspension bridge of lianas that would enable us to cross the river. And there too we should be able to get more porters, for those who had brought us this far would not cross on to the northern side of the Jimi. As far as they were concerned, that was hostile territory.

We reached Tumbungi that evening. The bridge was there all right – slumped in a shallow curve, limp, decrepit and entangled with an untidy web of supporting lianas that ran from the thick rope of twisted fibres forming the bridge's foot-way up to the tops of the trees on either side. But the place was hardly even a hamlet. There were just two huts and a shelter, crudely made from branches and leaves. And it was deserted – silent except for the buzz of insects, the gentle rippling of the water of the Jimi as it slid past the bank, and the faint creak of the suspension bridge swaying in the wind.

While we were counting the loads to make sure that all had arrived, Wawawi pointed to the opposite bank. A group of tiny half-naked people wearing huge dome-shaped hats were standing there silently watching us. Wawawi clambered across the bridge and encouraged them to follow him back to where Barry sat. Their spokesman was an elfin-like little man, about four and a half feet tall, wearing a particularly large pudding-basin hat garlanded with a long strip of brown fur from a tree-kangaroo's tail. Small pieces of bamboo, the size of toothpicks, were stuck through his nostrils and the beak of a hornbill hung from the back of his neck. He watched us with darting eyes and hesitant smiles that slowly began to displace his worried baffled expression. Communicating with him was not easy. Barry spoke in pidgin to Wawawi; Wawawi knew a language understood by one of the carriers we had recruited at Menjim; and the carrier, while not able to speak the pygmies' language, knew some intermediate language that the pygmy leader seemed to understand.

Barry asked for two things: food and help to carry our baggage. The pygmy leader nodded so vigorously that his pudding basin head-gear wobbled alarmingly and I feared it might tip off. In fact there was no danger of it doing so for, as I discovered later, it was irremovable. It consisted of hair clippings and mud that had been kneaded into, and was at one with, his growing hair. From these nods and slightly less transient smiles, it seemed that he agreed to both of Barry's requests and after an hour or so he and his companions made their way back across the swaying bridge into their own country.

He was as good as his word. Next day, forty or fifty small people, women as well as men, reappeared carrying bananas, plantains, breadfruit, yams and sugar cane. Not only that, one of the men carried a tame cockatoo on a stick on his shoulder. He happily exchanged it for three cakes of salt and we added it to our collection.

The people sat around in the clearing watching me as I went through my daily routine of cleaning and feeding our animal collection. The bird-cages were floored by trays covered in newspaper that I replaced every day so that their occupants as they sat on their perches, should not soil their tail feathers on their own droppings. As I took each piece out, the pygmies grabbed it eagerly and took it down to the edge of the river. There they carefully rinsed it in water and laid it out to dry. Later that day, they tore it into strips, wrapped some of their locally grown tobacco in it, and smoked it.

The day after that, the news of our presence had spread so far that a hundred pygmies appeared, prepared to carry our baggage. We crossed the river and started up into the mountains. We only had three days to get to McKinnon's camp before a plane would be flying over to look for Barry's signal.

We made it, but only just. Some of the porters were unhappy at going so far from their own country and simply put down their loads when no one was looking and vanished into the forest. Wawawi had managed to persuade those carriers that remained to take double loads for double pay, but we were close to having to abandon something.

Jim McKinnon, when at last we reached him, welcomed us effusively. His camp was a mess – a wooden shack roughly put together, furnished only with a battered table and an unmade camp bed, and strewn with opened and half-eaten tins of food. A few yards away he had dug long trenches down which he had diverted a small stream. This would provide him with the sluice in which he would wash the gravel to extract such gold as it might contain. He was convinced that, as he put it, 'she would come good.' If he could get in the equipment he needed, he was certain he would soon be a rich man.

He had not seen any other European for many months and now that some had appeared he was so excited and so unaccustomed to speaking his native tongue that he stuttered and his words tumbled over one another. He was also embarrassed and apologetic, as indeed he had cause to be. When he sent his request for Barry to come and validate his strip, he was relying on getting a roller to compact its surface. He had arranged for a plane to fly over and drop it – a standard way of getting in supplies. But for some reason the air-drop had not happened. By that time, Barry had already left Tabibuga and was uncontactable. The strip was still uncompacted and Jim knew as well as anyone that it was not firm enough for a plane to land on it. So he had cancelled the charter. Our haste, the extra days we had to spend on a diversion to get to him, the abandonment of any attempt to film birds of paradise had all been pointless.

That night, more pygmies vanished. Barry decided to leave some of his gear with Jim. He would reclaim it the next time he came through, though he had no idea when that might be. We could not delay for we only had four days left to cross the Bismarcks and get down to the Ramu valley and Aiome station before our charter was due to arrive. Without a radio we could not delay the booking. If we missed it, it might be weeks before we would be able to get another. So we left Jim waving a sad and stuttering farewell and set off for Aiome at as fast a pace as we could manage.

The following night, more porters disappeared. Now there were not enough to transport our gear, reduced though it was, even if everyone carried double loads. We would have to split. Charles and I, taking only our filming gear, our animals in their cages and food for just one day, would leave at dawn the following morning with as few of the remaining porters as was necessary. We would march as fast as possible and hope to get down to the Ramu and Aiome that night. We would then recruit Aiome men and send them back to bring out Barry and the remaining gear. If all went well there might just be enough time for him to rejoin us at Aiome and fly back with us on our charter for his long anticipated holiday in the Wahgi.

But things didn't go well. The following morning we left in good time and walked hard. We had one further tributary river to cross before the final descent into the Ramu. When we got to it we found that the crude bridge of branches and lianas that we had been told would be there had been partially washed away. It took us over three hours to improvise repairs. That night we ate the last of the food we had with us.

We marched into Aiome the following day at noon. The Patrol Officer gave us cold beer, chairs to sit on, and showers. In that order. He sent out messages asking for carriers and before evening several dozen had set off to try and reach Barry as quickly as possible. Our charter was due two days later.

The following day, revelling in Aiome's luxuries, strolling idly around its neatly clipped lawn with no need to walk, walk, walk as fast as possible, was heaven. But it was marred, for no matter how I worked it out, I did not see how Barry could possibly join us in time to catch the charter plane the next morning.

That evening, sitting on the Patrol Officer's veranda drinking still more of his cold beer, Charles cocked his head. He thought he had heard distant singing. He had. We looked across the airstrip to the silhouette of the mountain wall that rose from its far end. There we saw tiny pinpoints of light. They disappeared and the sound faded. Whoever it was must have left the open grassy slopes and entered the forest that covered the valley floor. And then we heard singing again, nearer and louder and bright lights appeared at the end of the airstrip. We rushed out to greet whoever it might be.

Barry was striding at the head of the column. Behind him, with the loads we had abandoned, walked not pygmies, not even the Aiome carriers we had sent to help him, but Tabibuga men, Marakas. Apparently, a group of them had arrived at the station with a collection of snakes they had caught for us. When they discovered that we had all left, they set out to follow us. They were so confident of themselves as warriors, so prepared to take on all-comers, that they had followed Barry's tracks, crossed the Jimi and finally caught up with him at McKinnon's camp.

Barry laughed.

'Just when I thought I had a couple of really good fool-proof murder charges to hang on them,' he said, 'the bastards turn up trumps like this.'

* * *

Our charter plane turned up on time and together with Barry we were back in the Wahgi valley in little more than an hour. Our Jimi jaunt had been quite productive. We had quite a large collection of snakes, a hornbill, the cockatoo I had traded from the pygmies, a cuscus – a snow-white, slow-moving climbing marsupial the size of a cat – and our three little dwarf fig parrots. But no birds of paradise either in the flesh or on film. Fred Shaw Mayer, however, had entirely unexpected and exceedingly good news for us. During the months that we had been away, Garai, one of the men working for him, had found a casuarina tree where a plumed Count Raggi's Bird had started to display. It was not a traditional dancing site. Instead of half a dozen or more birds dancing together competitively, there was just a single one. But it was appearing regularly every morning. So far Garai had not succumbed to the temptation to shoot it. Fred was not sure how long Garai's self-control would last.

We went out before dawn the following morning with Garai leading the way. As we neared the casuarina, a bird took off from its branches and flew off down the valley. It was too dark to see its colours but its silhouette, with clumps of plumes trailing from its flanks, was enough to tell me that it was a bird of paradise. We were too late. But I had seen which branch it was that the bird had left and so we were able to plan just where we might put the camera to get the best view of it.

The next morning we set out at 3.45 a.m. and were well concealed in our planned position while it was still dark. I had a parabolic dish, two feet across with a microphone at its centre which would focus the bird's calls – we did not yet have gun mikes with which to record distant sounds. The valley below was blanketed in cloud. The sky began to pale and up from the valley flew a plumed bird. It settled in the tree ahead exactly where we hoped it would, and began to preen. Its toilet completed, it called, but it seemed to be in no hurry to start its dance. That, in fact, was a mercy for there was still not enough light for Charles to film.

The sun bulged above the horizon and two unplumed birds arrived. There was no way of knowing from their appearance whether they were immature males or females but the fact that the plumed bird took little notice of them suggested that they were males. Had one of them been a female the plumed bird would suddenly have rushed at her and tried to copulate. Unplumed males were no competition so could be ignored. Then as sunlight began to spread across the valley, the male suddenly ducked his head and threw up his red plumes over his back. It shone on his golden head and glinted on his shimmering emerald throat.

Charles' camera whirred. It seemed to make an alarmingly loud noise, but the dancing bird, obsessed with his own performance, took no notice of it. Up and down he scuttled, shrieking passionately. Charles put a new magazine in the camera and continued filming. As the sky brightened, he was able to change his relatively wide-angled lens for a more powerful longer-focused one that would give him a closer shot. But I was getting anxious. The camera was so loud that I could not get any worthwhile recording of the bird's calls as long as he was running. I whispered urgently to him, asking him to pause for just a few seconds, but he was understandably so thrilled at getting, at last, the shots that had been our target for so long, that he was reluctant to do so. Finally when he had run out of film and had to reload a magazine anyway, the grinding noise of his camera stopped. I switched on my recorder. The bird called twice – and then opened his wings and glided down to the valley.

Fred Shaw Mayer had other good news for us too. Sir Edward Hallstrom had said that we might take back some of the immature birds from his aviaries as a gift to London Zoo. The group Fred assembled for us was rich indeed – as well as Count Raggi's, the species we had filmed, there were Magnificent, Superb, King, Princess Stephanie's Birds and –

best of all – a little male King of Saxony Bird, a species which had never before been seen in any zoo outside Australia.

Charles flew home carrying the undeveloped film with him. I took the birds and all the other animals to Rabaul. I crossed the South China Sea by cargo boat to Hong Kong and then flew to London. It was, the Zoo said, the most comprehensive and important collection of New Guinea species to reach the Zoo for many years.

* * *

Putting together our film of the bird of paradise's display was not easy. The shots were against the light so that the bird was little more than a silhouette and the fact that television – and therefore our film – was still in black and white meant that viewers could not see how magnificent the birds are in actuality. And there was a major problem with the sound. I had only managed to get a brief snatch of the male's call – one group of two quickly repeated notes – *wah-wah* – followed by one of three – *wah-wah-wah*. The film needed them not only during the dance itself but also to accompany the introductory shots which showed the cloud-filled valley at dawn. So I joined the beginning of the recording to its end, forming a loop that could be played continuously and faded up as needed. The sound track in its finished form, I thought, was quite atmospheric and certainly made a good backing to the limited footage we had managed to get.

Some appreciative letters arrived after the programme's transmission. One was from my old zoology professor who was an expert on bird song. After congratulating me on getting any coverage at all of the display, he made a particular point of how interesting the Raggi calls had been. Had I noticed, he asked, that the bird had only called in either groups of two, or groups of three and that it did so in strict rotation? It never gave two double calls following one another, nor two triple calls. He thought this so remarkable that he urged me to write an account of it for a scholarly journal. More than a little embarrassed, I had to reply and explain about recording loops. But his letter did make me realise just how inadequate our filming of these wonderful birds still was. Sometime I would have to do better. I had not yet got a film about birds of paradise out of my system.

8

Paraguay

I am not sure, now, why Charles and I decided that our next trip should be to Paraguay. Perhaps it was just that South American animals had been less seen on television than those from Africa. We ourselves, it is true, had been to Guiana five years earlier, but Paraguay was a thousand miles farther south, on the other side of the Amazon basin, and there we should be able to find many creatures that we had not filmed before. And Paraguay, for its size, is also a very varied country. It is bisected by the Paraguay River that runs virtually directly across it from north to south for five hundred miles. To the east of the river there is rain forest. To the west lies a flat plain, the Chaco, that floods during the wet season but is a parched cactus-ridden desert during the dry. And in the far south there are the beginnings of the grasslands that stretch down across Paraguay's border and ultimately become the flat pampas of Argentina. Each area, I reasoned, must have its own characteristic assemblage of animals, and I felt sure that each, somehow or other, would provide material for one or more programmes.

I managed to find someone in London who had influence with a great Anglo-Argentinian meat-packing company that owned several vast ranches in Paraguay. They were hugely helpful. One of its representatives would meet us at the airport to make sure there was no problem with customs; guest houses on the estancias, as the ranches are known, would be ours for the asking; and the company's VIP launch would be at our disposal to transport us up and down the rivers. If necessary we could even use one of the company's small aircraft. We were clearly in for the most luxurious *Zoo Quest* so far.

In Asuncion, the capital, one of the company's Paraguayan fixers took us round the shops to make sure that we were properly kitted out. *Sombreros*, of course, were essential. And we would certainly need *ponchos,* blankets with holes in the middle, to wear as cloaks. But that was just the beginning. The only way to travel in the Chaco, he said, was on horseback, so we should wear *bombachos*, pleated baggy pantaloons. And since the cactuses there was so thick, the *bombachos* would have to be protected by *piernera,* leather leggings. And we should certainly wear a *faja,* a strip of decoratively woven cloth about six inches wide and four feet long that is wound three times around the waist. I suggested mildly that maybe we might forego a *faja*, but our guide was insistent. Anyone on horseback had to wear one. It prevented the guts, as he put it, from banging about.

We emerged with the full set. 'I don't know,' Charles said, 'whether this lot will be any good in the Chaco, but I'm sure we would be winners at a fancy dress party.'

I had discovered before we had arrived that Paraguay has a delightful musical tradition. It is based on the country's individual version of the harps that were introduced to South America by the Spanish in the sixteenth century. I thought that, used properly, it could give the series an individual character. Happily, we found and recorded a splendid group of three harpists and five guitarists, who played all the standard Paraguayan dance tunes and folk songs with great panache. One melody was particularly catchy. It started with an imitation of the call of Paraguay's national bird, the bellbird. That, I decided, we should use as our signature tune. It would prepare people for seeing me in a *poncho*, a *sombrero*, *bombachos* and a *faja*.

* * *

Our first jaunt was into the rain forest. We recruited an interpreter, Sandy Wood, the Paraguayan-born son of Australian immigrants. He spoke Spanish, Guarani the local Indian language, and English with a broad Australian accent that sounded very strange coming from the lips of someone who had never left South America. He said he knew just the sort of forest we were looking for. We should sail up the Paraguay River for a hundred miles and then turn east up the Jejui, one of its major tributaries. He knew because he had once worked up there as a logger.

It sounded a good idea. The company generously said that we could have the use of their launch, the *Cassel*, for two or three weeks. We went down to the docks to inspect her. She was a thirty-foot-long diesel-powered cabin cruiser. Her galley and refrigerator were ready stocked with luxury foods, including considerable quantities of all the varied kinds of the company's tinned meats. A small high-powered speedboat was tied to her stern. That could swoosh us away on short trips up smaller streams if anything caught our fancy.

We would live on board. Each of us could have our own cabin. There was plenty of space for Charles to spread out his equipment. The main saloon had netting to keep out mosquitoes and other biting insects which, we had been told, were the only unpleasantness we were likely to encounter.

The captain of the *Cassel* was a short pot-bellied man who wore impenetrably dark glasses and a huge bell-shaped straw hat with a brim that was turned down well below his eyebrows. This, together with our limited Spanish, made it rather difficult to establish a rapport with him. Sandy helped us to explain our plan. The Captain was not impressed. None of his previous VIP passengers had ever wanted to leave the main river. He himself had never been up the Jejui and he had no ambition to do so. We tried to explain that we wanted to go there in order to get as far away as possible from human habitation. That seemed to horrify him so we tactfully discarded that as a persuasive argument. In the end, he agreed with a shrug of his shoulders, to go up and have a look at the Jejui, but it was clear that he was not going to be an ally.

* * *

Once installed on board, we set off northwards up the Rio Paraguay. Charles, Sandy and I settled down for a spell of the easy life. The Captain, still in familiar waters, drove the *Cassel* up the river at considerable speed, leaving lesser craft bobbing about in our wake. The only hazards he had to negotiate were rafts of *camelote*, a floating weed that could be so thick that the *Cassel* had difficulty in cutting through it. It was the cause of our first disaster. On the second afternoon, Charles and I were dozing in the cabin, when the engine suddenly stopped. We went on deck and saw the cushions and the wooden bench from the speedboat sailing away down-river. The speedboat itself was still attached to the *Cassel*'s stern, but it was submerged. The Captain had made a recklessly fast turn to avoid a patch of weed, and the speedboat had capsized in the wake.

It turned out that Charles and I were the only people on board who could swim so it was up to us to retrieve the speedboat. Fortunately the river here was not very deep. Even so, it took us more than two hours to drag the submerged boat into the shallows and bale it out. But the seats and their cushions were lost. We set off again, rather more soberly.

The next day we turned east up the Jejui. The Captain, who had driven with such panache up the main river, now proceeded very slowly indeed. The Jejui meandered a great deal and he approached each bend with increasing caution. If it looked to be at all sharp, he reconnoitred in the speedboat before committing the *Cassel*. On the evening of the fourth day, he returned from one of these inspections looking even more dismal than normal. The *Cassel* was too big to get round it, he said. We could go no further. It was already late, we would moor for the night and tomorrow we would return down- river. We cajoled, entreated, argued, but he was adamant. It was not our boat, so we could not overrule him.

We climbed into our bunks that night very depressed. It seemed that we might have to slink back to Asuncion having completely wasted a week. While we were discussing the situation, we heard the noise of an outboard engine and went on deck to see who it was so bravely travelling up-river in the dark. Sandy knew him. He was a logger named Cajo who was on his way with three axe-men and a load of stores to start felling trees up a smaller tributary, the Curuguati. That, said Sandy, was exactly where he had thought we should go. He knew of a timber-man and his wife who had a small homestead up there at a place called Ihrevu-qua. He was sure we could stay there.

We hatched a plan. Cajo's underpowered launch already had a substantial load, but he could find room for our stores, our drum of fuel and our baggage. He could take it up to Ihrevu-qua and drop it off there. We would follow in the morning in the speedboat and catch up with him. After Cajo had dropped the gear at Ihrevu-qua he would then go on further up-river looking for a suitable patch of forest for logging. He might stay for a couple of weeks, depending on what he found. Having shown the axemen what he wanted them to do, he would then return. On the journey back his launch would be practically empty so he could pick us up from Ihrevu-qua with all our gear and take us down to rejoin the *Cassel* which would be waiting for us on the Rio Paraguay. We shook hands on it. Quickly we assembled our stores and transferred them to Cajo's launch. He started his engine and disappeared into the night. We went to bed feeling greatly relieved at having found such a happy and fortuitous solution to our problem.

The following morning we realised that the plan, hatched at midnight in desperation, was quite daft. The speedboat was not at all suitable as our main form of transport. It had no shelter for its passengers. Having lost its seat, it had nowhere comfortable for anyone to sit. Effectively, it only had two speeds – either impossibly slow or very fast indeed with its bows lifted and planing on the surface, which was not really a sensible way to proceed on a narrow river where there might be snags and submerged logs that could rip out its bottom. What would happen if we broke down before we caught up with Cajo, since we only had minimal food with us and no camping gear apart from hammocks? Suppose Cajo found so much to do that he decided to stay longer than a couple of weeks, how would we get back? And how far could we go before we ran out of fuel? Our drum of fuel had gone with Cajo. All we had were two full tanks, one on the engine and one spare. But the sun was shining and the thought of exploring the river and the forest up-river was an exciting one. In any case, much of our gear had gone ahead. We were already committed. It was, at any rate, a relief to wave goodbye to the Captain.

* * *

That day we enjoyed ourselves. The Jejui was still broad enough for us to travel at speed without too much risk and we roared upstream, watching cormorants flapping their way up the river ahead of us, and toucans and parrots flying across it in the sky above. We caught up with Cajo in mid-morning, waved happily and went on past him.

Ihrevu-qua – the name meant, rather unattractively, 'vulture's hole' – was rather farther upstream than Sandy had remembered. We still had not reached it by nightfall, so we slung our hammocks in the forest and ate most of the food we had brought with us for our supper. In the middle of the night, we heard Cajo's launch labouring noisily past us; but he didn't stop.

The next day was not so good. As we approached a particularly acute bend, Sandy, who was driving very fast indeed, had to swing the steering wheel so swiftly that he broke the wire cable connecting the wheel to the rudder. That took a couple of hours to fix, two hours that seemed all the longer with the knowledge that Cajo and our gear was ahead and still steaming steadily farther and farther away from us. The skies started to cloud over. Soon after midday it began to rain. We had nothing to shield ourselves from it. Soon it was a torrential downpour, the pelting drops blistering the river surface. You get very cold indeed if you are wet through and travelling at speed, but that was better than travelling at snail's pace. Again and again we rounded a bend in the river and peered into the lashing rain without seeing any break in the forest lining the banks ahead. It was almost dark before we spotted the clearing.

A man and a woman, hearing us arrive, came out in the drenching rain to welcome us. They were Sandy's friends, Nennito and his wife. We sheltered in their tiny hut, but there was only room in it for one extra person to sleep. It seemed best that Sandy should be the one to do so. Charles and I, said Nennito, could sleep in an even smaller hut that served as a store. As we opened its door, a couple of bats flew out. Others – presumably as reluctant as we were to go out in the rain – stayed roosting in the rafters and craned their heads round to look at us as we went in. The smell inside was dreadful. It came from putrefying salt beef in three great jars that stood against one wall beneath a rough wooden shelf. There was only room to sling one hammock. Charles did so and curled up in it. I stretched out on the shelf above the stinking meat jars.

Lying there in the dark, I thought I could hear a curious rustling noise even above the steady swish of the rain on the thatched roof above. I switched on my torch. The mud and wattle wall within an inch of my ear was covered with a glistening moving veil of cockroaches that had come up from the rotting beef in the jars beneath. As I played my torch over them, they all turned as one and the veil sank beneath the shelf, like a projector screen rolling back into its tube. When it had disappeared, I turned off my torch. Within seconds the rustle started again and the torch showed me that the cockroaches were already climbing up again. After playing this game two or three times, I gave up. There was nowhere else for me to go. It was better than sitting out in the rain. Marginally.

Ihrevu-qua was certainly a paradise for an entomologist. It was not just the cockroaches. I have never, before or since, been anywhere that had a greater number or variety of insects that sting. They worked in relays. Mosquitoes took the morning shift. There were several sorts. The most vicious had distinctive white heads and were so aggressive that we ate our breakfast sitting in the smoke of the wood fire on which we had cooked it, hoping vainly that it would keep them away. In mid-morning they went off duty and retreated to roosts beneath the trees by the riverside. Their responsibilities were taken over by *mbaragui*, large flies like bluebottles, which when they stabbed you with their proboscis left a little scarlet spot of blood beneath the skin. If you were quick, you stood a chance of getting your own back by swatting them. But our main persecutors did not give us even that satisfaction. They were *polverines,* tiny black flies the size of dust particles, so small in fact, that it was difficult to believe they were living creatures – until you started itching because they had bitten you. Mosquito netting was no defence against them. They drifted straight through it. Insect repellent deterred them not at all. They harried us from the afternoon onwards into the evening and then throughout the night until such time as dawn brought back the white-headed mosquitoes.

But insects, as well as being our tormentors, also brought unforgettable glory to Ihrevu-qua. There were butterflies. After a heavy rainstorm, when the sky was washed clear and the sun beat so strongly on the rocks by the riverside that they were painful to a naked foot, butterflies appeared in thousands. Showers, squalls, eddies, blizzards of them. They came in such numbers you could hardly see through them to the other side of the clearing. They drifted down from the tall surrounding trees like ticker tape during a triumphal procession in New York. I knew of no way by which I could make even the vaguest estimate of their numbers. I started collecting some of those that swirled round our hut and gave up when I had over ninety different species. They were not large. Some were pure sulphur yellow, some a rich orange. One kind had a single red stripe across its fore wings and glinting blue on the hind wings. One had a delicate zig-zag border round its outer wing margin. Another Sandy called *ochenta-y-ocho*, for it carried 88 written on its underside in elegant black numerals.

Down by the river, where the water lapped on a narrow sandy beach, there were other, bigger ones. These were swallowtails, the rear of their wings being elongated into black commas. There were half a dozen or more species of these, some yellow patterned with black, some pure black, rich as velvet, and inscribed with spots of deep carmine. Each species, when choosing a place on which to settle, seemed attracted by its own pattern, with the result that each formed its own platoon on the parade ground of the beach.

They settled in ranks, close to one another, their upright wings a-quiver, and then uncurled their probosces that were coiled like watch-springs beneath their heads, to probe the sand and suck up water. As they drank, they absorbed the river water's dissolved salts and then squirted what remained from the rear of their abdomens. If we approached cautiously, we could sit beside them. Then others would come and settle on our hands and faces to sip the even saltier beverage of our perspiration. It was a delightful sensation to feel their thread-like probosces playing over our skin, but it was only achieved at a price, for sweaty skin is an even greater temptation to mosquitoes than it is to butterflies. We were lucky if we got a tickle without getting a bite.

The biggest butterflies of all lived in the forest. These were the famous electric-blue butterflies of South America, the morphos. They preferred to fly through relatively clear spaces rather than flutter around among the branches and would flap in a leisurely way along the clear logging avenues that Nennito had cut in the forest. Catching one was not easy. At first I gave chase, but one futile swipe from my net and the morpho would immediately change its flight style. Instead of flapping lazily, it would suddenly put on a turn of speed and fly up straight and fast into the branches, far beyond my reach. For some illogical reason, it had not occurred to me before that an insect might be able to see things far ahead and recognise them sufficiently well to react to them just as a bird might, but I soon learned that a morpho many yards distant would swerve to avoid me if I made any movement with my net. To be successful, I had to stand motionless with net poised, like a cricketer facing a fast bowler, and only make a move when the insect was right upon me.

But that was not the most efficient way to either catch or film them. Morphos have feeding preferences that are in distressing contrast to their beauty. They like the smell of urine, dung and putrescence. A highly-smelling bait based on any one of these would bring them down from the tree tops and give us a real chance to marvel at their glory – even though we might have to hold our noses.

My journal is full of names of birds we saw in the forest – surucu trogon, urraca jay, rufous-headed mot-mot, toco toucan, blue-fronted Amazon parrot, red and yellow macaw – but we had very few images of any of these on film. The forest was, for the most part, far too dark for filming and Charles' lenses, though the most powerful then available, were still not good enough to give us reasonable close-ups of the forest creatures that we were able to watch through binoculars.

As the days passed we began to make side trips in the speedboat up smaller streams or walk to an Indian village some miles away in the forest. But we did not dare to stay away from Ihrevu-qua for long. If we missed Cajo on his way down, he might think we had left and carry straight on down-river, in which case we might be marooned for weeks.

Eventually, after a fortnight, Cajo did reappear. With some relief we clambered aboard his launch, loaded all our gear and hitched the speedboat to its stern. With a relatively light load and travelling with the current he made much better speed than he had achieved on his way up. Two days later we were back on the Paraguay River. I would not have believed it possible that I should be glad to see the Captain again, but I was.

* * *

Our trip to the southern grasslands was altogether more successful – and much more comfortable. The company's plane took us down to an estancia in northern Argentina which had once been managed by a Scotsman who was an enthusiastic naturalist. He had left one part of it completely wild and his successors had kept it that way.

It was rich with birds. The major problem facing any bird on the pampas is finding a safe place for its eggs. There are hardly any trees or bushes in which to build nests. Burrowing owls solve the problem by using holes in the ground. Their name is somewhat misleading, for although they can burrow they prefer to take over holes that have been made by rodents, and during the day they sit glaring in a proprietorial way alongside the entrance. The oven bird builds a nest out of mud. It is as big as a football with a domed shape somewhat like the oven used by local people. The entrance is a vertical slit with a side-wall immediately within that acts like a baffle so that it is not possible for a paw – or a hand – to grope inside and reach the eggs. Wattled plovers don't make a nest of any kind. They simply lay their four eggs in a shallow depression on the ground and rely on their near-perfect camouflage to cloak them in invisibility. The rhea, South America's equivalent of the ostrich, also lays its eggs on the ground but it is so big and strong that it is able to chase away anything that threatens it.

A rhea's nest is a truly extraordinary sight. Each egg is twelve times the size of a chicken's egg and the nest we found contained thirty of them. We built a hide beside it and filmed the way in which this huge clutch was produced.

It is the male rhea that builds and tends the nest. He has a harem of a dozen or so females who live in his territory, cropping the grass. He courts all of them in turn, waving his wings like a fan dancer until the female, beguiled, sinks down to the ground so that he may mount her. When her egg, a few days later, is ready to be laid, he solicitously escorts her to his nest and paces up and down on guard beside her until she has deposited her egg and is ready to leave. After each member of his harem has delivered her contribution they drift away, leaving him to incubate the whole clutch.

* * *

We also wanted to go to the Chaco in the west of the country. There we should be able to find those quintessential South American animals, armadillos. The London Zoo was hoping that we would bring back a small collection of different armadillo species, and I was glad to try to do so for I felt sure that they would be engaging performers in the studio sequences that were still an integral and valuable component of *Zoo Quest* series. But travel in the Chaco was not going to be easy. Had we come a few months earlier, it would have been even worse, for then most of the Chaco had been underwater. Now, in September, the floods had drained away leaving behind wide muddy plains that had baked hard in the sun. There were a few swamps where the water still lingered, and isolated patches of cactus and thorn bush growing on small elevated patches of ground that only a few weeks earlier had been islands. There were no roads for motorised vehicles of any kind.

Our helpful meat-packing company allowed us to use their small plane to get to one of the more remote estancias and there we were able to borrow a wagon drawn by two oxen. That would carry the gear – and any armadillos we might catch. We ourselves would travel on horseback. We set off heading west towards the Pilcomajo River which forms Paraguay's western border with Argentina.

There were at least half a dozen different species of armadillo to be found there. In Guarani, they are known as *tatu*. After several people had reacted oddly to our explaining that we hoped to chase all kinds of tatu, we discovered that *tatu* was also a slang expression for girls. An English equivalent might be 'crumpet'. When we cleared up any misunderstandings on this point, people seemed to find it even stranger that we should be chasing four-legged *tatu* rather than two-legged ones.

Armadillos, as their Spanish-English name suggests, have an armour of horny plates, broad ones around the shoulder and the rump, with a varying number of narrow bands between, around the waist. The commonest species has nine such waist-bands. It is about the size of a large rabbit and much hunted by the local people who relish its flesh. The Indians we met, working as cowhands on the estancias, were able to catch them any time we wanted, so we delayed acquiring this kind until just before the end of our trip.

The six-banded armadillo was somewhat similar but hairier and rarer and has a taste for carrion. The seven-banded favours insects such as ants and termites. It has distinctive long ears and is called by the locals *mulita* meaning little donkey. All these were formidable diggers. If we failed to catch them before they disappeared into a hole, we gave up the chase, for they could dig much faster than we could. Even if we groped in the burrow and managed to grab their tail, they would brace their feet, so that their back pressed against the top of the tunnel, and in that position they were virtually impossible to extract.

The most charming of them was the three-banded. No bigger than a grapefruit, it can, unlike the other species, roll up into an armoured ball. A triangular plate on its head fits neatly beside a similar-sized one on its tail so that together they form a rectangle. Once curled and shut tight, few things can harm a three-banded armadillo. Leave it alone and sit still and a crack will appear. A tiny glittering eye peers out apprehensively. If it likes what it sees, the grapefruit splits apart and the armadillo trots away, running on the tips of its claws.

Three-banded were easy to catch for their only defence is to curl up. They would become alarmed at the sound and vibration of our approaching footsteps and we would find them lying around underneath bushes and beside cactus, like fallen fruit, waiting for danger to pass them by. We would pick them up and put them for safe keeping in a cloth bag until such time that we could give them bigger living space in the bottom of the ox wagon. Most animals dealt with in this way, snakes, small birds and mammals, will lie still in the darkness. The little three-banded however, uncurling in a bag, ran, just as if it were out in the open, and if we did not keep an eye out, we would see the bag rolling away across the landscape with the little *tatu* trotting away in the darkness within.

The biggest of them all, however, was a different matter. The people called it *tatu carreta* – the 'cart' armadillo. 'Cart' is perhaps something of an exaggeration, but the animal is certainly the size of a small wheelbarrow. One of those would be a great prize, for London Zoo had never exhibited one and they were very rarely seen in captivity anywhere. We asked about them everywhere we went. People on the estancia where we had landed told us that if we wanted to find one, we should go to the far west of the Chaco. In any case, they said, we ought to go that way because the *patron* of one of these remote and wild estancias was an Englishman who was very proud of his nationality and constantly complained that Paraguay was a barbarian country. There was no one in the Chaco to whom he could speak in his native tongue, the richest and most beautiful language in the world, whose literature gave him such limitless pleasure. We would be bound to be well received. I asked his name. 'Horchay Heeleth,' I was told. It did not sound very English to me – until I realised that this was a Hispanic rendering of the name 'George Giles'. I asked how we should find our way there. 'Just ride into the setting sun,' they said.

So we did. I had been somewhat apprehensive at the prospect of travelling on horseback for days on end, but we soon got used to it. Our mounts did not expect us to bounce up and down as is *de rigueur* in Britain and no matter what gait our horse adopted, we sat stolidly in saddles well padded with sheepskins. Our steeds, in any case, were hardly mettlesome. The only times they showed signs of exerting themselves were when we passed close to one of the drying swamps, for swarms of biting insects rose from them making a fearsome drone that was audible twenty yards away. When our horses heard it, they steadfastly ignored any suggestion we tried to give them by way of the reins that they should go any closer.

Chaco horses were expected to find all the food they needed in the thorn bush. If they were not required for some time, the estancieros simply turned them loose to fend for themselves. Those we had borrowed were tired after a week or so, and at one of the smaller estancias we visited, the *patron* offered to provide us with two more, to lessen the strain on the two we owned. I explained that we did not plan to return that way, so would not be able to return them. But that was no problem. We could buy them, he said, and then turn them loose when we had finished with them. Since their price in guarani converted to sterling was only about five pounds it seemed an acceptable arrangement.

After several days' travel, we saw far away on the western horizon what I thought must be George Giles' estancia. As we neared, I could see through my binoculars that someone was reclining in a hammock in the shade of its veranda. The figure did not stir as we approached. It was not until we had dismounted and tied up our horses to the hitching post, that he swung his legs out of his hammock.

I held out my hand and explained that I was English. We had come, I said, because we had been told that he too was English and yearned to be able to speak his mother tongue.

'Bugger me,' he said. And that sadly proved to be the only phrase he was able to recall of the language of Milton and Shakespeare.

* * *

George confirmed that *tatu carreta* were indeed occasionally found in his area. One of the Indians who worked for him took us out the next day to look for one. We found the burrows they had made when digging for termites. They were formidable excavations, big enough for me to wriggle into, but none, alas, as far as I could discover was occupied. Nor did we find any that seemed to be recent. So we moved on.

We camped at night in the thorn bush patches. Clearing space with my machete for my hammock one evening, I carelessly slashed at a palm stem and a six-inch thorn stabbed into the side of my hand and broke off. Charles, on the grounds that he had once had ambitions to pursue a medical career (though he had failed the initial examinations), took charge and found our first aid kit. He inspected the wound and tried unsuccessfully to grip the end of the spine, buried in the flesh, with a pair of forceps. He would have to operate, he said, severely. He found a scalpel. I lay in my hammock, looking away, but I could not prevent myself from flinching as Charles started his work. This craven behaviour on my part, he said, made his task impossible. I had to lie still. If I could not do so, then I needed sedation. Morphine was the thing. This seemed to me to be a little extreme but I submitted and Charles gave me an injection in my buttock.

I have little recollection of the dreams that followed except that they were delicious. I awoke to find my hand wrapped in so many bandages that it was the size of a football. Charles however was a little contrite. The spine was still in there. He hadn't after all, been able to steel himself to cut into my flesh with the scalpel sufficiently far to reach it. Perhaps it was as well that he had abandoned his medical career. Over the next few days my hand continued to throb and swell. When at last we got back to Asuncion I went to a doctor. By now the hand was greatly swollen. The doctor nicked the wound with a scalpel and the thorn, driven by the pressure within, rose of its own accord out of my flesh like toothpaste from a tube.

We now had a very varied group of creatures – tegu lizards, young maned wolves, savannah foxes, pygmy owls, parrots, caiman, and coatimundis – but the pride of the collection were the armadillos: three-banded, six-banded, seven-banded and nine-banded. But not, sadly, a giant.

We took the whole collection down to Buenos Aires by air freighter. The airline allowed us to billet them in an empty hangar while we waited for another to fly us to London. Keeping all the animals fed and properly cleaned did not give us much free time, but I heard that an Englishman and his wife had just arrived in the city at the beginning of their own animal collecting expedition. A little detective work located him and I spoke to him by telephone. It proved to be Gerald Durrell and his first wife Jacqui. I had not met either before. Gerald, young and still unbearded with a hank of hair flopping over his face, was as funny as he was in the books which he had just started to write. He had not yet started his own ground-breaking zoo in Jersey but over a lot of cheap Chilean wine, he told me of his plans for the zoo of his dreams. I showed him our armadillos and told him what we had learned about keeping them – how, for example you could keep the little three-banded free of diarrhoea by mixing soil with its food. And I lamented the fact that we hadn't seen a *tatu carreta*. I was, I confess, more than a little envious when Jacqui said that they had already got one. It was waiting for them in a town in northern Argentina, she said. I later discovered that they were just as unlucky as we had been. The man who said he had one, in reality only meant that he was sure he could catch one. In the event neither he nor they managed to do so and I felt a little relieved that such an experienced and accomplished animal collector as Gerry had done no better than we had.

* * *

When we got back to London, my first task, as always, was to try and bring some order into a folder full of crumpled scraps of paper and indecipherable scribbles that had accumulated over the previous three months and try to turn them into receipts that would give some support to my expenses. Achieving the right totals did, it is true, require some imagination, but I did my best to make the finished document as accurate as I could. Nonetheless, soon after I had sent it off, there was the inevitable telephone call.

'Mr Attenborough?' It was a lady and she spoke with the formality still proper in telephone conversations in the 1950s.

'David Attenborough speaking.'

'Accounts Department here. I would like to query an entry in your expenses. One thousand guarani for the purchase of two horses. Surely that should be for their *hire?*'

I recalled the transaction in the Chaco very clearly.

'No,' I said, 'We had to buy them.'

'But in that case,' the voice said primly, 'they are the Corporation's property. What did you do with them?'

'Madam,' I said sepulchrally, 'we ate them.'

* * *

While we were still working on the films, I walked through the scruffy entrance hall of Lime Grove studios and saw John Betjeman, the connoisseur of Victoriana and future Poet Laureate. He was slumped on a sofa, wearing his usual battered trilby hat, smoking abstractedly, with cigarette ash falling unheeded on his waistcoat. He brightened as he saw me.

'My dear nephew!' he said.

He had invented this relationship when I had first recruited him for *Where on Earth*, the doomed travel quiz I had devised some years earlier, on the grounds that he knew my father and had once stayed at our house. Since then we had corresponded about various programme projects, and while always addressing me in this way, he had signed himself as, variously, 'Jan Trebetjeman, the noted Cornish nationalist,' 'John Quetjeman (Manx),' or 'X Frank Zanzibar (Bishop).'

'Uncle John,' I said, sitting down beside him, 'How lovely to see you.'

'Tell me,' he said earnestly, 'What thrilling place have you just come back from?'

'Paraguay.'

'How wonderful! One of the greatest ambitions of my life has always been to go to...'

His voice trailed away. '...remind me, what is the capital of Paraguay?'

'Asuncion.'

'Yes, Asuncion, of course. What an exciting city that must be. Tell me,' he leaned towards me, eyes sparkling. 'Do they have...*trams*?'

Sadly *Zoo Quest to Paraguay*, if he ever saw it, must have disappointed him on this score. Whether viewers were as engaged by armadillos as I was, I cannot tell, but on one thing nearly everyone was agreed – the music. Paraguayan harp music had hardly been heard before in Britain. Today the Corporation doubtless would have rushed out a disc of it, but then the BBC had no commercial arm. There was, however, a Paraguayan trio, who had just started recording for a European company at the time. They wore even more decorative ponchos, sombreros and *fajas* than the ones we had been equipped with and they played many of the same folk tunes we had recorded. They themselves may well have enjoyed catching glimpses of the four-legged *tatus* of their country on British television. But they must have been even more delighted when viewers bought their records in considerable numbers to be reminded of the sight of a little three-banded armadillo trotting like a clockwork toy through the cactus of the Chaco.

9

Solving Sound

Creating sound tracks to go with our films could be a complicated business. Our cameras made a loud grinding noise and we still had no method of linking our recorders to them so that the two could run in perfect synchrony. We therefore usually filmed first and recorded later, as I had done with our bird of paradise sequence. The sounds were then transferred from magnetic tape to 16mm magnetic film. In the cutting room, picture and sound were edited on separate rolls that ran in parallel, so we were able to 'lay' a sound in such a way that it roughly fitted the action that theoretically was creating it. As long as this was something simple, like a lion roaring and not as complicated as someone talking, the technique produced a tolerable result.

Sometimes, however, we failed to get any recording whatsoever of the animal we had filmed. But that was seldom a major problem. The BBC's Natural History Unit already had a vast library of animal sounds and it would usually have a recording that would suit. Its collection of bird song was the biggest in the world. The fact that in the finished film, a bird's trills and whistles did not precisely coincide with the image of its throbbing throat or the movements of its beak, did not seem to trouble anyone. We made sure, of course, that the recorded calls and songs we used belonged to the particular species we showed in picture, for we knew well enough that if there was a mismatch a hundred fanatical ornithologists were waiting, pen-poised, to demonstrate how knowledgeable they were by writing outraged letters to us.

I had had the job of recording the sound on all the *Zoo Quest* expeditions so far, so I had only myself to blame if I returned from a trip with inadequate recordings. But in between, as Head of the Travel and Exploration Unit, I produced a series of programmes called *Travellers' Tales* and later *Adventure* with material brought back by travellers and explorers for whom filming and recording had been a side-line. Their film shots could be scrappy and they hardly ever bothered to record incidental sounds, such as canoes being paddled, insects singing in a tropical forest, or the flappings made by a cloud of bats. Only too often we had to cobble together a sound track from whatever recordings we could find in the recorded sound library.

Creating a sound track – dubbing – was done in special dubbing theatres. The dubbing mixer sat in a darkened theatre behind an enormous desk covered with a hundred or so knobs and faders, while the picture for which he had to produce a track was projected on a screen in front of him.

He had several different sound sources from which to compile it. If we were only able to give him a minute or so of some important sound, he would have the magnetic film carrying it joined into a loop which could run continuously round and round, as I had done with the bird of paradise calls. He could then simply fade it up whenever the need arose. We dealt with human chatter in the same way. I always asked film-making travellers to record a few minutes of the people they met talking among themselves. Chatter is very distinctive. An African crowd sounds very different from an Italian or an Indian one, even if you cannot distinguish a single word in any of them.

One film in the *Travellers' Tales* series concerned Bushmen. Their speech is particularly distinctive for they use a great number of clicks. Our film-making traveller had brought back a short recording of it. There was not enough to cover some of the scenes, so I asked the film editor to make a loop of it.

He was so late in coming to the dubbing theatre that I telephoned his cutting room. He was full of apologies. There had been some technical problems. When at last he appeared, he explained. The traveller's recorder, he said, had developed some strange fault, for the chatter recording was spattered throughout by strange electronic clicks. He had had to cut each one out physically and there were so many that there was barely enough left of the recording to make a loop. And then there was another delay while he went back to the cutting room and gave the Bushmen their clicks back.

The mixer could also use sounds from 78 rpm vinyl discs. The General Effects Gramophone Library could provide him with a vast range of sounds – claps of thunder, slams of doors, dogs barking, hinges creaking, outboard engines chugging. An experienced gram-swinger in charge of a bank of four turntables would mark the precise groove that produced the required noise with a yellow wax pencil and then lower the needle on to it so that it sounded at precisely the right moment.

Men marching were a special problem. The sound of their footsteps had to be at exactly the right speed to be convincing. A disc could be induced to produce that by adjusting the speed of the turntable even though doing so inevitably raised or lowered the pitch of the sound. Other repetitive sounds could be created in less obvious ways. The squelches made by an intrepid explorer walking through a swamp could be improvised by a disc of a donkey chewing hay, played at a quarter speed with the volume control being swiftly turned up and down in time with his feet. Silence, oddly perhaps, was also tricky. A total absence of sound might give viewers the impression that their sets had developed a fault. So for the awesome silence of the central Sahara we would add a recording of the Victoria Falls played very slowly and extremely softly.

Many noises, however, were best made physically in the dubbing theatre and many people were amazingly skilled at producing them. The most famous of all was a lady known as Beryl the Boot. She could improvise footsteps of every conceivable kind, with her footwear and metal trays of gravel, sand, coconut shells or – if the steps were in snow – custard powder in a silk stocking squeezed with the right rhythm. The sonic illusions she could create with nothing more than the London telephone directory beggared belief.

Animals eating were among her most widely appreciated and convincing performances. Such effects were in greater demand than you might suppose. It is not difficult for a cameraman to get a close-up shot of, say, an owl sitting high on the branch of a tree, ripping apart the body of a vole. To get a suitable accompanying sound, a microphone would have to be really close to the bird's beak and that is seldom if ever possible. In such a situation, Beryl the Boot was in her element, leaning close to the microphone, with her eye cocked at the screen showing a picture of the owl tucking into its meal, as she masticated something particularly chewy herself.

If all else failed, music could fill the gap. In the 1950s our programme budgets were so small that we could not afford to commission specially composed pieces for our films. Instead we used discs of works by obscure composers which had been specially recorded by BBC Orchestras and could be used in our sound tracks without any extra charge. We would describe the pictures to the Recorded Music Librarian, John Carter, a man of the most recondite musical tastes, and he would suggest works by composers few of us, back then, had ever heard of. Looking back now, I am astonished at seeing who they were. Telemann, Vivaldi and Berlioz at that time were very seldom played or recorded commercially and their works solved many of our problems. I remember using Bartok's Music for Strings, Percussion and Celesta to accompany the pounce of a praying mantis and thinking that by doing so I was being daringly avant-garde.

Cueing all these ingredients and mixing them together at the right levels without ugly unnatural bumps required great skill. The operation became the more nerve-racking because the whole confection had to be recorded on magnetic film in ten-minute stretches. A single mistimed footstep nine minutes into the reel would mean that the whole sequence would have to be started again from the beginning.

Bob Saunders was one of the great dubbing virtuosi. Every working day, he sat in the dark crouched behind his huge mixing desk in the Television Film Studios, shoulders hunched, chain-smoking, taking his cues from the footage counter numbers projected at the bottom of the screen, twisting knobs, shouting commands to the gram-swingers and gesturing to Beryl the Boot. He was as tense and as emotionally involved in the balance and accuracy of timing as many an orchestral conductor.

When we were in full production with a series of travel films, we had a weekly booking with Bob in his theatre. In the morning I would speak a commentary that I had written. In the afternoon we would prepare an effects and music track and then mix the two into the final version. At lunchtime we drank and ate together in The Red Lion, the pub opposite the studios. It was almost the only time that Bob saw the light of day. He particularly loved the travel programmes. He said they reminded him of what sunshine looked like. What wouldn't he give to get out of the daily darkness of his dubbing theatre into the wide open spaces.

10

A Quest for Custom

Soon after we got back from Paraguay, a letter landed on my desk from an anthropologist, Jim Spillius who, with his wife Elizabeth, was working on the remote island of Tonga in the south Pacific. He was passing on to me an invitation from Tonga's ruler, Queen Salote, to visit her kingdom and film one of its most important ceremonials.

In the 1950s, everyone in Britain knew who Queen Salote of Tonga was. They might have found it hard to pinpoint the position of her kingdom on the map, but they knew her. She had taken part in the procession for Queen Elizabeth's Coronation in 1953. On the way back from Westminster Abbey it had started to rain. At first it drizzled. Then it poured. Most of the heads of state and other diplomatic and royal dignitaries, understandably, had the hoods of their carriages put up. But not Queen Salote. She sat unprotected, beaming and waving while the rain drenched her. The sodden crowds lining London's streets cheered enthusiastically and millions at home watching the Coronation, British television's first smash-hit, were delighted.

Some said that the reason she had travelled unprotected from the rain was that the hood of her carriage was stuck. The fact that she did not seem to mind would have been enough to make people warm to her. But the truth was even more endearing. An attendant had, in fact, offered to put it up, but she had refused to let him do so. She said she wanted to share the joy of the day with the crowds.

The Tongan ceremony that she wanted to have filmed was one in which she herself took part. It was called the *taumafa kava* and it was so sacred that no European hitherto had been allowed to witness it. If, however, the BBC would make such a film, the production team would be entertained throughout their stay and be given every help to film any other aspect of Tongan life they wanted. Furthermore, she would even allow some or all of the Royal Kava Ceremony to be included in these programmes.

I had no doubt that the Queen's personality was still so intriguing that viewers would be delighted to see something of her in her own kingdom.

Tonga was still a little known island. Few people visited it for then it had no airfield and no hotel to accommodate them even if they got there. On the other hand, the *taumafa kava* ceremony, as described by Spillius, did not sound thrillingly spectacular. Fundamentally, it consisted of preparing and consuming the island's traditional drink, kava. I did not see how that, even embellished with a lot of background material about the island, could sustain more than a single programme. I would have to devise a series of at least half a dozen episodes to justify the expense of going to such a distant part of the globe.

I could hardly suggest that we should make a *Zoo Quest*, for the islands of that part of the Pacific are so small and so scattered that they have very few wild animals to go questing for. On the other hand the human inhabitants in neighbouring island groups practised all kinds of spectacular rituals. The notion of a quest had worked well zoologically. Perhaps it might do so, anthropologically. Why not make my next trip a quest for strange Pacific customs of which Queen Salote's *taumafa kava* would be just one? My bosses agreed.

Charles Lagus had decided that after five *Zoo Quests* in four years, he would like to spend a little time at home. He had been the first cameraman to film for the BBC on 16mm stock, but now, thanks to its cheapness and many technical improvements in both cameras and telecine machines, many productions were using it and there were several experienced cameramen who might succeed him. I asked his assistant, Geoff Mulligan, if he would like to come. He jumped at the chance.

* * *

We flew to Fiji and there boarded a ship that called at Tonga on its regular circuit around the islands. When we moored alongside the quay at Nuku'alofa, the island's capital, Jim Spillius was there to meet us. With him stood the Queen's special representative, Ve'ehala, the Keeper of the Palace Records.

Ve'ehala had been deputed by Queen Salote to look after us throughout our stay. He was small and stout with close-cropped black frizzy hair and an irrepressible gurgling laugh that usually lasted so long that he had seldom got to the end of it before he had to draw breath – and that turned his continuing gurgles into falsetto squeaks which were quite alarming until you got used to them. As befitted the custodian of the island's traditions, he was a noted performer on the Tongan nose-flute and he habitually wore formal Tongan dress. This consisted of a cloth kilt, a *vala,* with a mat woven from pandanus leaves wrapped around it and tied at the waist with string. Mats are extremely important in Tonga. Some are very ancient and are among the most valuable things that a person can own. Those worn around the waist vary. For everyday wear, Ve'ehala usually put on quite a narrow one with a few short straps dangling from it, as a token to indicate how long a proper one might be. On important occasions, however, and always when seeking an audience with the Queen, he would wear the full-size version, a huge stiff tube that extended from the middle of his shins to half way up his chest and was secured by a thick girdle of plaited coconut fibre. This, of course, was very cumbersome when surrounded by European style furnishings, such as chairs, so in those circumstances he might carefully step out of it, leaving it standing upright in a corner.

My first job was to work out with Ve'ehala and the Spilliuses exactly how we would film the *taumafa kava*. We sat together in Ve'ehala's office, close by the Palace. It had little in it apart from a desk and a telephone that linked him to other offices in the Palace and the town. There was also one book. Significantly, that turned out to be *Burke's Peerage*. He told us with a heavy sigh that much of his time was taken up with replying to correspondents who wrote to the island in great numbers asking for examples of its stamps. But he also had to deal with all aspects of royal protocol and preparations for the forthcoming ceremony were demanding his full attention. Philatelists would have to take second place for a month or so.

Kava is made by crushing the roots of a kind of pepper plant and mixing the pulp with water. It has a watery milky-grey appearance and a slightly antiseptic taste. There is no alcohol in it but it is said to have a slightly anaesthetising effect if you drink enough of it. In my limited experience it seems to rob you of the use of your legs, but that may have been because I mostly drank it while sitting cross-legged on the floor and after a little time in that position, my legs seemed to lose all sensation anyway.

It has its own rituals. It should be served in a coconut cup. You should receive it in silence, drain it in a single draught and then give thanks by clapping with cupped hands and muttering '*Malo, malo*'. It is an essential accompaniment for every kind of meeting. Friends when they call on one another habitually drink kava together. Business discussions inevitably start with sips of it and feasts of any kind would be unthinkable without it. The *taumafa kava* ceremony is the most sacred occasion of all, for then the Queen, who in the minds of most of her subjects was semi-divine, would drink kava with her people.

The Spilliuses and Ve'ehala explained the details of what would happen. All the chiefs and noblemen of the kingdom would assemble on the *mala'e*, the open grass-covered ground beside the Palace, and sit down in a huge circle, a hundred yards across, facing inwards. The Queen would take her place in a specially erected pavilion on one side. Each village in her kingdom would pay homage to her by presenting gifts which would be displayed in the centre of the ring. Then kava would be mixed in an immense wooden bowl. The Queen would drink some and after her, each nobleman, in turn, would be given a cupful. The ceremony would last four or five hours. It was imperative, I was told, that the finished film contained close-ups of every single one of the hundred or so nobles as they drank.

It did not sound very exciting, but then I told myself that while I would film all of it for the Queen's film, I need only show a small sample on British television. But there was a snag. The whole event was very sacred – *tambu*. Since we would be the first white people ever to be present on such an occasion, we should remain as unobtrusive as possible. Preferably, we should not move at all, but if it was absolutely essential to do so, we must not stand upright but shuffle about in a crouching position. Above all, we must not, once the ceremony had started, step inside the ring. That was an act of desecration that would deeply offend those taking part.

I explained that it was not possible to make the comprehensive film the Queen wanted under such limitations. Wherever we sat outside the ring, half of the participants would have their backs to us, while those who faced us on the other side of the circle would be so far away that they would barely be recognisable. Ve'ehala was horrified by this revelation. If even one nobleman was omitted from the film, he said, it would be a disaster.

For three days we sat discussing the problem. We worked out at which point we might move the camera to another position. We settled the embarrassing question of which of the participants were so junior that if for some reason a pause in filming was unavoidable, their absence might be glossed over. And we agreed as to who was so hugely powerful and important that whatever happened he had to be included. The Spilliuses, as anthropologists, had their own queries about the details of the ceremony itself. Different noblemen had to be consulted on exactly what should happen. Ve'ehala bustled back and forth to the Palace, carrying written proposals for the Queen's consideration. Elizabeth started to compile a detailed script which defined who should say what, do what, and in what order. Each suggestion had to be ratified – or dismissed – by the Queen. As the script grew longer and longer, and more and more detailed, I realised that the film was not going to be a record of a spontaneous happening as I had imagined, but a visual translation of Elizabeth's script. There was even a suggestion that if in the event, there was some deviation from what the script specified, then the film should be edited to match the script rather than show what actually happened. It wasn't my idea of anthropological filming.

In between these long sessions, and while awaiting decisions on the proposals they generated, we filmed around the island. We were entertained with feasts of roast sucking pigs and breadfruit cooked in earth ovens which we ate sitting cross-legged on pandanus mats with garlands of frangipani flowers, *leis*, piled up around our necks while villagers sang and danced for us. We visited the royal tombs, square two-tiered terraces covered with a gravel of white broken coral, and the *Ha'amonga*, Tonga's famous ancient monument, two huge rectangular blocks of coral-limestone standing twelve feet high with a third placed across their tops to form an archway. We recorded the whole process of making bark-cloth in which girls with mallets, sitting in a line behind a log, beat strips of the inner bark of a kind of mulberry tree until the strips become many times their original width and can be glued together. We went down to the coast where the Pacific breakers surge on to the low coral-limestone cliffs and spout up through blow-holes as huge fountains. And we filmed in and around the Palace.

The Palace overlooked Tonga's lagoon. It was a two-storeyed white-painted gabled building made of wooden clapboard with a red corrugated iron roof that had been put up in the late nineteenth century by New Zealand builders. It was not the centre of the town geographically, but socially it certainly was. A dozen or so people lived there permanently – officials of various kinds and servants – but over twice that number spent all their days there. It appeared that Tongan traditional hospitality was such that almost anyone could move in to the Palace provided they could ingratiate themselves with one of the residents. Chiefs from outlying villages called there to pay their respects to the Queen. Members of the Government constantly arrived to seek decisions on legal points from Palace officials, and processions of people from all over the islands queued up to leave gifts of homage.

In one corridor, a small band played almost continuously. They sang with the warm harmonies of South Seas popular music, accompanying themselves on guitars and mandolines. The Queen herself was very fond of music and had composed several songs which were, of course, in her band's repertoire, but their favourite song was 'You are My Sunshine', a cheerful number that had reached the island during World War II and had never lost its popularity in Tonga. We spent a lot of time with the band, either waiting for yet another meeting with some Palace official to get decisions on further *taumafa kava* problems, or simply for the pleasure of listening to them and drinking kava. They even gave us Tongan names. Mine was Ratu Tavita – Tavita being the Polynesian version of David and Ratu meaning roughly, Chief.

Eventually, the day of the *taumafa kava* arrived. Barefooted Tongan policemen wearing their best uniform *valas* and khaki bush-ranger hats, were stationed around the *mala'e* to make sure that other European residents on the island were kept away. Even though the proceedings were to be broadcast eventually on television, foreigners – British advisors to the Government, traders and representatives of commercial firms – were not yet allowed to witness what went on, much to the irritation of some of them.

Noblemen began to assemble. Heated arguments broke out about who should be placed where and Ve'ehala, resplendent in an enormous mat, had to be summoned to try and sort things out. One particularly senior and physically massive aristocrat became alarmingly incensed and refused to sit in the position that Ve'ehala said was his proper one. As things threatened to turn ugly, a policeman stalked across and handed the nobleman a note. He read it and sat down meekly in the position he had complained about. I discovered later that the note had come from the Queen who had been watching what was going on through binoculars from the Palace balcony a quarter of a mile away.

Geoff and I established ourselves beneath one of the large Norfolk Island pines that lined one side of the *mala'e*. Before things became too *tambu*, I walked into the middle of the circle and planted a microphone, rather optimistically, on the ground and then laid out a long cable from it to the recorder that I would keep beside me as I sat next to Geoff.

The ceremony began with long lines of villagers bringing gifts, singing in harmony as they came. What each group presented was specified by tradition. There were cooked chickens, fish and mandioca, whole roasted pigs with their livers skewered to their chests, pandanus mats and rolls of bark cloth, some a hundred yards long. Most important of all, there were a dozen or so large uprooted kava plants. As each category of gift arrived, a palace official pointed to it and counted it in a loud voice, so that everyone should know exactly what tributes had been paid and by whom. Another official sat down on the side opposite the Queen's pavilion behind a wooden bowl, five feet across, in which the kava would be mixed.

Everything and everyone was now in place, ready for the Queen.

She appeared in the Palace garden and walked to take her seat in her little pavilion. Around her waist she wore a mat that was said to be some five hundred years old and was famous throughout the island. She looked truly regal – tall, erect and dignified. Roots from the biggest kava plant were ceremonially crushed, taken to the kava bowl and mixed with water.

Then the serving began. The first cup was presented to the Queen. After she had drunk, the master of ceremonies, an elder named Motu'apuaka, called the name of one of the nobles and a cup was taken to him. Inconveniently for us, since the drinking had to be done in order of social precedence, Motu'apuaka did not work his way round the circle in an orderly fashion, but moved from one side to another, so Geoff had to continually change lenses. I was able to warn him of where the server was going next by having the agreed seating plan on my knees and following the announcements from Elizabeth's script. To my surprise, the words spoken by Motu'apuaka exactly matched what she had written – and then I saw that he too had a copy of the same script half concealed in front of him.

The following day, we had a post-mortem with Ve'ehala and the Spilliuses. Some incidents we had missed altogether but we had already planned how we would edit the picture to cover such gaps. Some commands and responses we had recorded in sound only, but we knew what pictures we could use to back them. Other speeches were so distant that they were barely audible on my recording. That, however, we could deal with by arranging for Motu'apuaka to re-record the whole script on a later occasion when I would be able to hold a microphone within a reasonable distance of his lips. I could not, of course, understand Tongan and the Spilliuses were fearful that somehow commands or responses would get misplaced when we put the whole film together. Although they were not due to return to Britain for several months, they were insistent, as the Queen's advisors, that the editing of the final full version must wait until they could oversee it. Eventually I had to agree to this, but with the proviso that I would be able to copy enough from the master negative of the film to enable me to produce a five-minute condensed version of the ceremony that we needed for our own programme.

At the end of our stay, we were given a party. During the three weeks on the island we had made many friends and the celebration was an enthusiastic one. The band from the Palace played non-stop. There was beer and dancing and songs. Frangipani *leis* were loaded on our necks. At midnight, in spite of all the commotion, someone heard the telephone ring. It was a message from the Queen. She had heard that the party was going well and recommended that it should continue for at least another hour. The following morning, exhausted and sorry to leave, we sailed away.

* * *

Tonga's *taumafa kava* was an example of an ancient ritual that had been tidied up, and codified and maintained as a mechanism to support the social order and the status quo. Indeed, we had been recruited to assist in that process. But there were many other kinds of custom for us to film in the south-west Pacific. Some were in decay, others being distorted to attract tourists, and there were some that had germinated recently and were only just coming into bloom. The rarest kind, however, was that shy and unobtrusive variety that was ancient but still survived unnoticed in remote and little visited parts. One of those we found in the small island of Vanua Mbalavu that lies with half a dozen others of lesser size in a small group half way between Tonga and Fiji.

We stayed in the little village of Lomaloma. Looking back now, across forty years, it seems an almost absurdly naïve model for a Polynesian paradise, straight from the drawing board of a Hollywood designer. Its houses were thatched – scarcely a corrugated iron roof to be seen. Neatly trimmed grass sward covered the ground between them. Scarlet hibiscus and purple bougainvillea grew freely everywhere. The air was loaded with the heavy scent of frangipani and the trade-winds rustled steadily and caressingly through the feathery leaves of the palm trees that lined a blue crystal lagoon. I can hardly believe it was so, yet it was.

Two men of our own age had been seconded to us by the authorities in Suva, the cosmopolitan modern capital of Fiji, to act as our guides. Both had family connections – always invaluable and irresistible in Polynesia – in Lomaloma. We had brought gifts with us, as prescribed by custom, including the still essential whale's teeth which we had managed to buy in the Government store. The headman, the *mbuli*, welcomed us and allocated a house for us to live in. We slept on the silky-smooth pandanus matted floor. We went down to the lagoon most mornings to swim with the men and help catch fish for our meals. In the evenings, we drank kava and exchanged songs. We found it hard – very hard – to remember that we had a film to make.

We had come to the island to record a little known ritualised fishing ceremony. It took place in a small shallow lake, nestling in the low hills in the centre of the island, which contained large freshwater fish with a particularly delicious taste. If the right rituals were performed, these fish were said to give themselves up, leaping right out of the water into the people's hands. The lake was under the protection of a tribal priest. He had not allowed fish to be taken from it for many years but now, after pressure from people in Lomaloma and several other villages along the coast, it was to happen again.

Half the population of Lomaloma went up to the lake and camped by its shore. We went with them. The priest was already there and each group as they arrived presented him with a gift of kava, here known as *yanggona*. When everyone had assembled, he announced his instructions. They were precise. That evening everybody, without exception, had to go and swim in the lake. No clothing of any kind could be worn, except for skirts made from the leaves of a particular plant that grew in the nearby bush. Everyone had to be anointed with the fragrant coconut massage oil, scented with crushed flower buds, that every village made. The lake would punish anyone who ignored these requirements by biting their skin. People must then swim, two by two, supporting themselves on specially cut logs, throughout the night. At no point could the lake be totally deserted. In the morning, if the rules had been scrupulously followed, the fish would give themselves up.

Our friends in Lomaloma needed no encouragement. Neither did we. We anointed one another. We put on leaf skirts. And we swam and sang as we drifted back and forth across the lake. After an hour or so, we came out, drank a little kava, ate a little of the pork and chicken that was roasting over the fire – and then went back to swim a little more. By the early hours of the morning, the lake's secret was a secret no more. A gentle whiff of sulphuretted hydrogen began to waft over the water. The lake was shallow. Leaves falling into it from the trees on the surrounding banks drifted to the bottom and decayed to form a deep ooze. So a great number of people, swimming continuously for hour after hour, stirred the mud, releasing the gas and making the waters slightly acidic.

By the early hours, the increasing acidity of the water caused the fish to swim near the surface. They were about two feet long and it wasn't difficult to catch one by the tail. But the time to take them had not yet come. The priest would say when that was, and we must wait for him to do so. By dawn more people were in the camp, drinking or dozing, than swimming in the lake but, as the priest had ordained, the lake waters had never been totally deserted and as the sun rose and warmed us, enthusiasm returned. In mid-morning, the priest gave another command. This time all those still left in the camp rushed down to the lake and plunged in. The commotion was such that almost immediately the lake was alive with fish leaping from the surface. Some were speared by the men. Others leaped out so close to the swimmers that they could be grabbed. A few near the edge jumped right out of the water and landed on the shore. Soon a couple of hundred of them were lying on the bank ready to be shared out and cooked over the fires. And once again, there was more feasting and more happy singing.

The advantages of ritualising such an event and putting it under the control of a priest were plain. The lake, being comparatively small, could have been easily fished out if there was no limitation. The method used to catch the fish required a considerable number of swimmers and some authority was needed to assemble and coordinate them. The rules requiring you to anoint your body with oil prevented the acid created by stirring up the mud from irritating your skin. And wearing nothing other than a leaf skirt was also obvious sense. It was, after all, a party.

* * *

Farther still to the west of Fiji we filmed very different events – in Vanuatu. At the time of our visit, Europeans still called the islands the New Hebrides, the name Captain Cook had given them nearly two hundred years earlier. Here we were among Melanesian people who differed from Polynesians both physically and temperamentally. Skins were darker, hair frizzier and customs based less on pleasure and delight and more on physical bravery, even punishment.

On the southern tip of the island of Pentecost in the northern part of Vanuatu, the people indulged in land-diving. In 1959, the outside world had only recently heard of the custom. Photographs of it had been published but it had never been filmed and when we asked if we might do so, the villagers readily agreed to put on a performance.

The ceremony was to take place on a steep hillside just outside their village. One tall tree had been trimmed of most of its branches and enmeshed with scaffolding to form a tower. It stood about a hundred feet high. The slope running down from it had been cleared of all vegetation and dug over to form a soft landing ground. When we arrived, men were constructing diving boards that projected from the front of the tower. The lowest were some twenty feet above the ground, the highest right at the top. There were forty or fifty in all. Each consisted of two planks supported by two vertical struts running between the outer end of the platform and the main body of the tower. A pair of long forest vines, secured to poles in the centre of the tower, ran along each platform and dangled down the front. These would be hauled up and tied to the ankles of the men who would jump from the boards. Their length was obviously crucial. Too short and a diver would be left dangling upside down in mid-air. Too long and he would hit the ground and break his neck. You might have thought that every diver would want to have a good look at the pair he was going to use, but the villagers assured us that that particular detail was entrusted to the builders alone. Those who had decided to dive would not bother to check.

The whole village turned out on the day of the jump. Women and men who were not going to dive formed lines at the foot of the tower and started to dance. Then young men, scarcely more than boys, clambered up the scaffolding and took their places, one after the other, on the lower boards. Each had an assistant who tied the lianas around his ankles. That done, the diver stood on the end of the board, summoning up his courage while he gestured with his arms out to the side or above his head, like an Olympic high diver. He then tipped forward into the air. He was still falling, travelling down and outwards, away from the tower, when the vines tightened. The strain broke the lashings that secured the struts at the ends of the board he had left so that it collapsed downwards, thereby absorbing some of the shock and slowing his fall. When the vines could stretch no further, the diver was jerked backwards to land on the soft soil.

As men climbed higher and higher to the platforms, so the dancers at the bottom of the towers stamped and chanted with increasing excitement. After an hour or so only one platform, the highest, remained unused. The man who climbed up to that was brave indeed. He stood on the far end of it and hesitated for a long time. He called out. He raised and lowered his arms and threw down a red hibiscus flower. Then with arms outstretched, he toppled forward in a graceful swallow dive. He landed like all his predecessors on his back, unharmed and jubilant. Everybody cheered.

The villagers had asked us for a substantial sum of money for permission to film this performance and we certainly did not begrudge it. But how did the ceremony start? Was it originally a rite of passage, a test of bravery that a youth had to undergo in order to demonstrate that he had become a man? We could not discover. Was it now being performed more as a sport? Perhaps. Would it become before long a regular tourist spectacle and a source of income? In due course, it did.

* * *

Other rituals, however, in the south of Vanuatu were certainly not for tourists. Indeed quite the converse. They were specifically anti-European, part of a whole category of new religions called cargo cults that have flared up over the past century in many parts of the Pacific, from Tahiti to the east and New Guinea to the west. When we were in New Guinea, a missionary had explained them to me like this.

'Cargo' is the pidgin word for factory-made objects that are brought to the Pacific by plane or ship. When the islanders first saw things such as metal cutlery, glass tumblers, or plastic pens, they were mystified. Such objects plainly could not have been made by any of the technologies with which they were familiar. How could you chip or plait or weave a radio or a rifle? And if such things were not man-made, it followed that they must have come from the gods. The islanders asked about these new gods, and some of the white newcomers were only too happy to talk about them and enrol the islanders as worshippers. But the new religion let the people down. The share of the cargo that they gained as a result was tiny. So they took no more notice of what the missionaries said. They then asked the traders how to get a share of the cargo. The traders said they should work in the plantations and earn money with which they could buy what they wanted at the stores that belonged to the traders. Many did so, but however hard they laboured they were never able to earn enough to buy anything but the most trivial bits of the cargo.

The colonists must have some secret way of getting the cargo that they were not revealing. The islanders watched them closely and saw that much of what these newcomers did served no practical function. They ate their food in strange ways, sitting on small platforms with bigger, taller ones in front of them covered with white cloth. They sat and shuffled papers about. They persuaded some of the local people to dress up in identical clothes to stamp up and down with exaggerated movements. Perhaps all these were the rituals of a secret religion by which they induced the gods to send material riches to them alone. So the islanders started to perform such actions themselves. At the same time, they ceased to co-operate with the colonists, refusing to work for them and rejecting their money.

It is hardly likely that any one individual islander thought out sequential arguments in such a way. The rationale is too pat, too obviously neat. The driving force doubtless came from a more sub-conscious un-articulated reaction to what people saw going on around them. Nonetheless, it accounts for the many common features of all these movements – all of them anti-European yet incorporating imitations of European actions, all urging people to abandon the teachings of the white missionaries, all claiming to be ways of persuading the gods to send material goods not to the colonisers but to the colonised.

Many of the cults personified the supernatural being who would ultimately bring this cargo. In Tanna, his name was John Frum. European planters living on the island first heard talk of him in 1940. It was said that when he arrived, there would be a great cataclysm. The mountains would fall flat and the white people expelled. John would bring his own money with him, coins stamped with the image of a coconut, and the white man's money would be worthless. There was a run on the stores with people getting rid of the old money before it totally lost its value. An area of bush was cleared to serve as a landing strip on which John Frum would land in his plane, and huge sheds were built alongside to receive the cargo that he would bring with him.

The colonial government tried to repress the movement by force. The cargo sheds were burned down, the airstrip destroyed, and the leaders of the cult imprisoned. But this only drove the movement underground and made it more difficult to control or monitor. Few people thought it had been eradicated. In the 1950s it surfaced again. This time people said that John Frum had instructed them to form an army. They made uniforms for themselves out of bark cloth and imitation rifles from bamboo. Platoons of them, wearing red singlets, a colour sacred to John Frum, started drilling in the same way that they had seen Government police doing. They began to march round the villages, ordering the people to support them and give them food. Once again, the Government sent down armed police and arrested the leaders. But they could not imprison whole villages and there were still a large number of people in the south and east of the island who wore the scarlet singlets of John Frum's army and were preparing for the apocalypse that would signal his arrival and the beginning of his rule. This time, the Government decided that as long as John Frum's soldiers did not threaten others with violence, it would leave them alone.

One planter still lived down there, an Australian named Bob Paul. When he had first tried to start his own plantations on the island, he had been given a very bad time by the long-established planters, even to the extent of one of them trying to prevent him from loading his copra at gun point. This led the local people to treat him differently from other European settlers and now he was the only one who still could persuade them to work on his plantation. With him as our ambassador we hoped to meet some of the members of the cult.

Bob did not look like someone who would stand out alone against a hostile population, yet we knew he had done so. He was quietly spoken and undemonstrative, a tall lanky man in his forties with sandy hair and a small moustache. He took us to see Yahuwey, the volcano that lies close to the east coast, in the heart of John Frum country. As he drove us down the dirt tracks through plantations of coconut palms, he raised his hand to acknowledge the few people we passed on the road, in a way that was neither effusive nor patrician. The men, many wearing the red shirts, the women mostly carrying loads in nets slung from their foreheads, looked back at us impassively.

Volcanoes are very varied. They can be sullen, threatening, ominous. Some can be overwhelmingly beautiful, especially at night, when the rivers of lava streaming from them glow with the richest red you have ever seen. Others are straightforwardly ugly, their lava a tide of irregular blocks, creaking and tumbling as the whole flow inches forward like waste from an industrial furnace. But Yahuwey had an atmosphere I have not encountered on any other. It was eerie.

From a distance it seemed no more than a low dome with a dirty yellow-brown mushroom of smoke rising from it, twisting and ballooning as fresh spurts of gas were injected into it from below. We could hear muffled explosions like distant thunder. When we were a mile or so from it, we came to a grey dune of ash that had invaded the lush bush of tree ferns and stilted pandanus like an overflow from the spoil tip of a mine. From there on, a plain of naked sterile ash stretched in a gently rising slope towards the crater.

It was eerie because lines of red-painted sticks, as tall and as widely separated as fence posts had been stuck in the ash. But there was no fence connecting them. In one place, we found a gate, again painted red, serviceably hinged, with a wooden arch connecting its two posts. Yet again there was no fence on either side. I was reminded of those ceremonial arches that stand in European cities with the traffic swirling around them, that are only opened on the most important of ceremonial occasions when some high personage, as a symbolic act of distinction and honour, will be allowed to drive through them. We trudged past them, along the margin of a small lake that lay in a fold of the ash dunes and up towards the crater. On its lip, stood a cross, nearly seven feet tall, its red paint peeling from it, corroded by the acrid fumes that swirled up from the vents deep in the crater. We looked down to the vents themselves, six hundred feet beneath us – until there was a sudden detonation of ear-splitting volume and a cannonade of blocks fired from below sent us sprinting downwards.

Beside a track not far from Yahuwey we came across a wayside shrine, protected by a small thatched roof. A human figure, with white hands and face and a white belt around its red body, stood upright with arms outstretched, its right leg bent at the knee and pointing backwards. Presumably this was an image of John Frum himself. He was flanked on one side by an enigmatic animal figure that looked like a rat with wings surrounded by a small fence-like cage, and on the other by a model of a four-engined aeroplane with huge wheels. We presumed that this represented the transport that would bring in the cargo but what the other figures meant we could neither guess nor discover.

We spent several days driving round the island, making ourselves visible, avoiding doing anything or making any contact that might suggest we were allied to officialdom, letting it be known that we represented the outside world which wanted to hear all about John Frum.

Bob arranged for us to meet Sam, one of the senior men of the movement who had once been a teacher in the Presbyterian-run school. He had been selected for that job by the missionary because he had been a particularly good student and spoke excellent English. He talked to us willingly but quietly and modestly, his eyes fixed to the ground. He himself had never seen John, he said, but his brother had – nineteen years ago. John was a tall white man who wore shoes, so he was clearly not a Tannese. Nonetheless, he spoke the Tanna language, not English. Sam knew that John would come soon, bringing cargo because he had appeared to several men and had promised them all that he would do so. Nineteen years was a long time ago, I said, and John had not yet kept his promise. Wasn't that a long time to wait? Sam lifted his eyes from the ground and looked straight at me in a solemn but challenging way.

'If you can wait two thousand years for Jesus Christ to come, and he no come,' he said, 'then I can wait more than nineteen years for John.'

The Friday after we arrived, we went down to the village of Sulphur Bay, the centre of the cult and the home of Nambas, its current leader and prophet and the man who had led John Frum's Army when it had made its march round the island. Bob Paul thought that the people would talk more freely if we went there without him.

Friday was the day sacred to John when his followers assembled in some numbers. When we drove into the village, there were several dozen men, all in red singlets, sitting around beneath a huge banyan tree in the centre of the village. Not far away stood a bamboo mast, thirty feet high with a red cross on its top and a small protective fence around its base. A crowd gathered around us and a tall grey-haired man stepped forward to introduce himself. This was Nambas.

'Me savvy you will come,' he said loudly so that all should hear him. 'John Frum 'e speak me two weeks ago.'

It was hardly surprising that he was expecting us. We had been doing our best to spread the word that we wanted to hear about John Frum ever since we had arrived.

Had Nambas seen him when he told him this? No, John had sent the message by radio, and Nambas pointed to the bamboo mast.

We had heard about this radio. Apparently, its messages were received by an old woman who wrapped electrical flex around her waist and went into a trance behind a screen in Nambas' hut, close by the mast. No one could understand the sounds she made, except for Nambas who interpreted them.

I asked if we might see the radio.

Nambas said that was not possible. John had said that white men must not see it.

I asked why gates had been built on the slopes of the volcano.

'Because man stop inside volcano,' he said. 'Many man belong John Frum. Red man, brown man, white man, man belong Tanna, man belong South America, all stop 'long volcano and bring cargo.'

Had Nambas seen John? Many, many times.

What did he look like?

'He look like you,' said Nambas, jabbing his finger at me and leaning forward with burning eyes.

'When will John come?' I asked.

' 'e no say when. But 'e come,' replied Nambas with quiet confidence; and the crowd around us, who had been listening intently, grunted their agreement.

That evening, a group of men with guitars, mandolins and drums made from tin cans started to play. Women wearing long grass skirts began to dance and soon everyone was strutting and jigging in an awkward gawky fashion quite unlike any other dances we had seen in the islands. The music and the song they sang was neither a traditional Tannese chant, nor a version of the Pacific pop music that incessantly blared from the traders' stores. The followers of John Frum belonged to neither world.

* * *

There were two sequels to our Pacific trip, one happy, the other very much less so.

I returned from the Pacific in time for a family Christmas and then had to start editing the programmes, which had already been scheduled for transmission the following April. The Spillius' return, however, was repeatedly delayed and in the end I had to complete the last in the series, about Tonga, without their advice. But I had given my word that I would not start work on the full-length film of the *taumafa kava* until they were able to supervise the whole process in person. By the following September, nine months after my own return, they still had not come back and I had to leave on another Quest – to Madagascar. So the Tongan rushes, still unedited, were sent to a film vault to await my – and the Spillius' – return.

I got back from Madagascar in December. There had been a disaster. That autumn there had been torrential rainstorms. Several of the BBC's film-vaults had been flooded. Among the many cans of film that had been destroyed were the Tongan rushes. No rescue or repair was possible. The catastrophe was total. I wrote to Ve'ehala to explain as best I could. I had already sent him a copy of the television version. He had thought very well of it, and had great hopes for the long version. There was nothing I could do but apologise and to ask him to convey my deepest regrets to Queen Salote. The disaster had been beyond my control, but I was greatly dismayed that we had failed, after all that effort, to produce the film document that we had promised.

The other happier sequel also involved Tonga. One evening, some months after I had returned from the Pacific, the telephone rang. 'Ratu Tavita!' said a voice. It was Isaia, the leader of the Tongan Palace band.

He told me his entire group had decided that the time had come to broaden the audience for their music. They knew for sure that since I worked for BBC Television and was their friend that I would arrange everything if they came to London. So they had worked their passage to Britain on a tourist ship. Now they were here and ready to appear on television just as soon as I gave the word. They understood, of course, that it might not be that very night. Next week would do perfectly well.

I explained that I had nothing to do with musical programmes. Getting on to television was very difficult. There were agents who knew how to handle these problems. I could find the name of the best and most trustworthy. Isaia explained patiently that the band had already considered the possibility of an agent, but didn't like the idea. They thought it would be better if I fixed things for them myself.

I had to do something. The next morning, I rang up Gordon Watkins, a friend who worked as a director on *Tonight*, a topical early evening programme that went out every weekday. During the afternoons, before the evening's live broadcast, the production team sometimes used the otherwise empty studio to try out new ideas and performers. Would he let my Tongan friends come in and sing a couple of numbers to camera? Nothing, I knew, would come of it, but it would at least show that I had done *something*. Gordon gallantly agreed.

That afternoon Isaia and his friends arrived in Lime Grove looking very uncomfortable in ill-fitting slacks and sports jackets, with plastic flower *leis* and cloth *valas* in a canvas bag. I took them up to the studio. They had given some thought to what they might play, they told me, and decided that probably their best number was 'You are My Sunshine.' I persuaded them that British viewers would also like to hear one of Queen Salote's compositions so they agreed to play one of those as well. The cameras moved back and forth in front of them as they did so and their pictures appeared on the studio monitors. They were delighted.

As I saw them into taxis back to their hotel, I impressed on them once more that having an audition was a long way from appearing in a transmitted programme. 'Don't worry, Ratu Tavita,' they kept saying. 'We trust you.'

At six o'clock that evening Gordon telephoned me at home. He was in something of a panic. His calypso singer, who regularly provided a musical end to the programme, and who was due to appear live in an hour or so's time, had been struck dumb by laryngitis. Gordon's problem would be solved if he could end the programme with 'You are My Sunshine', which had been recorded as was usual with auditions. However, he had no contract with my Tongan friends and it was strictly against the rules to put performers on air without one. Who was their agent? 'Don't worry,' I said, 'we'll sort out the fee afterwards.' I rang the Tongans and told them the good news. They were grateful but not in the least surprised. 'We knew you would arrange everything,' they said.

They then disappeared. I heard nothing more from them until several months later when the book I had written about our Pacific journey was published. A celebratory dinner was arranged by the publishers at a West End Polynesian-style restaurant. As we walked down the stairs through a cluster of plastic coconut palms, with fishing nets draped on the walls studded with dusty dried starfish, the sound of 'You are My Sunshine' floated up to me.

And there they were. They had the same plastic *leis*, the same cloth *valas*, but every one of them had put on at least a couple of stone in weight. They were clearly flourishing. We stood in the doorway listening and Isaia caught sight of me. He stopped playing in mid-phrase. 'Ratu Tavita' he roared. The rest of the band stopped as well and they all rushed across the restaurant floor to embrace me. And then, as we stood with arms around one another's shoulders, we all cried. The Palace corridor in Nuku'alofa seemed a long way away.

11

Radio Excursions

Radio also had its own programmes about exploration and producers there, noticing that I was covering a similar field in television and therefore knew some of their contributors, invited me on occasion to act as an interviewer. Few of the interviewees they chose, however, were natural talkers. Most considered it bad form to admit that they had encountered danger of any kind, except that which came from their own ineptitude. They were even uneasy at suggesting that they had ever experienced any discomfort. Upper lips nowhere came stiffer.

Bill Tilman was the archetype. He had accompanied Eric Shipton in 1934 on one of the great episodes in the European exploration of the Himalayas – the reconnaissance of Nanda Devi and the approaches to Everest. The two men had lived 'off the country' – that is to say, they had subsisted almost entirely on roasted barley, *tsampa*, and shared a single small tent throughout the entire journey. It was said that after a month or so of this, Shipton had said, rather shyly, to Tilman, 'Now that we know one another pretty well, do you think perhaps we might stop calling one another by our surnames?' To which Tilman had replied 'Are you suggesting that I should call you Eric? I'm afraid I couldn't do that. I should feel such a bloody fool.'

When Tilman became too old to continue mountaineering at the very highest level, he took to sailing small boats in the roughest seas he could find. He refused to have any kind of electrical gadget on board, and cooked only the simplest food in a tiny galley. No one, they say, ever sailed with him twice.

He wrote a book about one of these voyages in the Antarctic Ocean and Radio asked me to interview him. Considering that I was trying to give him publicity about his book, he was hardly forthcoming.

'I understand you had one companion on this voyage.'

'Ysss,' he said. He was a habitual pipe smoker and spoke most of the time through clenched teeth.

'How did you find him?'

'The way one usually does, of course. Advertised in the Personal

Column of *The Times*.'

'And what qualifications did he have?'

'He'd sailed across the Atlantic on the *Queen Mary*, playing the double bass in their dance band. If he could live through that I reckoned he could live through anything.'

The target of his voyage had been the Crozet Islands. Getting him to agree that reaching them had been to any degree difficult was almost impossible.

'The Crozets are quite far south, aren't they? Well down towards Antarctica.'

'Ysss.'

'And down there in the Roaring Forties, the seas are pretty rough I imagine.'

'Ysss.'

'Navigation, I suppose, that far south is quite difficult.'

'Tricky.'

'And the Crozets are quite small, aren't they?'

'Ysss.'

This was like getting blood from a stone.

'So here you are, the two of you, in a small boat with a more or less permanent gale in your sails blowing you westwards, in mountainous seas, navigating with nothing more than a sextant, trying to find landfall on a group of tiny islands. You could easily have missed them, couldn't you?'

'Ysss.'

'What would you have done then?'

'Gorn round again.'

What he would have gone round again, of course, was the globe.

* * *

After each of our own overseas trips I took a selection of our recordings to Broadcasting House. In those days, the Corporation saw it as part of its public service duty to create and care for all kinds of archives. It held historic sound recordings of musicians, writers and politicians that dated back to the 1920s. It had systematically assembled one of the most comprehensive of all collections of bird song and other natural sounds. It commissioned folk-lorists to make carefully planned surveys of British folk music without being concerned whether or not what they recorded was likely to be used in programmes in the immediate future. Music from other cultures was looked after by a wonderfully enthusiastic and formidably intellectual lady named Madeau Stewart. Nothing was too esoteric

Recording a frog chorus in Sierra Leone. The recorder was the first portable battery-driven model. With one hand holding a torch to keep check on recording levels, the other had to hold both the speaker to the ear and the large microphone. Smoking a cigarette, as so many did in those days, may or may not have kept away mosquitoes but certainly complicated the procedure.

Jack Lester and Alf Woods trying to persuade a newly caught *Picathartes* to feed.

Charles Lagus filming a column of marching ants in Sierra Leone during our 1954 *Zoo Quest*. His clockwork camera only took 100 foot rolls.

△
The first Komodo dragon we saw. He was a little over nine feet long.

◁ Sabran – unquenchably cheerful and greatly skilled in handling animals.

Feeding Charlie the orangutan, on the *Kruwing*.

New Guinea highlanders dancing in Minj. Their bodies have been anointed with pig fat and soot. The plumes of different birds of paradise adorn their head-dresses and are worn through their noses: King of Saxony's Bird, Count Raggi's Bird, the Superb Bird and Princess Stephanie's Bird.

△ The daily butterfly blizzard at Ihrevu-qua on the Rio Jejui.

◁ Heliconid butterflies settled on my hand to drink my sweat.

◁ Preparing to swim in the sacred lake of Masomo. One of the Lomaloma villagers anoints me with massage oil to ensure that the lake does not 'bite' me.

After a night-long swimming party, the Masomo fish begin to give themselves up. ▷

A land-diver on Pentecost Island hurtles towards the ground with vines tied to his ankles. As they tighten, the struts supporting the outer end of his diving board will collapse and the board will hinge downwards, so absorbing some of the shock on his legs.

for her. She would urge anyone who was about to visit a remote area to record whatever music they heard. Few refused her. I was an obvious target and over the years I had managed to contribute to her archive recordings of Balinese gamelans, African drums and Tongan nose-flutes.

One March, she asked me to compile a programme for which she had managed to get a special slot on Radio Three. I chose to talk about the Sheba Islands.

The Shebas, or Looflirpa to give them the name their own people use, had been discovered back in 1567 by Mendana who sailed westwards into the Pacific from the coast of South America. Two years earlier he had been the first European to reach the Solomon Islands and was greatly disappointed to discover that they contained no gold. His backer, Sarmenito da Gamba, however, persuaded Mendana to set out again two years later and renew the search. This time Mendana sailed further north and so reached the group that he called the Sheba Islands, after Solomon's consort. He gave the name of his own flagship, *Todos Santos*, to the main one. After his visit the islands were, in effect, lost and few European voyagers succeeded in locating them again. Even today, they are little visited for they lie far from any other shipping route. Nonetheless they are of great interest to any naturalist.

The Shebas have much in common with the Galapagos. There, as is well known, finches that accidentally reached the archipelago, subsequently evolved into different species on different islands. In the Shebas, however, the crucial immigrant was some sort of mouse which arrived there, perhaps as an involuntary passenger on a raft of floating vegetation. A German naturalist, Helmut Winderup, was the first to document these species in the early twentieth century when the Shebas, along with the Bismarcks, were part of the German Pacific Empire. One of the most interesting of these rodents Winderup called *Mus minimus*. It is smaller even than the little English harvest mouse and has evolved to fill the ecological niche occupied elsewhere by wood-boring beetles. Its young, which are born hairless, are regarded by the Shebans as great delicacies and traditionally are swallowed alive. Their calls, as you might expect from such a small creature, are extremely high pitched and the recording I played of them in the programme was probably only audible to younger people whose ears are so much more sensitive to high frequencies than those of their elders. I certainly couldn't hear it myself.

Another Sheban species of *Mus*, in the absence of any grass-eating competitors, has evolved in the opposite direction. It inhabits the grassy slopes of the great volcano that dominates the main island of Todos

Santos. It is not quite as big as the pig-sized capybara of mainland South America, but not far short of it. Winderup called it *Mus maximus*, but its popular name is the Sheban guinea pig and its calls are remarkably like those of a cow.

Perhaps the most enchanting of these island species is *Mus dendrophyllus*, the Sheban puff-tail. This lives in trees and with no predatory birds to threaten it, is active during the day. Individuals are therefore able to communicate with one another visually, and do so by waving their tails over their heads. Each island has its own species with its own characteristic colour so there is, for example, the russet puff-tail and the cream puff-tail, each living on its own individual island. Their fur for the most part is extremely soft, but on the tail it is mingled with individual filaments that have been thickened into hollow rattles. These make a most characteristic and haunting sound. Since each species has spines of slightly different diameter and length, it is possible to identify the island a puff-tail comes from entirely by the pitch of the sound, as I was able to demonstrate with recordings.

The Looflirpan people are also of great interest. They have a complex musical culture. They use hollow branches that have been bored down the middle by *Mus minimus* as musical instruments, cutting them into sections of varying length to make a kind of panpipe. In one specially sacred area, volcanic gases are continuously expelled through fumaroles. For festivals, the people erect particularly large pipes over them so that the island itself, as it were, produces its own music.

The people also erect standing stones and Todos Santos in particular is dotted with great numbers of them. At first sight their arrangement appears to be haphazard. But not so. Winderup was the first to notice that they were all connected by straight lines. That is to say, if he looked through a hole in the middle of one, he could see another lying in a direct line from it along a straight compass bearing. Indeed the simple straight line is the dominant element in Sheban material culture. One might be tempted to suppose that this is a primitive condition, that in decorative terms Sheban art has not yet evolved beyond straight lines. Later workers, however, are inclined to think that the reverse is the case and that the Sheban visual language is, in fact, extremely refined. Whereas, for example, the spears and sceptres found in New Guinea or the Solomons are decorated with extremely complex curvilinear designs, the Shebans have gone far beyond this. They have so refined, distilled and purified their art that now it is devoid of any ornament whatsoever. The standing stones, it is thought, were once erected in long avenues but the Shebans, with their taste for minimalism, eliminated the intermediate ones and retained only those at the beginning and end of the lines.

I ended the talk by saying how appropriate it was that the programme should be broadcast on this particular day, for it was the precise date of Mendana's discovery of Looflirpa, four hundred years ago – April the First.

A number of letters arrived a few days after the broadcast, asking for details as to how to visit the Shebas and for some references to Dr Winderup's taxonomic works on the genus *Mus*. I replied to them all and gently suggested that they looked again at the date on which the talk was given.

12

Lions and Lemurs

I was running out of continents. I had filmed in the tropical forests of Africa, South America and south-east Asia. I suppose I could have chosen to go next to Antarctica, but I have always preferred being over-heated to being frozen stiff. Where in the tropics could we go where the animals would be unlike anything we had shown in the previous four *Zoo Quests*? Leafing through an atlas with this question in mind, my eye lit on Madagascar. At first sight, it looks no more than a chip from the east coast of Africa. That, indeed, it is. But the island is immense, a thousand miles long. And it split away from Africa at a crucial time in evolutionary history – after the expansion of the reptiles, but very early in the development of mammals. Since then the few animals that it inherited from its super-continental past have evolved in isolation and have given rise to a whole range of creatures that are unique.

The most remarkable and varied of its mammals are its lemurs. Lemurs are primates, primitive relatives of monkeys and apes. The smallest of them look very like the bush babies of Africa, but the biggest, which are as large as monkeys, are unlike anything found elsewhere. Its reptiles are odd too. The island, apparently, was the centre of chameleon evolution and has more species of them than any other country. It has spectacular day-flying moths and unusually handsome tortoises. Even the people are surprising, for many are not African in origin, as you might assume, but are descended from Asiatic people who arrived in Madagascar about a thousand years ago. Clearly there was a great deal there for another *Zoo Quest*.

I started to do a little research in the library of the Zoological Society. Most of the books I found were in French, for although the London Missionary Society had sent its members to work in Madagascar in the early nineteenth century, the French were also very active in the island and by the end of the century they had incorporated it into their empire. So it was French scientists who had published most about the island's natural history. The Malagasy had won their independence in 1960, but the French influence remained very strong and French is the most widely spoken European language on the island.

Detailed first-hand information about lemurs, however, was hard to find. One species, the ring-tail, is certainly very well known for it makes a charming pet and thrives in captivity. But I was interested in the bigger species and articles about them were illustrated either with dull drawings dating from the nineteenth century or photographs of what were, only too plainly, stuffed specimens. This was because very few of these species had ever been seen alive in Europe. They have specialised diets and are very difficult if not impossible to keep alive for long in zoos. If there were no still photographs of the living creatures, there could hardly be film of them either. That made them a very alluring prospect as far as I was concerned. Madagascar it would be.

Geoff Mulligan, after our trip in the Pacific, was eager to come on another. Two weeks before we were due to leave, I was summoned by Leonard Miall, who was still Head of Talks Department. Billy Collins, the London publisher, had just sent him the proofs of a book, soon to be published, about a lioness that had been reared by the wife of an African game warden. That, in itself, was hardly exciting or unusual. But what made this book unique was the fact that the lioness, once she was adult, had been taught by her human foster parents to hunt so that she could take her proper place in the wild. There she had not only caught her own prey but had mated with a wild lion. And when her cubs had been born, she had returned to her foster parents, bringing her cubs with her. The author's name was Joy Adamson and the book's title was *Born Free*. Billy Collins had said that if a BBC camera team were to visit the Adamsons' camp it would be given privileged access to film this sensational story. 'Why not call in on your way to Madagascar?' said Leonard. Why not indeed?

* * *

Geoff and I flew to Nairobi. There we chartered a small plane to take us up to the Adamson's camp in the wild country in the far north of Kenya. The plane bumped down on a tiny newly-cleared earth strip. Joy was awaiting us alongside it with a Land Rover. She was in her fifties, tall and slim with a forehead that was slightly too broad and a chin that was slightly too prominent for her to be described as beautiful, but she was nonetheless strikingly handsome. She was dressed in khaki shorts, with a loose khaki top and a sun helmet. Rather to my surprise she spoke with a faint but unmistakable German accent. She was, as she told me later, an Austrian baroness by birth. She had bad news. A strange lioness had appeared in the area and had tried to oust Elsa. A few nights earlier there had been a fight and Elsa had been wounded. She had now disappeared and

Joy did not know where she had got to. It seemed that we might have no star for our film.

Joy gave orders to her African staff to collect our baggage in a way that was simultaneously imperious and coquettish. Back at camp, she introduced us to her husband George. He was as taciturn as she was loquacious. Bare-chested with a goatee beard, he wore only shorts and sandals. We never saw him dressed in any other way. Neither did we often see him without a pipe in his mouth and whenever he left camp, he always carried a rifle on his shoulder. He was the Senior Game Warden in charge of this part of Kenya, though he was now on the verge of retirement. It was he who, a year or so earlier, had shot a lioness and had brought her three orphaned cubs back for Joy to rear. Two of them, as they grew up, had been sent to European zoos. The third was the missing Elsa.

The Adamsons had been camped here for over a year. Joy had a large tent surrounded by a boma, a tall and impenetrable hedge of thorn branches. George had a smaller one beside it, but outside the boma. The African staff had their own quarters behind. Geoff and I, for safety, would sleep in the enclosed back of the vehicles, one of us in Joy's Land Rover and the other in the small lorry used for ferrying supplies out to the camp from a small town several hours' drive away.

It was very hot. We had been travelling all the previous night and were now very tired. After a brief meal, Joy said she was going to take a siesta. We decided to do the same. I took a camp bed down to the edge of a small river fifty yards away and put it down on a sand bank in the shade of a leafy fig tree. I was asleep within seconds.

I was woken by a powerful and repellent stench of bad breath. I opened my eyes. Fur, snaggled with saliva, was just a few inches above my face. It was the underside of the jaw of a lioness that was leaning over me. Had I sat up with a jerk I would have bumped into her chin. While I lay there rigid, wondering what on earth to do, I heard Joy calling out in delight, 'Elsa, my darling Elsa.' The lioness turned away from me and strolled lazily across to meet Joy who was running over towards her.

'*Jinja mbusi*,' Joy shouted as she came. That is Swahili for 'Kill a goat'.

Elsa looked up at Joy and brushed the side of her head against Joy's cheek. Joy fell to her knees to embrace her ecstatically around the neck and the two then lay down beside one another.

Joy tenderly examined the lioness, talking to her softly. Elsa had indeed got some wounds, but they were not deep or serious. Joy went to her tent and returned with some antibiotic powder with which she dressed Elsa's injuries and solicitously smacked any tsetse flies that settled on her.

While these endearments were going on, I heard the sudden plaintive bleating of a goat. A few minutes later an African appeared from behind Joy's boma, dragging the carcass of a goat, its throat newly cut. He chained it to a tree stump in front of Joy's tent. Elsa got to her feet, went over to the carcass and settled down to rip it apart.

It seemed sense not to start filming that afternoon. Elsa might need to get accustomed to our presence. In fact, however, she seemed to take very little notice of us. Perhaps it would be more accurate to say that we ourselves needed to get accustomed to having a full-grown lioness strolling about the place before we started thinking about how and what we were going to film. While Elsa continued with her meal, the first of her cubs appeared. This, George told us, was Jespah, a young male and the boldest of the three. He was the size of a large Alsatian dog. The Adamsons had made no attempt to tame or discipline him and now he was quite beyond their control. Neither did he have the fear of fire or of human beings that a wild lion might have. It was clear, even then, that as he got older and bigger he might cause serious problems. A second male cub followed him. The third, a female Joy called Little Elsa, did not turn up until after dark.

We dined that evening in a big mess tent within Joy's boma. Joy, full of vivacity with eyes flashing, was thrilled to have Elsa and her family back. But she was still worried about the presence of the rival lioness. George, she said, would have to go out on the following day to shoot her. George said little. Joy became more vehement. Eventually such harsh things were being said between husband and wife that, out of embarrassment, I got up from the table and went out towards the Land Rover where I was to sleep. Joy came running after me.

'I know you think it is terrible for us to quarrel,' she said. 'But George is so lazy – and I love Elsa more than any man.'

We started to film the next day. Jespah made life difficult. His favourite game was to swipe at our legs as we walked. Joy wore leather leggings to protect herself, for Jespah on occasion neglected to retract his claws and could easily draw blood. The best we could do was to keep a sharp eye out for him and try to dodge. The other two cubs were rather more tentative and contented themselves with waiting in ambush to burst out on us as we passed.

Joy herself didn't make matters any easier. She was even more protective of Elsa than Jespah, and credited her with all kinds of human feelings for which I found it difficult to find any evidence. One afternoon, we followed Elsa as she stalked up a bare rock kopje and lay down in one of her favourite positions from where she could survey the surrounding bush. We walked up quietly towards her. In the distance, an elephant squealed and Elsa lifted her head.

'Come down at once,' commanded Joy from the bottom of the rock. 'Can't you see she is bored with you? You are taking liberties. You must not let her think that you are visiting her just to make ciné.'

Our stay was not turning out as I had expected. I had naively thought we would be able to film an idyll. Instead, violence lay beneath the surface wherever we looked. Goats, cooped up in a boma behind Joy's tent, would bleat in terror as Elsa stalked by. The half-grown cubs were already sufficiently powerful to be dangerous should they wish to mount an attack and the African staff treated them with extreme caution. The fact that the cubs had none of the in-built suspicion of humanity that restrains many animals and that they were accustomed to wandering fearlessly about a camp, boded ill. The rows between George and Joy continued to smoulder beneath the surface and only too frequently flared into open quarrels. The brusqueness with which she told him of her wishes was exceeded only by the imperious way with which she dealt with her African staff.

Soon after we left, even before the film we had shot was shown on television, Elsa contracted tick fever and died. The three cubs, still at large, started to raid the herds belonging to local people and the new Chief Game Warden, George's successor, ordered them to be shot. George managed to trap them before that could happen and took them down to the Serengeti where he released them. George and Joy stayed in the Serengeti for some time thereafter. One night, a young male lion walked into Joy's tent and left without harming her. On another occasion she woke to find a lioness drinking from a bowl beside her bed. They might or might not have been Elsa's cubs. Soon after that, however, a European man on safari was seized by a lion, dragged out of his tent and killed. Tracks of two males were found and the Game Warden shot them both. Whether any of these were Elsa's cubs no one could be sure, but they were never heard of again.

Joy's book, *Born Free*, was turned into a feature film. That too had a violent aftermath. George trained several young lions to play the part of Elsa at various stages of her life. Having lost their natural caution with human beings, they also became dangerous. One of them attacked and killed one of George's assistants and George himself had to shoot it. Joy some years later was murdered by one of her African staff. George was killed when shiftas, local bandits, attacked his camp. Elsa's story, at first sight so touching and tender, had started with violence when George had shot her mother. So it ended as it had begun.

* * *

From the Adamsons' camp, Geoff and I flew back to Nairobi and then on to Madagascar. I already knew that the Merina, the people living in the centre of the island, had an Asiatic origin, and indeed the capital, Tananarive, even from the air, looked very un-African. It was surrounded by extensive flooded rice fields, just like an Asiatic city. The traditional buildings, that then still dominated it, were not low and thatched but two-storeyed with tiled roofs. The people in the airport were straight-haired, fine-featured and wore brightly coloured cloaks with broad-brimmed hats.

The routine of trailing round offices seeking permits was, however, much the same as anywhere else, though this time we had to do it in French. The English community welcomed us hospitably and entertained us nightly. As usual, we sought advice from people who were interested in the local natural history and soon heard of the wife of a long-established English resident. She was a White Russian by birth and was said to be a little eccentric, but everyone told us that she was a passionate animal lover. We were delighted when she sent us a formal invitation to dinner.

She proved to be small, somewhat withered, dressed in a number of voluminous cloak-like garments, one on top of the other, and given to large and dramatic gestures. 'Welcome,' she said huskily and waved us into a huge living room. A wide shelf ran all the way round it. It was lined with a spectacular collection of crystals. Here was an obvious topic with which to break the conversational ice. I admired a double-ended quartz crystal, which pleased her, and moved on to a large pillar of a black mineral. Could this be a tourmaline? It was. I raked through my memories of mineralogy lectures at Cambridge and became bolder. Surely this wonderful specimen of mica, with its pinkish tinge, must be the rare variety known as lepidolite. For a moment I feared that I had overdone it, but she was greatly impressed.

'You are exactly the man I have been waiting to meet,' she said. 'People here are so ignorant. You know about these things.'

I demurred, modestly.

'Wait!' She turned on her heel, swept away into an adjoining room and returned within seconds, holding in front of her an object swathed in cotton-wool.

'You are the only person here who would realise how rare this is.'

Fearing that the limits of my expertise were about to be exposed, I somewhat apprehensively unwrapped it. Inside, I found a long quartz crystal with two much smaller ones attached to its base. All three were covered with haematite, an iron-rich mineral with a smooth black surface that not uncommonly encrusts crystals of various kinds.

'You do know what it is, don't you?' she whispered.

'Well, yes,' I said nervously, 'It's three quartz crystals covered with mammiferous haematite.'

I sensed that this was not the right answer.

'You are wrong,' she said firmly. 'Those are the fossilised private parts of a pygmy. Why on earth would you think they are quartz crystals?'

'Well,' I said , 'they are hexagonal. They have six sides and a pointed end.'

She snatched the specimen away from me and wrapped it up again.

'And how do you know,' she said haughtily, 'that a pygmy's private parts don't get six sides when they fossilise?'

To which question I failed to find an answer.

* * *

We acquired a Land Rover; we recruited an expert Malagasy helper from the Scientific Institute in Tananarive, which also ran the city's zoo; and we travelled around the island for the next three months. No matter where we looked, we found things that had never been filmed before.

Down in the deserts of the south, even the vegetation looked extraordinary. There were forests of tall unbranched stems forty feet tall and a foot or so thick, studded with spines and rows of small circular leaves. Some, at their tip, carried tassels of withered brown flowers. They were *Didierea*, Madagascar's equivalent of the cacti that grow in American deserts. It was in these surreal thickets that we saw a sifaka for the first time.

I could hardly believe that such a beautiful creature could be so little known. Until a few months previously, I had never even heard its name. It was the size of a large monkey and it was clinging to the top of a pillar of *Didierea* in a very un-monkey-like posture, its torso vertical, its hands grasping the stem in front of its chest and its hind legs drawn up beneath it. It had a black cap and face, but otherwise its long silky fur was a pure and wonderful white. It stared down at us with blazing yellow eyes. Then it took fright. It straightened its hind legs and soared through the air with its long white tail streaming out behind it, to land on another *Didierea* stem a good twenty feet away. The *Didierea*'s formidable spines seemed to cause it no problems. It landed with a thump and did not even shuffle its feet or adjust its hands. Even now, I do not understand how sifakas can do such a thing without drawing blood. I can only assume that they have palms and soles like leather.

The power needed for such giant leaps comes from the sifaka's extraordinarily long legs. They are almost equal to half the animal's total length and are very much longer than its arms. Such proportions make it very difficult to run four-footedly as all monkeys can. So on the infrequent occasions that a sifaka comes down to the ground, it stands upright and gets about by skipping with its feet together in a series of fey balletic leaps. I recalled the awkward, graceless pictures that I had seen in the Zoological Society's library. If ever an animal has been libelled by its image in natural history texts, I decided, it was this enchanting creature.

There was another attraction for us down in these spiny forests of the south. It was here that we hoped to find the remains of one of Madagascar's most extraordinary earlier inhabitants. Stories about it were circulating in Europe long before anyone there had heard of Madagascar itself. Crusaders back from the Holy Land in the twelfth century relayed the legends of Sinbad the Sailor who had travelled beyond the Arabian Gulf and down the eastern shores of Africa. He and his crew had come across an egg as big as a house. One of his men had accidentally cracked it and in revenge the bird that had laid it, an enormous creature called the roc or rukh, appeared in the sky and eventually sank Sinbad's ship by bombing it with boulders. Marco Polo, a century later, heard similar stories of a bird in these regions that was so immense that it fed on elephants which it killed by picking them up with its talons and dropping them.

In the middle of the seventeenth century, a French traveller who had spent some years in Madagascar brought back more sober and believable reports of a giant bird. He said that it lived in the south of the island and was flightless, like an ostrich. Then, in the nineteenth century, material evidence of such a creature was discovered – gigantic eggs.

These eggs were not the size that one might expect from a bird that could lift an elephant. Nonetheless they were the biggest egg that anyone, then or now, had ever seen – about the size and shape of a rugby football. Soon after, bones of the creature that must have laid such things were found. It was a flightless bird like an ostrich and stood about ten feet high. Scientists named it *Aepyornis*, the elephant bird. Ten feet is not the greatest height of any bird – the extinct moas of New Zealand were taller, but the elephant bird was very stocky and heavier than any moa. Some scientists estimate that it weighed around a thousand pounds. Most agree that it was the heaviest bird that ever existed. Occasionally its enormous eggs are still found intact among the sands of the southern Madagascan deserts. I didn't really suppose that we could make a dramatic action-packed sequence about an extinct fossil bird. Nonetheless, I yearned for the chance to look for fragments of the eggs of such a marvel.

We had been told in Tananarive that bits of its eggshell were not uncommon. They could be picked up, people said, in and around the *Didierea* forests. I tramped through the dunes, staring at the ground. I scrabbled in the low dusty cliffs of dried-up streams. I crawled on my hands and knees, sifting through the sand. Eventually, I found something. It was an irregular, slightly curved plate, an inch across, dull white on the inside of the curve and pale yellow with a distinct grain on the other. I had no doubt that it was a piece of an enormous egg. Another hour or so and I found two more bits. I took them back to camp in triumph to show Geoff.

That evening, a herd of goats wandered into our camp, followed by a small boy. Still full of enthusiasm, I showed him my finds. He looked at them blankly.

'More?' I said.

I tried my schoolboy French. '*Encore*?'

I waved a bank note at him. He showed no sign of understanding what I was trying to convey. As his goats had found little to eat in our camp, they drifted away and he trailed after them.

The following morning, a tall dignified woman wearing a sari-like costume and carrying a basket on her head walked up to our tent. She took down the basket and emptied on to the ground several hundred pieces of eggshell exactly like mine. She held out her hand for payment. Had I been able to make myself understood to the goatherd, I would now have been in danger of bankruptcy for I would have been committed to pay for every single piece brought to me. As it was I was able, without too much dishonour, to give a reward for the whole basketful which she seemed to think was wholly adequate. But it seemed painfully clear that my eyesight was as deficient as my French.

That point was made only too clearly as the day wore on. People arrived at hourly intervals with more basketfuls. A great tip of fragments accumulated beside my tent. I tried to explain that now I had enough. But too late. The reward we were now paying for a basket-load was still good enough for people to be keen to earn such easy money. Basketful after basketful arrived and was tipped out in front of me. Eventually, I managed to explain that we had enough small pieces. Now all we needed, I said, were really big ones.

The day after that, the goatherd reappeared, carrying something tied up in a grubby cloth. He undid it and laid out in front of me a dozen or so huge fragments. I quickly saw that several would fit together. Maybe more would. Whether or not he had found all the pieces in one place or whether someone in his village had once discovered a complete egg and had deliberately broken it open to see if there was anything inside, I could not discover. But we were overjoyed and I paid the lad what I think he considered, from the brief flicker of his smile, a princely reward. That afternoon, Geoff and I stuck the bits together with tape. But for one tiny triangular piece that was missing, it was a complete egg.

* * *

The elephant bird was not the only Malagasy creature that may have given rise to European legends. When I had started research into the island's natural history in preparations for the trip, I read of the biggest of all living lemurs and the only one without a tail. It was called the indri. The only picture I could find of it was a photograph of a stuffed specimen. It had been mounted standing upright leaning against a branch. It looked faintly absurd, like a man with a dog's head. And I thought of the dog-headed man of European legend. Marco Polo had told stories of that too. He called it the *Cynocephalus*. Even as late as the sixteenth century, naturalists included it in their encyclopedias. Aldrovandus, one of the great Italian scholars, had even illustrated one. I looked it up. It did look very much like that poor stuffed indri. Even if there were no connection between indri and *Cynocephalus* – and who could prove it, one way or the other? – the real animal would clearly make a remarkable sequence, if only we could find it.

Indris live in rain forest. Madagascar has only a small patch of it left, running in a broken strip parallel to the eastern coast and about a hundred miles or so inland. We established ourselves at Perinet, where there was a largely empty hotel that had been built to cater for passengers travelling on the railway that ran down from the capital Tananarive to the east coast port of Tamatave.

We soon discovered that the name indri was not a local Malagasy name. That seemed odd. It did not seem to have any Greek or Latin meaning. If it wasn't a local name, what was it? It turned out that the first naturalist to have described the species, a Frenchman called Sonnerat, was walking one day in the rain forest, when his local guides pointed to a large lemur sitting high in the trees. '*Indri, indri*,' they said. Sonnerat, understandably, wrote down a description of the animal they had shown him and noted that it was called 'indri'. Unfortunately, the word is an expostulation in the local language which means simply, 'Look at that.' However, such are the rules of zoological nomenclature that the animal is still known in the scientific world as *Indri* and that too has become its popular name. The local people, however, call it *babakoto*.

The presence of indris in the forests around Perinet was easy to establish. Walking there on our first day, early in the morning, the sounds of the forest were suddenly drowned by a chorus of unearthly howls. They rose and sank and rose again like wailing sirens. I searched the canopy with binoculars but couldn't see the animal that was making them. I started to creep as quietly as I could into the thicker bush by the side of the track, but the calls stopped abruptly. A few seconds later, a branch some distance ahead shook as some big creature left it. Getting the clear view we needed for a respectable film shot was not going to be easy.

There was one obvious trick we could try. The calls of the indri made it clear that the animal was territorial. Birds also make territorial calls. Recording their songs and playing them back to them usually gives a male bird the impression that another male has invaded its territory and it will quickly seek it out to try and expel it and in doing so make itself visible. Maybe the same trick would work for the indri.

There was no difficulty in recording the calls. The animals performed morning and evening with great regularity. And when, one day after their performance of choral evensong, we played them a reprise, the male who owned the territory responded with great indignation. He leaped magnificently from trunk to trunk towards us until he was in a perfect position for Geoff and his camera. He looked neither absurd nor freakish. He was as skilful and as accomplished a forest athlete as you could wish to see and for us it was a particular excitement. It was another filming first.

* * *

We spent nearly all our time filming, but I had to remember that the series I was making was still called, like its five predecessors, *Zoo Quest* and I was committed to producing programmes in which film was interspersed with sequences during which I would show animals, live, in the studio. So I had to think about collecting some. I could not and should not collect sifakas, indris, or any other of the large lemurs for they were, very properly, protected. Even if I had been given permission I would not have been justified in catching any for they would be unlikely to survive for long in captivity.

There was, however, one exception. Mouse-lemurs are the smallest in the family. They are, literally, mouse-sized and they look very like their African cousins, the bush babies, except that they are even smaller. Indeed, they are the smallest primate in the world. They are not in the least fussy about what they eat – fruit, insects, leaves, flowers, they will take them all. Nor were they difficult to find for they are the most abundant of all the lemurs in Madagascar. They live in tree holes and local people knew just where to find them and how to catch them. We soon had as many as we could properly look after.

There were also other creatures that could justifiably be caught to take back to the London Zoo. Malagasy tortoises are particularly handsome. They have bright yellow stripes radiating from the centre point of each plate on their chocolate brown shells. They also have the habit of wandering slowly across roads. Once they get into the middle, they seem liable to forget which way they are going and to settle down to think about the matter. Collecting half a dozen of them, therefore, involved little more than swerving the Land Rover so as not to run over them and stopping every now and then to pick them up.

The Zoo had also asked us to bring back some of Madagascar's snakes. Happily for me, the species they were most anxious to obtain was a non-poisonous constrictor, the Malagasy tree boa. Doing so, however, turned out to pose an unexpected problem. Some of the Malagasy people believe that these particular snakes are the living incarnations of their ancestors. If they come across one, lying on the forest floor, they will inspect it very carefully. Which of their dead relatives could it be? A scar on its head, a wart on its flank, a particularly dark colour or an unusually sluggish disposition could be a clue linking it to an ancestor who had a similar physical character. They will then ask the animal a few questions. If the snake, in reply, moves its head from side to side, then that is taken to mean that it is saying yes. Since most boas frequently move their heads from side to side, it may not be long before a positive identification is established. The snake is then reverently picked up and taken to the house in the nearby village which was once the home of the person of whom it is the incarnation. There it is offered milk or even given a sacrificed chicken and there it will stay until it chooses to leave.

Clearly we did not want to offend anyone by picking up their grandmother and putting her into a bag. Fortunately, however, the belief is not a universal one and we made sure that before we started on a snake hunt, none of the local people would be upset. The Malagasy tree boa is also mysterious from a scientific point of view. Boas and pythons look very like one another. Both are constrictors. Both have relics of their hind limb bones buried in their flanks. Pythons are found throughout Africa and Asia and lay eggs. Boas give birth to live young and are restricted to South America. It would seem more than likely, therefore, that a Malagasy constrictor would be a close relation of the pythons of nearby Africa. But oddly, it is the only boa to live outside the Americas. What quirk in Madagascar's geological or biological history led to this odd distribution no one has yet worked out.

Chameleons were also high on our list – and not surprisingly. Madagascar has the most spectacular and varied species in the world. The biggest, a garishly coloured monster with a helmet on its head and two long scaly blades on its nose, can grow to two feet long. Chameleons are famously able to change their colours, but the range achieved by this species is particularly splendid. Turquoise flanks, yellow stripes, green blotches, orange eye-turrets, all these are well within its range. Picking one up, however, does require a certain nerve. No chameleon has the ability to run fast. An impatient prance is about the best they can manage. If you pursue one it is likely to turn round to confront you, hissing ferociously. It gapes alarmingly, exposing a sulphur yellow lining to its mouth and then sucks in air, inflating itself so that it increases visibly in size. You can, if you are feeling brave, reach over and grab it by the neck. But then there may be something of a struggle, for its toes have a powerful grasp and are difficult to detach. The best way of collecting one is to put a stick in its path so that, as it stalks away, it walks inadvertently on to your stick and you can remove it and drop it into a bag.

The local people, hardly surprisingly, credited chameleons not only with a bite that leads to an immediate and agonising death but believed them to have a baleful magical aura that would bring disaster to anyone who interfered with them. That, at any rate, we were able to turn to our advantage. Half way through our trip, thieves broke one of the windows of our Land Rover and stole some of our stores and clothing. Thereafter it was impossible to lock the truck – something of a disadvantage when one considers the value of the filming gear we had to leave in it. Our biggest and most bad-tempered chameleon came to our aid. We regularly left it loose, prowling over our pile of baggage in the back. No one interfered with our car again.

* * *

At the end of our trip, we had to visit the Scientific Institute in Tananarive to get permits to export our chameleons, boas, tortoises and mouse lemurs. The authorities generously added a pair of ring-tailed lemurs to the collection as a gift to the London Zoo. Ring-tails were abundant, breeding well in their own zoo and they had many more than they really needed.They also presented us with a ruffed lemur. This is a particularly beautiful and spectacular species with long silky fur, part black part white. It is related to the ring-tail and has hind and forelegs of about the same size. So, unlike indris or sifakas, it is essentially a four-footed runner rather than a two-legged leaper. It is also, like the ring-tail, not fussy about its diet. This particular one was an elderly female. They said she was long past breeding age and they were having difficulty in finding space for her in their small zoo. So she came with us too and it was quite a big collection that we eventually delivered to the London Zoo.

They all did well in their new home. The mouse lemurs were the first of their species that the Zoo had ever possessed and so were of special interest scientifically. They started breeding almost immediately and the colony survived for many generations. One of the tree boas proved that it had nothing whatever to do with any of those common, egg-laying African pythons by unexpectedly giving birth to three lively ochre-coloured babies. The ruffed lemur also surprised everyone. She was so gentle and handsome that the Zoo gave her an enclosure to herself. They also discovered that Paris Zoo had a breeding group and an elderly male who seemed rather left out of things. So he was sent to join our female in her London retirement home. To everyone's astonishment and against all prediction, the two geriatrics thought so highly of one another that they bred and produced twins. So all in all, the creatures we brought back to the Zoo flourished.

* * *

Handing over our collection of animals to the London Zoo after each of the *Zoo Quest* trips was always a great relief. Keeping them all properly fed and cleaned took a lot of time every day. Even so, I was particularly sad to part with some of them and occasionally I became so attached to one or two that before they went to their ultimate destination in the Zoo, I took them home. There was no suitable enclosure for the ring-tails, so they became part of our domestic menagerie.

I had been keeping animals for as long as I could remember. There were always dogs in our household when I was a boy. They were Irish setters and I was, of course, fond of them as mute but pattable friends, game players and occasionally comforters. But I was equally interested in the unpattable – sticklebacks, frogs and grass snakes, all of which I was able to collect from countryside within bicycle range of home.

We lived in Leicester where my father was Principal of the University

College, a recently founded institution supported by local philanthropists, which under his care and leadership was to grow and become the city's University. It was housed in a large Victorian building which had been the municipal Lunatic Asylum, a fact not unnoticed by local critics. Our house, which had been built I believe for the Superintendent, had a large greenhouse attached to it which we dignified with the name of conservatory. It was there that I had tanks in which I installed my captures.

As I grew older I also began to keep tropical fish there. Guppies fascinated me. They were indefatigable breeders. I watched astonished as the little brightly coloured males darted around the larger more stately and soberly coloured females, firing off invisibly small bullets of sperm at the female's genital opening with their pistol-like anal fins. I checked each female daily to see if she was developing the black triangle on her underside that would show that she was pregnant, and when she did, watched it darken until at last minuscule babies, smaller than tea leaves, dropped out of her and took shelter in the leaves of the water plants. I also kept gouramis. They had sky-blue bodies with two large black spots on their flanks and they also bred readily. They blew nests of bubbles under the floating duckweed in the corner of their tank. Then the male wrapped his body transversely around the female, and the two performed barrel rolls, sinking down through the water and simultaneously expelling sperm and ova which floated up into the bubble nest. The male, flushed bright with his success, then stationed himself beneath it and fiercely drove away any other inhabitants of the tank that approached his corner.

These dramas never staled for me. I maintained my tanks throughout my schooldays and only abandoned them when I left home to go to university. When Jane and I married in 1950 and set up home, I managed to persuade her that part of my tiny salary as an apprentice publisher should be spent –'invested', I suspect, was the word I used – in a tank in which I would breed white cloud mountain minnows. They were the fish of the moment in the aquarist circles of the 1950s. I was sure I would be able to persuade them to breed on a virtually industrial scale and that I could then earn enough by selling them to the local aquarist shop to pay for the tanks and the purchase of stock. Sadly, the white clouds never obliged me, except in the most desultory way. Nonetheless they established a place for tropical fish in our house which remained occupied for the next twenty years.

So after each *Zoo Quest* trip, our menagerie in Richmond expanded slightly. The tanks and vivaria spread from the dining room into the hall. In them lived small jewel-like tropical frogs, bizarre stick insects and chameleons. From one trip I brought back a clod of earth the size of a large grapefruit that I had dug from a dried-up swamp in Africa. It looked totally inanimate. Our two children, Robert and Susan, sat and watched as I dropped it into a tank of water. Bubbles streamed from a small hole in one side. As the water permeated the clay, bits of it fell away and exposed an object covered in a wrinkled parchment. This too softened in the water. Suddenly it shuddered. A split appeared. It widened and a long black sausage-like creature removed its tail from around its head and unfolded itself. It was a lungfish. It lived with us for many years, growing prodigiously, and later starred in a programme I made about the evolution of amphibians.

After a few years, we decided to devote an entire room on the ground floor to animals. One side was divided off by a screen of wire netting. Within, a window could be opened giving access to a long enclosure that stretched down to the far end of the garden. Here, at various times, lived parrots and monkeys, lemurs and pythons, armadillos and bush babies. In Borneo, during our Indonesian trip, I acquired so many hanging parakeets that I was able to retain a little group of six. They lived in a large cage in the dining room. The white cockatoo I had bartered from the pygmies in New Guinea also came home. She was very like the more common Australian species except that she was not quite so large and had a bright blue ring around each eye. She spent a lot of her time during the summer on the wisteria that grew over the back of the house. She had damaged her wing as a young bird so that she could not fully extend it. She could not, therefore, fly. When we gave her a bath by spraying her with the garden hose, she would stretch out her wings as far as she could, visibly delighting in the sensation, and squawk vociferously. But she never got used to British butterflies. Unaccountably, they seemed to terrify her. If a tortoiseshell or a cabbage white fluttered up to the wisteria flowers, she would let out a scream that could be heard a hundred yards away down the road.

Theoretically, these creatures were there for the interest and education of the children; formally, I was the one who was in charge of them; practically, needless to say, it was Jane who cared for them. She was so gentle, so instinctively understanding of their natures, that there were few that did not respond to her. Recognition of her skills among our zoologist friends became such that she was often asked to look after sick animals. Even the London Zoo sought her help. A baby gibbon had been sent to them. He was far too young to have been separated from his mother and he had chronic diarrhoea. They doubted whether he could survive. What he needed to save his life was tender-loving-care. No one in the Zoo could provide that continuously. Could Jane? She immediately saw what was required. She made a cloth sling so that Sammy, as we named the little gibbon, could remain attached to her throughout the day. If he was removed, he howled so miserably and so loudly that she allowed him to stay with her, no matter what she was doing.

Sammy's vocal repertoire was wide. If he was thwarted in some way, he would yell in outrage, but he also conversed in a range of gentle grunts and he and Jane would murmur reassuringly to one another as they moved around the house. If the telephone rang, Jane would necessarily have to answer it with Sammy on her hip. He soon realised that on such occasions some other entity was contributing to the conversation and he would lean over and belch gently into the mouthpiece. Jane said that she never knew whether to explain to her caller that she had a gibbon on her hip (which she thought might sound a little pretentious) or whether simply to murmur, 'I beg your pardon.'

We also looked after a small woolly monkey named William. He had been acquired by a producer in Children's Television who wanted him to make regular appearances in a new series. We were to care for him between times. Every week, a chauffeur-driven car called for him to take him to the studios for his performance. William developed a particular attachment to me and was very jealous. He was also very intelligent and learned to manipulate the catch on the door of the outside enclosure if it was left unlocked for even a short time. He would then scamper off to the end of the garden and sit truculently in the laburnums. But it was easy to get him back. All I had to do was to sit with Susan, who was five years old at the time, on my knee and give her a cuddle. William would immediately rush back and squeeze himself between us.

Our biggest success were our bush babies. These engaging creatures are primitive relations of the monkeys, small furry creatures with big wide eyes, continuously twitching membranous ears, long furry tails and very human-like grasping hands. They make enchanting pets, particularly for those people who have to be away from home during the day, for bush babies are nocturnal, spending all their days curled up asleep and only becoming active in the evenings. Then they emerge and soon become sufficiently tame to be allowed out in a room, where they leap entertainingly from curtain to picture rail to mantelpiece. They do, however, have a habit that makes them less than perfect as domestic pets. They urinate on their hands, rubbing them together enthusiastically as they do so. Then they deliberately lay trails of smelly sticky hand-prints all along their favoured routes. If they have the freedom of a suburban living room, the results can be awkward. For the first few weeks all seems well, but eventually, the householder realises that his home is acquiring a curious and pervasive smell. Getting rid of it not only involves getting rid of the bush babies, but may require the redecoration of an entire room and the renewal of all its soft furnishings.

Some pet-keepers keep their animals scrupulously clean at all times and regularly swab out their animals' accommodation with antiseptic. This, of course, does not suit the bush babies who will then sedulously set about re-creating the perfumed ambience they like best. We adopted a different policy for a pair that I picked up in Nairobi on the way back from our Madagascar trip. I suspended a long hollow log from the ceiling of the enclosure in the animal room in which they immediately made themselves at home by thoroughly anointing it with urine. The room thus quickly acquired the atmosphere in which they felt most relaxed and settled.

Bush babies are fond of fruit but they relish, above everything else, mealworms. These are the larvae of a small black beetle and can easily be produced in quantity by keeping adult beetles in old biscuit tins and providing their larvae with loaves of stale bread to nibble. Many cage birds love mealworms. So do many fish. But bush babies are addicts. I have never succeeded in giving a bush baby so many at a single sitting that it would refuse to eat more. Our two flourished and were soon so tame that they would leave their enclosure and leap on to my shoulder to take mealworms from my fingers.

Eventually, to our delight, they produced an infant. The female moved it from place to place in their enclosure, carrying it in her teeth. Slowly the colony grew. At one time there were ten of them, so many that we had to give some of the offspring to the Zoo. Eventually over two dozen babies were born there. It is true that sometimes guests who came to dinner were moved to enquire about the strange perfume they detected in the hall; and that mealworm beetles would occasionally appear crawling over the carpets, but that was a small price to pay for the pleasure of watching the little colony in the evenings chasing one another around their enclosure and nursing their offspring.

* * *

I imagine that guests to our house were not particularly surprised that someone who appeared on television, year after year, collecting animals in the tropics should keep a few at home. But, in truth, I had only started collecting animals for London Zoo by default. Had it not been for Jack

Lester's illness, I would never have done so. I was much more interested in making films about the ways animals behave than I was in catching them. The *Zoo Quest* format was beginning to look increasingly antiquated. And by now it was no longer necessary to have animals in the studio in order to show them in close-up. Now we had camera lenses and highly sensitive film stocks that enabled us to get such shots out in the wild. I certainly hoped to continue making trips to the tropics. But from now on, I wanted to make programmes entirely on film. *Zoo Quests* had come to an end.

13

The Top End of Australia

In the early 1960s, 16mm filming technology made an important advance. A way was at last found to link tape recorders and 16mm cameras so that both worked on the same time code and we could film and record simultaneously. Portable recorders also suddenly became much smaller. Glass valves that had been their crucial components until then were replaced by much smaller more efficient devices called transistors. Padded jackets were made to muffle camera noise. Sound now acquired a new importance. It was no longer something a director like me might record when he wasn't doing anything else. The next trip I made ought to have a properly qualified recordist. I asked Bob Saunders, who had mixed the sound tracks for so many of our *Zoo Quest* and travel programmes, if he would like to come with us. He did not hesitate. He had been trying to get out of the dubbing theatre for years.

How should we use this new capability? One thing was immediately obvious. The studio inserts that had been an essential element in *Zoo Quests* were no longer necessary. If I were to introduce the programme I could speak to the camera directly and more effectively from wherever it was we were filming. So could other people. No longer need Indians in the Amazonian rain forest, pygmies in the Congo and Dyaks in Borneo appear on a television screen apparently performing in some kind of dumb show accompanied by a background of indeterminate chatter. Now they could be seen and heard talking, arguing and singing, just like everyone else. So I decided that our next *Quest* should shift its emphasis. Now we could feature people as well as animals. Where should we go to exploit this new-found ability? One place, as far as I was concerned, was far ahead of the others. Australia.

Mention outback Australia and most people, I believed, thought of desert. But in the far north of the continent lies a country that even Australians at the time knew little about. It is nearer to Indonesia's capital Jakarta, than it is to Sydney. It is closer to the equator than is Fiji. Far from being a parched desert, it is drenched with rain through much of the year. There lived groups of Aborigines who still organised their lives in a traditional way. And there too we would also be able to find a rich range of animals – marsupials like wallabies and bandicoots, gorgeous birds and bizarre reptiles. Today a national park complete with luxury hotels has been established there and given the name of Kakadu, after the Aboriginal people whose homeland it is. At that time, however, the whole area was referred to vaguely, if at all, as the Top End.

Charles, having spent a year at home with his wife and young daughter, wanted to come on another trip. He would be cameraman. And for the first time, we would have a recordist – Bob Saunders.

* * *

We based ourselves at a disused logging camp at a place called Nourlangie about fifty miles from the north coast, close to the South Alligator River and from there drove into the bush country to camp wherever we wished. It was the dry season and there was no chance of rain so we did not bother with tents. Our camp beds were set up beneath gum trees or tree ferns with no protection or cover other than mosquito nets suspended above them. This did not really suit Bob. It turned out that he had never camped before. In fact he had never travelled outside Europe. He found it difficult to settle down to sleep without a roof over his head. The meals we improvised did not suit him either and he became particularly concerned about his bowels. 'I haven't *been*,' he told me confidentially, 'for five days.' I had a pill for him that would deal with that. But it didn't. Nor did a second pill the day after. Bob lost his chirpiness. A pained, worried look began to haunt his face. 'All those poisons building up inside,' he explained. 'It can't be good for you.'

This went on for three more days. On the morning of the fourth he seemed sunk in gloom as we searched in the bush for something to film. I was scanning the dusty ground hoping to recognise the tracks of something interesting when I found beneath a tree a huge pile of dung.

I knelt beside it.

'Buffalo,' I said, doing my best to sound expert and intrepid.

'Well you can bet your bottom dollar,' hissed Bob, clutching my arm, 'that it's not *mine*.'

Hoofprints nearby confirmed, had there been any doubt, that the droppings had indeed come from a buffalo. These animals were common in the Top End at the time. They were water buffalo from Asia, descended from those that had been brought here during the nineteenth century to pull wagons and ploughs and provide milk and meat for the military settlements the British had established here on the northern coast. When these settlements were abandoned, the buffalo were turned loose.

In the wild, they flourished. There were no other grazing animals of their size to compete with them. Before long there were great herds of them. And they were very dangerous. Bulls could weigh as much as three quarters of a ton and were liable to charge if they encountered human beings on foot. Soon after their release they had been hunted for the sake of their hides which made excellent leather but by the middle of the twentieth century there was not much profit in it. An optimistic pioneer, Allan Stewart, had taken a lease on Nourlangie and was trying to persuade big game hunters from southern Australia to come up and shoot buffaloes as sport. But putting bullets into these great lumbering beasts was scarcely difficult and there had been few takers. Nor did buffalo make particularly dramatic subjects for us but we dutifully filmed them as they plodded their way around the swamps. In hindsight we should have filmed them in more detail than we did. Now such film would have considerable historical interest for soon after our visit, the authorities decided to exterminate the buffalo. They were a foreign species and were upsetting the ecological balance of this part of the country. Teams were sent in to shoot them systematically. Today there are none.

There were, however, many species of native animals to occupy us. The lagoons were as rich in waterfowl as any in the world. Egrets, ducks, pelicans, ibis, pygmy geese and lily-trotters. Most of these we had seen elsewhere but the lagoons of the Top End had one splendour that was unique – vast flocks of magpie geese.

These strange black and white birds live only in Australia and New Guinea. They are primitive members of the goose family and so odd that some experts in these matters think they should be given a family of their own. They have a slightly comical pyramidal bump on the top of the head and toes that are only half webbed. Once you could have seen them all over Australia. But the systematic draining of swamps in the southern part of the continent robbed them of their feeding grounds and before long they could only be found in the Top End. They made spectacular subjects for our camera as they fed in assemblies several thousand strong, blanketing the surface of the lagoons.

There were not many Aboriginal people living around Nourlangie. Two or three were camped nearby hoping to get employment as guides or trackers. But there was abundant evidence that once this land had been home to a substantial population. The rocks, here and there, among the precipices and tumbled boulders, were emblazoned with paintings. We soon learned to predict where they might be. An overhang that could serve as a shelter was an obvious place to look, for people might well have spent time there to escape from the rain or the sun. In any case, designs painted on the back wall of such shelters, protected from the weather were more likely to have survived than those painted in more exposed positions.

Another indication of the presence of paintings nearby were small cup-shaped depressions on the tops of flat rocks. These seem to have served as mortars in which the artists ground their pigments. Prominent landmarks were also likely sites. Aborigines today see their own origins in the landscape around them and few conspicuous landmarks are without significance in their sagas of creation. A waterhole may be the place where the great serpent emerged before arching across the sky as a rainbow. A strangely shaped rock could be the petrified remains of a club thrown by the ancestral dingo at the first two human spirits. The artists who painted the rocks, whoever they were, seemed to have viewed the landscape in a similar way and celebrated its more important and sacred features by decorating the rocks around them.

Some paintings were naturalistic. They seemed to be a century or so old because of the subjects they represented – a flint-lock pistol, a sabre, a musket, a steam ship with smoke pouring out of its twin funnels. Others were much more ancient. A few were scratched in outline – skeletal figures racing across the rock carrying spears and spear-throwers. These, according to local stories, were self-portraits of *mimis*, shy gentle spirits that lived in crevices in the cliffs. They were so frail and delicate that the slightest wind would crumple them and so thin that they could slip into the narrowest crack if anyone approached. So no one had ever seen a *mimi*. In any case, the Nourlangie Aborigines said, these pictures were certainly not drawn by anyone living there in recent times.

Supernatural beings were also portrayed in a different style. The most spectacular of these we found at the foot of a monumental rock called Anbangbang that rose six hundred feet above the scrub. Immense blocks of stone, as big as houses, that had sheared from the cliff in primeval times lay around its base. In one place its face leaned outwards forming a shallow overhang and there we saw on the wall strange human figures, almost life-sized, painted in red and white ochres. Their white faces were featureless and surmounted by large headdresses. Their bodies were painted with dots and cross-hatched designs. Three of them were female, with prominent breasts splayed outwards and extending to their elbows. The lower parts of their thighs were indistinct for they had been worn away by buffaloes which had used the cliff as a scratching place.

In later years, this group came to be regarded as one of the special treasures of Kakadu. It is illustrated in guide books and was selected for reproduction on postage stamps. Our photographs of it seem to have been the first taken. But the paintings today are slightly different. They have been retouched and other designs added, including a huge human-like figure that looms above all the rest with arms and legs spread wide. It seems that they were originally painted by Old Najombolmi, one of the trackers who worked at Nourlangie, and that during the year after our visit, he had modified and added to them.

Many designs we could not film for they had been painted in places where there was not enough light – at the back of shelters or on the ceilings of overhangs. We had a small hand-lamp powered by a rechargeable battery which was enough to illuminate individual figures but it did not have the spread needed to show the full extent of many of these galleries. At one site, however, we did have a chance of filming a great frieze in its entirety. Ubirr lay seventy miles east of our base at Nourlangie. Its rock, a coarse sandstone, is horizontally stratified and at its western end, one immense leaf juts out from the main mass. The back wall of this wide deep shelter is decorated by a procession of giant barramundi, a local fish that is much relished for food. Among these swim fork-tailed catfish, long-necked turtles and goannas. All are painted as though in X-ray. The artists drew not just what they saw when they looked at such creatures but what they knew was there within them – backbones, livers and bunches of muscle.

For most of the day, this spectacular frieze was in heavy shade, but just before sunset, the near-horizontal rays of the setting sun reached into the depths of the shelter and illuminated the back wall. In order to make plain the spectacular dimensions of the barramundi paintings and to exploit our newly acquired ability to record in synchronous sound, we decided that I should be filmed as I walked along the entire length of the wall, talking about the individual designs which in places were painted on top of one another in such profusion that they were difficult to disentangle. But there was only about ten minutes between the wall becoming fully illuminated and falling into darkness as the sun dropped below the horizon. And we only had one evening left to get the shot, for we had been away from Nourlangie for over a week and our supplies of water were running out.

I rehearsed my words, walking along the frieze muttering to myself and trying to decide which of the designs needed explanation. Charles set up his camera so that he could pan along with me as I walked. Bob put his recorder alongside with his microphone on a long rod so that he could move it to catch my words. The sunlight crept inwards along the shelter floor and climbed slowly up the wall at the back. Minute by minute it rose higher and higher. The effect was magical. It was as though a proscenium curtain was being slowly raised in a theatre to reveal a magnificent painted backdrop.

At last the whole height of the frieze was lit. I cleared my throat and started my walk. Before I had taken two paces, Bob called a halt. His recorder had suddenly developed a high-pitched electronic howl. Out came his screwdriver. Within a minute or so, the machine's entrails were exposed and he was prodding around inside. The seconds ticked by. The sun was visibly dropping. 'Bloody transistors!' he said. These were the new components in our wonderful synchronous recording system about which he had been so enthusiastic. Then he remembered something in their technical specifications suggesting that they did not work properly above a certain temperature. He fanned them with his hat. Slowly the howl in his headphones faded. He jerked up his thumb; Charles pressed the camera button; and I set off once more along the ledge describing the paintings. I managed to do so without an intolerable verbal stumble. That was just as well. Within seconds of my finishing, the sun dropped behind a cloud and the paintings vanished into darkness.

* * *

In recent years, archaeologists have found ways of dating paintings such as those we had filmed at Ubirr and Anbangbang. Their results were so startling that at first many archaeologists refused to believe them. Some paintings were 45,000 years old, older, indeed, than most if not all of the spectacular cave paintings of France and Spain which are in many ways so similar. No one can ever know for sure why either was made. Even at Nourlangie, where some at least seemed to be fairly recent, no one could tell us. Old Najombolmi, who had painted those at Anbangbang, had not even admitted that he had done so when we were filming them. But farther east, in Arnhem Land, there were still people who had not yet abandoned their traditional way of life. They still travelled continuously and found in their land all they needed. And they still painted. There, I thought, we might be able to find out why they were so addicted to doing so.

A few missions had established settlements on the coast, each hoping to convert the people to their own brand of Christianity. They had encouraged the people to produce bark paintings in order to sell them to collectors farther south and swell mission funds. But since the missions were also urging the artists to abandon their traditional beliefs, such establishments were hardly the most promising places to go in order to discover the Aborigines' traditional motives for painting. However, the Government at that time was building a new settlement close to the mouth of the Liverpool River at Maningrida. Its purpose was to provide education and medical help. It was not, primarily at any rate, to convert them to Christianity. There, I thought, we might find out a little more about the Aborigines' zest for painting.

* * *

No one had ever reached Maningrida by land. The only way for visitors to get there was by ship from Darwin two hundred and fifty miles away to the west, or by air. We flew in. Maningrida, then, had just half a dozen buildings – a school, a hospital, communal kitchens, a store and accommodation for the European staff. But more were being constructed. A small cargo ship was moored on the jetty by the river and European bricklayers and carpenters were energetically carrying ashore bags of cement, building machinery and drums of petrol. It was cripplingly hot and I marvelled at the enthusiasm with which the men were working in the heat of the day. Mick Ivory, the Superintendent, explained. 'This is the first ship to call for over three weeks,' he said. 'The boys have been out of beer for ten days and I told the shippers to put the beer at the very bottom of the hold. So the blokes have to shift twenty tons of stores before they can get at it.'

The Aborigines were camped on the outskirts of this embryo settlement, living in low shelters improvised from bark, branches and scraps of cloth. Some were dressed in the tattered remains of western clothes but most wore nothing but a square of cloth passed between the legs and knotted at the hips. There were two groups. The Gunavidji occupied the strip of land along the coast and lived largely from the sea, the men fishing for barramundi and turtle, the women collecting shellfish along the shore. The Burada, living on the inland side, were hunters, who by tradition caught kangaroo and bandicoot and gathered seeds and the swollen roots of the desert plants. Both had been attracted here by the flour, sugar, tea and other food that the station could provide, but the two groups were unaccustomed to living so close to one another, or in such high densities, and quarrels were frequent. We were sipping tea on Mick Ivory's veranda when a particularly violent argument developed between two men a hundred yards away. Both carried spears with which they were threatening one another. Mick glanced round at them but otherwise took no notice. 'No worries,' he said. 'They've both got their trousers on. For some reason, they take them off if they are going to have a real set-to.'

Mick told us that there was one Burada man who was regarded locally as a particularly good painter. His name was Magani and we found him in a shelter standing somewhat by itself on the fringes of the scrappy eucalyptus bush. He sat square-shouldered and cross-legged with his knees touching the ground in a posture that few could assume unless they had sat that way from birth. Scar tissue, so swollen that it formed thick horizontal ridges, stretched across his chest. Ribbon-like scars also streaked down from his shoulders to his elbows and across his thighs. A lifetime of eating gritty food had worn his teeth down close to the gums but his smile, when it came, was impish and dazzling. He had already painted some barks that Mick Ivory had sold on his behalf through the station store. We explained that we had come to see how people painted and he accepted that, naturally, we would come to him since he knew he was the best.

He took the job of demonstrating his techniques very conscientiously and made quite sure that we did not miss any of the details. The bark he used came from the stringybark gum tree. He flattened it by throwing it, inner-side down, on to a small fire. This turned the sap that covered its inner surface into steam and made the whole sheet so pliable that after a few minutes he was able to take it off, press it flat on the ground and weight it down with stones. This was to be his canvas. For pigments he used ochres which he collected from special places in the bush. Red and yellow came from pebbles of iron-rich rocks, black from charcoal, and white from clay dug in the mangrove swamp. He also had a particularly deep red ochre which came from far away and which he hoarded in a little packet made from the folded bark of paperbark eucalyptus. For brushes he used a thin twig which he chewed in order to splay its fibres, and another, to which he had tied a few trailing animal hairs, for drawing particularly fine lines.

* * *

We visited Magani every day for a week or so to watch him working and to film. Another more solemn and morose man, Jarabili, was often with him. He occasionally added a figure to one side of the bark while Magani worked on the other. Because I once asked him the name of a bird in his own language, Jarabili concluded that I wanted to learn Burada and every time we sat down with him, he dictated a list of Burada words to me. Then when I arrived at the shelter the following day, he tested me to see how well I had learned them. He also tried to explain to me the relationships between his family and Magani's, how it was that they were intimately linked and yet had different rights and responsibilities. The full complexities of this, I have to admit, I found hard to follow.

Magani did not work intensively. He seldom added more than one or two designs a day. Nonetheless, the bark gradually filled with drawings. They were simple and diagrammatic but readily identifiable. Lizards were not just generalised lizards. One, from its proportions, was certainly a goanna. Another, with a broad flattened tail, was a gecko. A third kind had flaps flaring from either side of its throat. That was a frilled lizard. Fish could be identified as shark or barramundi or sting-ray. There were leaping kangaroos with paired footprints trailing behind them and hunters carrying spears and spear-throwers. A line of small oval creatures covered with short lines, were, Magani told me, cheeky buggers. 'Cheeky', in Aboriginal pidgin, means 'stinging' and I guessed that these must be caterpillars for several local kinds have long hairs with an unpleasant sting.

One design, somewhat more diagrammatic than most, dominated the centre of the bark he was painting for us. It was a long elliptical shape with stubby projections on either side and an interior filled with neatly drawn cross-hatched bands of white and yellow. Magani called it 'sugar-bag', the pidgin name for a hollow tree containing a bees' nest, full of honey. Close by, he drew a second elongated rectangle, also meticulously cross-hatched but lacking side projections. Beside this he added the figure of a man, with his head touching one end of the rectangle and his hands touching its side. Another man sat beside the first holding rhythm sticks. Since Magani did not use any particular perspective but showed figures in plan or profile according to how he fancied, it was difficult to know if these figures were standing or lying on their sides.

I asked Magani what the long rectangle represented. For the first time, he hesitated before he answered. Then he leaned forward.

'Yurlunggur,' he whispered.

'Why you talk so soft?' I whispered back.

' 'im secret. Woman, young boys might be hearin' me talk his name. Im b'long business.'

I knew enough pidgin by this time to understand that 'business' referred to sacred rituals.

'Me no woman, no boy,' I said. 'Orright *me* see 'im?'

Magani scratched his nose and looked at me hard.

'Orright,' he said.

The next day Magani led us away from the encampment for half a mile or so to part of the bush that seemed little frequented, for instead of being criss-crossed by a network of trails there was just a single indistinct track. It led to a rather bigger shelter than any I had seen nearer the station. Jarabili emerged from it. He and Magani conferred. Then the pair of them went into the shelter and came out carrying a hollow pole, eight feet long, trimmed smooth and coloured uniformly with red ochre. It was painted with a few figures of goannas.

'Yurlunggur.' said Magani.

He lay down and put his mouth to one end that was partially stopped with bees' wax.

'Im talk,' he said and then he blew, producing a deep booming note. Yurlunggur, whatever it might represent, was in material terms a giant sacred didgeridoo.

It was also unfinished. Magani explained that it was to be used in an important ritual in which he, Jarabili and many other Burada men would take part. Women would never be allowed to see such a thing but young men would be summoned to watch, for during it an episode from the story of the Creation would be re-enacted. That was how boys learned about their gods and ancestors.

Yurlunggur was the great serpent that during the mythical past, the Dream Time, emerged from a waterhole and swallowed the twin sisters who were to be the progenitors of the whole human race. After many adventures during which all the animals and plants of Burada country were brought into existence, Yurlunggur arched into the sky. He still occasionally appears today after storms as a rainbow and speaks with the voice of thunder.

Before Yurlunggur's ritual could be held, the whole length of his wooden embodiment had to be covered with paintings. Magani was not the only one who worked on it. Jarabili drew some figures. So did other men who occasionally emerged silently from the bush, took up twigs, and sat alongside Magani, to add some designs themselves. When each design was finished, there would be a pause. Jarabili would take up rhythm sticks and he and Magani would chant, 'singing in' the new addition.

We visited every day to watch progress but had to take care in approaching the shelter. The bush all around the Burada encampment seemed empty but it was, in fact, divided by invisible boundaries of considerable importance. Once, on my way to Magani's shelter, I saw a small brightly coloured parrot a little way from the path. As it flew away, I followed it to try and get a better view. A young boy was with me carrying the camera gear, but he was lagging behind. I called to him to catch up, but he seemed oddly reluctant. Then I saw, some twenty yards ahead of me, an old man sitting quietly by himself on a log. A long wooden object bound with twine lay across his knees. He was anointing it with sweat from his armpits. We had strayed into sacred territory and intruded on his privacy at a particularly intimate moment. The old man pushed his wooden object under the log on which he had been sitting and called the boy over to him. The boy came back to me and explained that he would have to pay a fine of several tins of tobacco for breaking the rules. Both he and the old man knew that I would pay, but if the lad had been older and had not been acting at the request of a white stranger, the penalty might have been a much more severe one.

Eventually Yurlunggur was finished. Each end was bound with a narrow ring of orange parrot feathers. At the mouthpiece end there were also two long tassels of downy white cockatoo feathers weighted with bees' wax. A procession of goannas ran along its length in between rectangles symbolising Yurlunggur, like the one I had seen on the bark Magani had painted for us. The time for the ritual had come.

* * *

The following day was a Saturday and work on the station had stopped for the weekend. In mid-morning, men began to drift in and sat down around Yurlunggur's shelter. Many I recognised. Some during the week worked at the sawmill, stacking planks. Others tended the vegetable gardens. The didgeridoo was out of sight, somewhere in the back. Those few men who had any kind of European clothing removed it. Each man had a woven tasselled dilly-bag suspended from a string hanging around his neck. Then they began to paint goannas on one another's chests. The one to be painted lay flat on his back, eyes closed as though in a trance, while the other carefully cross-hatched the design in yellow, red and white on the shining jet-black skin. A group of younger men sat quietly to one side. They were observers, not participants.

Yurlunggur's deep drone sounded inside the shelter. The goanna men lay in a line alongside one another. The huge painted didgeridoo emerged from the shelter. Magani, crawling and holding its mouthpiece end, was blowing down it. Another man, bent double, held the other end a foot or so from the ground. Magani did not have a goanna painted on his chest but nonetheless was splendidly decorated with red and white stripes on his face and a string of orange feathers wound round his head. An old man, sitting to one side, started to beat rhythm sticks and to sing. Yurlunggur, still sounding, moved along horizontally, a foot or so above the ground, passing over the prostrate goanna men. It paused and the goanna men came to life. Lying on their stomachs, they wriggled in pairs through the dust beneath Yurlunggur. When all had done so and they were once again lying alongside one another, the rhythm of the music suddenly changed. The goannas, electrified, reared on to their knees holding their dilly-bags in their teeth and knelt rigidly in front of Yurlunggur, with their muscles trembling.

Again and again this happened until it began to get dark. Yurlunggur, escorted by Magani and the old man, moved back into the shelter and fell silent. The goannas turned back into men and walked away to the main encampment, their emblems, rubbed and smeared with sweat, barely recognisable on their chests.

Later that evening, we sat smoking with Magani in his shelter close to the station. I tried to find out the meaning of what we had seen. The shelter in which Yurlunggur had been painted was the waterhole from which the rainbow snake had emerged during the Dreaming. When it discovered the ancestral goannas, it swallowed them all, and then spewed them out. More than that I could not establish. Why did not Magani have a goanna on his chest? Because goannas were Jarabili's dreaming. Magani had rain cloud dreaming. It was rain clouds that brought the rainbow snake. There were many other significances and implications which were difficult to comprehend. One thing, however, I had learned. Painting for Aborigines was not idle doodling stemming from a vague impulse to fill in blank spaces. It was a profoundly important part of the people's religious life.

I was worried lest what we had filmed should be so secret that to show it to others would offend Magani or those taking part. I explained to him that when I got back to 'my place', all kinds of people would see it – not just men but women and children. Magani thought this over.

'Which way your place?' he asked. 'Down this way?' and he pointed south.

'No. Long, long way, 'cross this way.' I pointed west.

Magani thought more.

'Orright,' he said. 'But you no show 'im in my place.'

We smoked some more.

I asked him what would happen to Yurlunggur now that the ceremonies were over. Apparently, it would probably be buried down by the river in a sand bank. It would certainly never be used again.

'Dayfit,' he said. 'You like Yurlunggur?' I nodded.

'Okay,' he continued, 'you take 'im.'

'I no *sell* 'im,' he added, ' 'im present.'

I thanked him very much.

'Blackfella custom,' added Magani, to avoid any misunderstanding, 'one man give present, nudder man give present back.'

I told him that that was a whitefella custom too.

Later that night I gave him all the tobacco we had. It seemed to be the thing I had that he most wanted. I hoped that my gift was as great in his eyes, as his was in mine.

The next night, under cover of darkness, Magani and Jarabili brought Yurlunggur to our house in the settlement wrapped in old newspapers. We added more coverings of paper and wound strips of sackcloth around it. At the end of our stay, a small charter plane came in to collect us and take us back to Darwin. There was no way in which we could sneak Yurlunggur out unobtrusively for it was so long that we had to remove the plane's passenger seats in order to get it inside. But none of our Aborigine friends commented on the strangely shaped addition to our baggage.

* * *

Aborigines are not the only people living in the remote parts of the Australian outback. A few Europeans also choose to do so. We heard of three who had settled in the derelict town of Borroloola, two hundred and fifty miles south of Maningrida where the MacArthur River flows into the Gulf of Carpentaria.

Back in the 1880s cattlemen were starting to move westwards from Queensland round the southern shores of the Gulf – that vast bite taken out of the middle of the northern coast of Australia – and then on into the semi-desert south of Arnhem Land. The MacArthur River was the last place where they could water their stock and they regularly camped on its banks. So a small settlement grew. By 1913 there were about fifty white people living there. The young town had two hotels, five stores, and a gaol. Its policeman must have been a man of literary tastes for he discovered that some federal regulation entitled a town to a library. He applied to Melbourne for it and after six months on the road it finally arrived in Borroloola.

But then artesian wells were drilled in the desert and the town lost its importance as a watering place. Roads were driven from the cities in the south right across the interior of the continent. Few people needed to go to Borroloola any more and the town died. One of the hotels collapsed. The wooden shacks that had been homes mouldered to dust and blew away. Apart from a store where you could get petrol, a teetering assemblage of corrugated iron that had been the second hotel, and a few bits of ancient unidentifiable machinery half-covered by the desert grass, there was little left.

Nonetheless, one man, Roger Jose, had stayed. He had arrived in 1916 and had discovered that the same legislation that entitled the town to a library, also provided an annual salary for a librarian. He got the job and he did not intend to let it go even though eventually there was no one there to read the books except himself. He lived in a five thousand gallon corrugated iron water tank that had once belonged to the still standing hotel. He had dismantled it and reassembled it, upside down, about a mile away in the desert. He had cut a small door in its side but had not gone so far as to provide it with any windows. I imagine it must have been blisteringly hot inside but Roger spent most of his time outside it, sitting on a stool under an improvised awning. He was a noble-looking man with a luxuriant white beard and he habitually wore a strange costume of his own design – a khaki shirt with its sleeves cut off at the shoulder, khaki trousers trimmed off just below the knee and a peaked cap woven from pandanus fibres shaped rather like a Foreign Legionnaires' kepi.

His post as librarian was not simply a device to get an unearned official wage – though it was certainly that. He had a genuine love of books and literature. He said he had read every one of the three thousand books he once had charge of. 'Just managed to keep ahead of the termites,' he said. But the termites had won in the end. We found the last surviving volume on the floor in the hotel. It was, I could see from the binding, *The Imitation of Christ* by Thomas à Kempis. But Borroloola termites, it seems, had a particular taste for printer's ink, for when I opened the front cover, I found that they had eaten every one of the holy man's words, leaving only the white margins of the pages intact. Roger mourned its loss. 'Now' he said, 'I just have to read anything I can find – jam labels, sauce bottles, anything.'

Words delighted him. He rolled them around his mouth and sucked them like sweets. When I asked him where he got his food he said 'Well, I would fain pursue the elusive marsupial, but I can't catch the bastards.' His habit was to ask any visitor who came his way if they knew the meaning of some particularly obscure word he had dredged up from his memory, just so that he could have the pleasure of explaining what it meant. A policeman visited Borroloola about once a month just to check up on things. He got this treatment every time he met Roger. Eventually it got him down and he decided to do something about it. Before setting out for Borroloola, he consulted a dictionary. When he arrived, Roger asked him if he knew what 'transubstantiation' meant.

'No, I bloody don't,' said the policeman, 'But here's one for you. What's a leotard?'

'I'm not sure,' said Roger, 'but I think I saw the skin of one once.'

'No, you bloody didn't,' said the policeman.

Roger gave up. 'Okay, what *is* a leotard?'

'I'm not going to bloody tell yer,' said the policeman and he got into his truck and drove off.

Three days later, the policeman was asleep on the veranda of the police house. He was woken by someone gripping him by his shirt and shaking him. Roger, with his strange hat and long beard, even though in silhouette, was unmistakable against the stars in the night sky.

'What's a bloody leotard?' hissed Roger passionately through clenched teeth.

The last keeper of the Borroloola hotel had never left either. Mull – his full name was Jack Mulholland – was an Irishman in his late fifties who still spoke with a hint of brogue. He sat on the veranda of his hotel as though preparing to welcome visitors, but business, he told me, was slack. He had only had three guests ever since he took the job. Inside, on the sagging floor, lay a dune of empty bottles. I asked him, conversationally, if he had ever considered tidying the place up.

'Tidiness,' he replied severely, 'is a disease of the mind.'

He had a motor car. It was difficult to say what its make was. It was more an anthology of parts of the different machines that had roared and coughed their way to the town over the decades and had expired there. Its radiator, according to the label, was from a Pontiac. Its rear wheels had wooden spokes and were entirely different from the front. Its cylinder block was immense with an elemental rectangular grandeur. Looking at the grass growing around its tyres, I nearly offended Mull by asking when it had last moved but rephrased my question, just in time.

'How often do you take her out?'

'I can start her any time you like,' he said defensively. 'Would you like a spin?'

Once he had spoken the thought, he was reluctant to let it go and although I politely said he was not to bother, he started to make preparations for an excursion. After long contemplation of the engine, he decided the water pump needed replacement, but he found one in another hulk and spent several hours filing grooves in it to make it fit. The following morning he offered to take us for a ride.

The car had to be started by hand, for it had been built before the advent of the electric starter. Unluckily, the starting handle had been lost, so

Jack put the engine into gear, jacked up the back axle until both rear wheels were off the ground, and started to heave them round. The engine let out a tremendous roar and the wheels revolved by themselves. Mull went round to the front, put the car into neutral, removed the jack from the back and invited us to join him on the front seat. We then made a lap of honour around the hotel.

Roger told us of an occasion some time back when the town was due to be visited by a police sergeant. Mull heard he was coming and decided to drive off for a few days. It was not that he had anything to hide. Just that he saw no point in taking risks and getting unnecessarily entangled with the law. No one heard any news of him for three weeks. Then one day, there was a distant unmistakable roar and Mull in his car appeared on the horizon. When he was half a mile from home, the engine stopped. He might have decided to repair it there and then. Alternatively he could have walked the remaining half mile to spend the night back under his own roof.

He did neither. He clambered out, lit a fire, put on a billy for tea, and spread his swag out in the shade beneath the car. He stayed there for three days pondering what might have gone wrong with the engine. Finally he opened the bonnet and cleaned the plugs. The engine started immediately and Mull majestically completed the journey back home.

He said he used the car to go prospecting – for gold, silver, lead, opal – but had never found anything payable. I said I thought this must be disappointing.

'Not at all,' he said with unexpected vehemence. 'It would break a man's heart if he did discover something. What else would there be to live for? Money's no good to you anyway.'

'It can make life easy and comfortable,' I suggested.

'If I've learned one thing in my life,' said Mull, 'it is that the measure of a man's riches is the fewness of his wants.'

The third of this trio, Jack Kitson, lived in a cabin several miles out of town. It was said that he sometimes sat for hours in the desert by himself playing a violin but we were warned not to visit him without an invitation. He had a violent temper and had been known to order unexpected visitors off his property with a shotgun. One morning, we saw a strange lorry filling up with petrol at the store. Mull was in conversation with its driver, a small thin man with steel-rimmed spectacles. He introduced us. It was Jack.

'These fellers,' said Mull, pointing at us, 'have been asking me why I came out here. Why do you reckon you came, Jack?'

Jack looked at us aggressively. 'I was sent out of England for England's good,' he said flatly.

I decided to change the subject.

'They tell me you play the violin,' I said.

Jack admitted that he did. I asked what kind of music he played.

'Scales mostly,' he said.

'What about tunes?'

'Wait on,' said Jack. 'Fritz Kreisler and fellers like that started when they were knee-high to a grasshopper. The violin's a tricky instrument. I've only been at it myself for seven years. I might have a crack at Handel's Largo next year but I'm in no hurry.'

He climbed into his lorry. 'Well, I can't waste my time jabbering to you blokes,' he said.

'I was wondering whether we might come and visit you some time,' I said tentatively.

'I don't think you had better,' he said. 'You can never tell what mood I'll be in.'

He put the lorry into gear and then leaned out of the window.

'If you don't bring any of them cameras and recording machines, I suppose it might be all right,' he said and drove off.

The following day, we took him at his word. His cabin stood on the edge of a crescent-shaped billabong on which sailed a flotilla of pelicans. Cockatoos, perched in the skeleton of a gum tree, flew off as we arrived. We could see Jack through the glassless windows, busy with something. He did not look up as we parked. We called g'day but he took no notice. After several minutes, he at last emerged and gestured to us to sit on boxes in the shadow of the eaves.

I asked him if he would show me his violin.

'I'd rather not,' he said.

We talked in a desultory way for a minute or so. Abruptly, he got to his feet and went over to a table improvised from a sheet of bark. He picked up an enamel mug and started to polish it with a cloth, even though it was clearly bone dry.

'People call us hatters, you know,' he said bitterly. 'Well they're right. Mull and Roger have been telling you they are happy, haven't they? Well they are not telling the truth. They are hatters just like me. A man comes out here for one reason or another, he stays on and before you know what has happened, he's got to a stage where he can't change his way of life even if he wanted to.'

He hung the mug on a nail.

'Time you fellas were going,' he said and he disappeared into his cabin and shut the door.

* * *

The films we brought back from Australia were, as I had intended, as much about human beings as they were about animals. The series was given the title *Quest Under Capricorn*. The word 'quest' indicated a link with the *Zoo Quest* series that had preceded it over the previous half dozen years, but in truth I had little justification for using it. The series certainly had nothing to do with zoos, and I would have found it hard to explain what we had been in quest of. It was the last time I used a title with either word in it.

14

Casting Around

Leonard Miall summoned me to his office. It was time, he said, for me to consider my future. I ought to be applying for some of the administrative jobs that were coming up. An Assistant Headship of something or other would give me valuable experience. 'After all,' he said, 'you won't want to be gallivanting around the world when you are fifty.' I was not convinced of this. However, I thanked him for his solicitude and promised him I would think about it.

There was something in what he said, I decided, though not for the reasons he gave. It was 1962 and I had been at the top of the producer ladder for some time. Looking around, I could see that people who arrived at ladder-tops and stayed there for long periods can easily go stale, or at any rate be dismissed in the estimation of others as having become so. On the other hand, I didn't want to abandon travelling, in spite of what Leonard had said. How then was I to remain fresh? The answer, it seemed to me, was to hop on to a different ladder – and then go on doing the same thing as before while wearing (to mix my metaphors) another hat. Perhaps the time had come for me to leave the BBC's staff and become a freelance – but to go on directing similar kinds of films. I also had another ambition. *Quest Under Capricorn* had made me aware of how little I knew of anthropology. Maybe I could earn enough to support my family by working for the BBC for six months and then for the remainder of the year, I could study for an anthropological degree.

I went to see Sir Raymond Firth, the great Polynesian scholar who was then head of the anthropological faculty at the London School of Economics. He very generously agreed that I might enrol in his department's lectures part-time with a view to ultimately writing a post-graduate thesis on the use of film in ethnographic research. The BBC considerately worked out a scheme whereby they took six months of my time every year for the next five years, provided I could offer them programme ideas that interested them and did not work for the opposition.

So I went back to university. I listened, jaw sagging, to a brilliant young anthropologist, Robin Fox, lecturing on the intricacies of kinship systems. He began by pointing out how many variables there could be in the formal relationships between male and female and the inheritance of property by their progeny. A man could live with his wife's family, or a woman with her husband's. Each could have single or multiple partners. A male child could look to his mother's brother for tuition and protection, or a girl to her father's sister. Property could be inherited along the male line or the female line. All children might share their parent's property equally, or the first-born might take all. He then drew charts mathematically permuting all such variables and demonstrated that there was no theoretical combination that had not been adopted by some society somewhere in the world. Matrilineal, patrilocal, polyandrous, polygynous – the polysyllables whirled round our heads. We dutifully drew kinship diagrams of algebraic complexity. Dr Fox's triumphant conclusion at the end of the course was that of all the kinship systems in the entire world, the most complex by far were those evolved by Australian Aborigines. No wonder I had found it so difficult to make head or tail of Magani's explanations.

I also attended Professor Firth's seminars. The right to do so was a considerable privilege but I was seriously out of my depth. The seminars were held in a lecture theatre with lines of seats, steeply raked and curved around a small desk in front of a blackboard. We were a mixed lot. Some were still in their 'teens and frighteningly bright. A few, including some colonial administrators (for in the 1960s we still had the vestiges of empire), were middle-aged students as I was. I chose to sit next to one of them, George Milner, a scholar who had worked in Fiji and compiled the definitive Fijian-English dictionary. On occasion he seemed – no doubt because of his innate modesty – to be almost as baffled as I was.

Each seminar started with some dazzlingly intelligent young graduate delivering a paper on a subject of his or her own choosing. At its conclusion, there would be a discussion, led by the Professor. His gimlet eye would sweep slowly round the tiered ranks deciding on which of the listeners would be selected to make the first comment. I was reminded of those war films (one of which had, indeed, featured my brother Richard) in which daring British officers trying to escape from a German prisoner of war camp were caught by – or managed to dodge – the pitiless beam of a searchlight sweeping round the barbed wire fence. George Milner and I, more canny maybe than our younger colleagues and less in need of making a public academic reputation, habitually sat on either side of a pillar behind which we could lean if the Professor's gaze came dangerously close.

On one occasion, George was just too late. The Professor caught him.

'Dr Milner,' he said, 'the paper has explained that the Azande believe that twins are birds. Now do you suppose that, in the Azande mind, they really *are* birds?'

I may have got the precise wording of this question marginally wrong after this long time. I remember George's answer, however, word for word, for I made a careful mental note feeling sure that it would come in very handy some time in the future.

'I think,' said George, after a considerable pause 'that the question, as phrased, is not meaningful.'

I bided my time. In due course, one of the younger members of the seminar delivered a paper on the behaviour of fighting dogs and the parallels it might provide with human aggressive displays. At last – a subject that I felt I knew something about. The dog research on which the paper was based had been done decades before, not by a zoologist but by a psychologist and its conclusions, as any animal behaviourist would have known by the 1960s, were hopelessly wrong. It was going to be my moment of glory. I carefully formulated a question in my mind that would not be too unkind to the reader of the paper but would at the same time in a magisterial way demolish the very foundations of his arguments.

The paper came to an end. I prepared myself for my maiden speech. Professor Firth did not immediately start his searchlight scan.

'Before I ask for comments,' he said, 'let me make one thing clear. We are an anthropological seminar. We deal with human beings. So let us hear no more of fighting dogs, talking budgerigars or any other animals.' My penetrating comments congealed on my lips.

Much of our studies at the LSE were concerned with the philosophical attitudes of the great names in anthropology – Evans Pritchard, Levi-Strauss, Malinowski (who had himself been professor at the LSE) and many others. The tribal peoples among whom they lived did not seem to receive quite as much attention. This approach did not suit me. I had imagined that anthropologists might try to be as objective and clinical in their observations of the human species as zoologists endeavoured to be when observing animals.

Even before enrolling in the seminar, I had begun to sketch a series of programmes that would look at human beings in this way. One programme was to examine human territoriality. I planned to take a half-wrecked old car and park it, perfectly legally, in a fashionable part of Mayfair where expensive eighteenth-century houses had front doors that opened directly on to the pavement. I would then arrange to hide a camera in a window on the opposite side of the street. I suspected that the house-owner, when he came out of his door, would look faintly irritated at seeing such a disreputable car in what he must surely consider his territory. If we could manage to park it there day after day, I guessed that he might eventually physically attack it. Later in the same programme I suggested concealing a camera in a hotel room to watch the way a new occupant would mark his sleeping territory, by putting his pyjamas on the bed, his shaving gear in the bathroom, his papers on the desk. I only sketched in the vaguest terms what we might do in the programme we would have to include that dealt with sexuality.

If we were to get really valid natural behaviour, it was essential, of course, that the people concerned should be totally unaware that they were being observed and filmed. And with that thought came the realisation that such spying on one's fellow human beings was an intolerable intrusion on their privacy. Programmes could not be made in such a way. Human beings are not, after all, the same as other animals and television should not treat them as though they were.

* * *

Soon after I had become a freelance, I walked into the bar in Lime Grove where people assembled to drink and gossip. Huw Wheldon was there. He had just been put in charge of all documentaries. He hailed me. Did I want to go to Japan? Yes, of course. Next week? Certainly, what was the story? He explained that the London Symphony Orchestra was going there on tour. It would be the first time a whole European orchestra had given concerts in Japan. Their manager had chartered a jet and there were four spare seats. It broke his heart to see them going to waste so he had offered them to the BBC in return for making a film about the whole event. There must be a story there somewhere, Huw said. I agreed. What it was, I couldn't be sure at the moment, but I felt certain that something would emerge as we went along.

The opportunity was, in any case, not one to be refused. It was not only that visiting Japan would be exciting. So would be the chance to see what went on behind the scenes in classical music. My cameraman was to be Ken Higgins. He had been a news cameraman during the war and had carried his ancient Newman Sinclair 35mm camera across Burma with the Chindits. Now he was one of the star staff cameramen. He had the reputation of chewing up inexperienced or indecisive directors and spitting out the pieces. I was not certain how far my experience of making films with monkeys was going to take me with him.

The long flight to Japan gave me the chance to meet the orchestra's players. They were as baffled as I was about the kind of film that should be made. Some suggested it might be a travelogue about Japan, intercut with short excerpts from their concerts. Others were apprehensive that it might be an embarrassing exposé of the swashbuckling life led by orchestral musicians when out on a tour. I suggested, somewhat tentatively, that we might film during rehearsals.

'You ought to get a few fireworks from one of the maestros,' said one of them, 'a bit of an argument about interpretation, that sort of thing. Several of them who are coming with us have short fuses. We could get one of them going in a rehearsal, if that would suit you.'

In truth I did not know what I wanted. Perhaps the film should contain bits of all these ingredients.

One thing was certain. We would need sizeable chunks of at least one concert performance. This was not going to be easy to film with only one camera. We would not be allowed to move around during the concert itself. The only solution, therefore, was to film the same piece during several performances, each time from a different point of view. Then, using one master sound-track, we could cut the pictures together. Fortunately, a number of works were being played several times so that this was possible. I decided that one of the major excerpts would come from the last movement of Tchaikovsky's Fourth Symphony which Antal Dorati was conducting.

The first concert was in Osaka. The Tchaikovsky symphony formed the second half. Ken and I sat in a box. He said that he would need prompting about what to film when, so in the relative privacy of the box I was able to whisper and point and all was well.

We had star conductors. Pierre Monteux, magisterial and imperturbable; Georg Solti, sometimes conducting with electrical energy, sometimes with no more than a rhythmic wriggle of one wrist; and Dorati occasionally baffling the orchestra with Englished versions of his native Hungarian, such as 'While the going is good you make mistake and then it is too late.' We filmed the principals of the orchestra giving master classes to Japanese instrumentalists and post-concert celebrations with Dorati and Solti, fellow-Hungarians, boisterously singing songs of their native country together at the piano and the whole orchestra responding with a virtuoso version of 'Old MacDonald Had a Farm.'

The final concert was to take place in Tokyo. We had now filmed the last movement of the Tchaikovsky symphony four times – concentrating in turn on the brass, the strings, the woodwind and the percussion. They would all cut together nicely. But one crucially important element was missing – the conductor, face-on. Sitting in the dark at the back of the hall, or in a box reserved for us, we and our camera so far had been relatively unobtrusive. But there was only one place from which to get proper face-on coverage of Maestro Dorati conducting and that was in the orchestra itself. To sit there would require the permission of the Japanese concert promoters.

The thought appalled them. There was much in-taking of breath and hissing in horror. Out of the question! For one thing our camera would make too much noise and if we sat in the orchestra with their microphones adjusted to capture its every sound, the noise of the camera directly in front of them would be heard on the radio relay. I assured them that our camera was one of the quietest in existence. Indeed, to all intents and purposes, it was totally silent. This was not strictly truthful but I felt able to exaggerate just slightly for I did not intend to start filming until the last few minutes of the last movement when every instrument of the orchestra was playing at full throttle. Under those circumstances, no one could possibly hear any camera. After much persuasion, permission was reluctantly given.

The symphony formed the second half of the concert. After the interval, Ken and I in our best suits silently filed on to the platform with the French horns and took our place alongside them. The camera, draped with all the cloth we could find to muffle its sound, stood alongside Ken like some unfamiliar addition to the percussionist's battery. Ken focused it on the podium and locked it off. All he then had to do was unobtrusively to reach up and press the start button. He wasn't sure of the right point to do this but I knew the symphony well, particularly after having sat through four performances of it so recently. So I would give him a nudge.

We sat with folded arms and appropriately solemn expressions through the exposition of the themes, and their elaboration and combination. The music slowly built up towards its final climax. The strings were playing furiously. The trombones brayed and the French horns alongside chorused an ear-splitting response. I gave Ken an unobtrusive nudge and he reached up and turned on the camera. Maestro Dorati was at his most athletic, urging the orchestra on to a huge climax. But then, instead of the repeated thunderous final chords I had expected, a sudden hush fell on the orchestra. An oboe played a long plaintive phrase over soft pizzicato from the strings – accompanied by the grinding noise of our camera. I had forgotten that the last section of the movement was repeated before it came to its final climactic bars. Ken and I sat there perspiring, pretending that we had nothing to do with the black shape audibly growling beside us. It seemed an age before its noise was once again drowned by that of the orchestra. Relief – but only a short-lived one, for now I had another worry. I had started the camera far too early. It was essential that we filmed the final bars and saw Maestro Dorati lay down his baton and turn to the audience. Would we run out of film before that happened? We reached the final bars and, with all the noise and the tumult of applause that followed them, Ken couldn't tell from the camera's sound whether it had run out or not.

Back in the cutting room in London I waited anxiously for that shot. Dorati laid down his baton and turned to the audience – and the film came to an end. It was a dead heat. But by that stage I was able to cut to a shot of the audience applauding, taken at a previous concert.

Now, however, I had a further problem. By the time the camera approached the very end of its magazine, the battery, close to exhaustion, was having to pull a heavy load of film. So it was running slightly slowly. That meant that when the film was shown at the proper speed, Maestro Dorati appeared to be whirling his arms like a demented windmill. But I had no other shot to use. In the editing room, I timed it so that he laid his baton down at exactly the same time as the music finished on the master sound track, but in consequence his last frantic gestures bore little relation to the rhythm of the music. Would the Maestro's agent think that the result was libellous?

I decided to invite several members of the orchestra to have a preview of the finished film. They showed every sign of enjoying it, nudging one another when one of them appeared in shot, and making ribald remarks about sections of the performances. At the end, I asked them if they had seen any musical faults. None. What about Dorati's gestures at the end, weren't they out of time with the music? 'Oh, don't take any notice of that,' they said. 'We don't. He always gets a little excited in the final bars.'

* * *

After my first six months at the LSE I had to think of a proposal that would appeal to the BBC for the six months I was to spend earning my living with them. Should it be about animals or people? Maybe I should try to move into other subjects that interested me such as archaeology or history? Or perhaps something concerned with the ecological problems that humanity was creating on the planet, an issue with which I was becoming increasingly involved in my non-broadcasting private existence. In the end, I came up with an idea that would combine them all – a journey down the Zambezi from source to mouth.

We would certainly see a lot of animals. Archaeologically, we could look at the ruins of Great Zimbabwe, a little way to the south in what was then Southern Rhodesia. Anthropologically, we could find subjects among peoples along the middle stretches of the river in Barotseland who still lived in traditional ways. Historically, we could tell the story of Livingstone's travels along the river that he was the first European to explore. And the recent building of the Kariba Dam half-way down would bring up issues of conservation.

Geoff Mulligan was once again the cameraman, but Bob Saunders could not come with us. The dubbing theatre had reclaimed him. His Australian trip, however, had affected him profoundly. He had come to enjoy travel in the wilder parts of the world so much that he determined that he too would earn his living in that way. Before long he was back on the road as a recordist and not long after that he started to direct very successful travel films himself, making some extremely arduous journeys in the rain forests of Brazil. In his place, we recruited another staff recordist, Bob Roberts.

On the Zambezi, it was not difficult to find dramatic wildlife. Elephants in parties several dozen strong came down to the river to drink and took little notice of us as we drifted past them in a boat. Hippo snuggled up close to one another in midstream so that you might have mistaken them for a small island, until one of the rounded grey rocks discharged twin jets of steam from its nostrils. Carmine bee-eaters nested in colonies on the river banks and above, great processions of pelican followed one another across the sky in a single file five hundred birds long.

The most unusual species we saw was a semi-aquatic antelope, the red lechwe. Its hooves are long and wide and splay outwards so that the animal's weight is spread on the sodden vegetation. Herds of them, tens of thousands strong, lived on the flooded plains around the Kafue River, a tributary of the Zambezi that joins it from the north. We watched them grazing in the swamps and then, when they took fright at our presence, filmed them bounding away with their horns laid back on their shoulders, making great splashing leaps like schools of porpoises.

The herds migrate with the rhythm of the river. As it floods during the rains, the lechwe move away from the main channel, grazing on the submerged grasses and reeds, sometimes wading up to their bellies. Then as the plain dries they return to the flats nearer the river.

The people of the Zambezi flood-plain move seasonally just as the lechwe do. As the waters retreat during the dry season so the Barotse up sticks and travel down to the plain, led in ceremonial procession by their ruler, the Litunga. We had timed our trip there to coincide with this migration and we went up to Limalunga, his rainy season residence on the high ground to the east of the river, to ask for permission to film the proceedings.

Limalunga was photogenically traditional – clusters of circular thatched houses, each surrounded by a neat palisade of reeds. The Litunga's palace stood in the centre of the settlement surrounded by a palisade much taller than any other, its reeds bound with lashings of a special pattern and supported by tall pointed posts, painted white at their tips. Both are decorative flourishes that are the prerogative of Barotse royalty. In the open space in front of its gate stood a flagpole. At one side, in the shade of a tall tree, an orchestra of drums and *sirimbi* – a kind of xylophone – played almost constantly throughout daylight hours. Men and women stood around in groups, chatting. Occasionally one or two of them would unself-consciously start to dance, turning to face the palace gate, raising their arms in the air and swaying their hips. The Litunga had recently returned from attending talks in London about Northern Rhodesia's imminent independence and the people were expressing their pleasure at having him back safely in their midst.

We requested an audience with this semi-divine personage in the proper traditional fashion. We sat in front of the gate, with legs together and to one side, and clapped our hands in a ceremonial salute called a *kandalela*. A court official, an *induna*, appeared. We told him who we were and what we wanted and in due course we were escorted through two courtyards to the palace itself, a two-storeyed European-style building.

The Litunga received us in a large room sparsely fitted out with one or two pieces of western furniture. He was sitting in a wooden armchair, wearing a smart European-style suit. He looked remarkably like Mr Khrushchev. On a sideboard behind him and hanging on the walls were framed photographs of himself and some of his predecessors with British royalty – Georges V and VI, Edward VIII and the present Queen and Prince Philip. The *induna*, before we went in, had questioned us closely about our respective roles and made up his mind about our relative seniority. He directed us to a line of chairs that extended to the left of the Litunga. I was seated next to the throne.

The following moments were a little sticky. I had always understood that when meeting British royalty, one never introduced a subject into the conversation but simply responded to whatever topic was raised. I did not know that in Barotseland, the reverse is the case. The Litunga waits for his visitors to initiate a conversational exchange and then responds. As a consequence, we both looked vacantly into mid-air for an embarrassing length of time. Eventually, I decided that, whatever the protocol, I ought to say why we had come. That broke the ice and things proceeded more easily thereafter.

The Litunga was happy, he said, that we should record something of court life and agreed that we might also film the *fuluhela*, his procession down to his dry season residence at Lealui. However, he could not say when this would be. It might be within a day or so – or weeks ahead. He had not yet made up his mind. We settled down to wait. During the next few days we filmed village scenes, the daily round within the palace which echoed to the sound of almost continuous *kandalelas*, and spent some happy times out on the flood-plain watching birds.

The Litunga was keen to know how we were getting on and I was summoned every now and then to give a report. He was particularly anxious that we should record his musicians and arranged for them to perform for us in his inner courtyard of the palace. The musicians prepared their instruments with care. Having in mind the commentary I would eventually have to write, I asked the *induna* what the substance was with which the senior drummer was so carefully anointing his drum skin. The *induna* conferred with the drummer. 'A mixture of Sunlight soap and Stork margarine,' he reported gravely.

To please the Litunga, we filmed the recital in rather more detail than we might otherwise have done. When the performance was finished and we were packing away our equipment, a message arrived from the depths of the palace. The Litunga had been listening, and had noticed several mistakes in the music. The recital was to be repeated and we would have to film it again, using up still more of our precious and limited film stock.

Eventually the time came for the royal departure. The Litunga's barge, the Nalikwanda, was moored in a canal some way from the palace. It was a sixty feet long dug-out canoe with, amidships, a domed cabin covered with white fabric like a huge cocoon. A smaller canoe, with a smaller cocoon in its middle, lay behind it. That would carry the Litunga's first wife. Beyond that, there was a fleet of more modestly sized canoes that would follow behind.

The royal chattels were brought down to the canal in trucks and loaded on to the Nalikwanda – two large safes, top hat boxes with fashionable West End names stamped on them in tarnished gold, cabin trunks, and a great number of cloth-wrapped bundles. Gourds full of cassava gruel were taken into the cocoon, to provide refreshment on the way. The national war drums, immense wooden hemispheres three feet across, were put in front of the cabin and a *sirimbi* hung at the back.

The Litunga emerged from the palace carrying a sacred eland-hair fly whisk and wearing a top hat and a grey tail suit as though he were on the way to Royal Ascot. Once the royal party was on board, the procession set off. The Nalikwanda was paddled by fifty men, many wearing lion-mane headdresses with leopard skins around their waists. Crowds lined the banks and ululated enthusiastically as the procession passed. In places the canal was so shallow that there was a risk that the Nalikwanda, so much bigger than any other canoe in the armada, might be grounded. But that had been foreseen. Some time earlier, a series of mud dams had been built across such shallower stretches so that somewhat deeper ponds had formed. The Nalikwanda sailed serenely towards the first dam. That was then broken by attendants and the barge was carried forward with the surge of water into the next stretch.

The voyage took over four hours and it was too dark to film by the time we arrived at Lealui. The following day there were further ceremonials to make it apparent to all that the king was properly installed in his dry season residence, ready to deliver any judgements that might be demanded of him and to preside over the welfare of his people. We filmed those as well. That evening I asked for a final audience so that we might thank him and make our farewells. The *induna* who escorted us in, squatted and started to deliver a clapping *kandalela* with particular gusto. The Litunga stopped him with a few sharp words and the *induna* retired, abashed. When he had gone the Litunga leaned towards me. 'Sometimes,' he said confidentially and a little wearily, 'royal salutes can make too much noise indoors.'

* * *

Limalunga was the site of David Livingstone's first mission station in Barotseland. He had been the first European to enter the country, in 1853, and had formed a firm friendship with its ruler. I had planned to devote a whole programme to Livingstone for no single person could have played a greater part in the history of the Zambezi valley than he. The river had obsessed him. He saw it as a highway into the heart of Africa along which trade and Christianity could travel, thus opening up the heart of darkness. He first set eyes on it when he came up from southern Africa and arrived at the village of Sesheke which stood on its banks. There, under a huge baobab, he set up a magic lantern that he had brought with him and showed an astonished crowd, many of whom had never seen a white face before, slides of Biblical subjects. The old chief of Sesheke, now blind, confirmed that we had come to the right place and told us that when he was a boy, there were old people in the village who still remembered Livingstone making his camp there. Sadly, the baobab itself had blown down a year earlier. Talking to camera, I did my best to try and conjure up a picture of Livingstone giving his sermon but it was hard going.

I had a better subject eighty miles farther on down-river, for there Livingstone made his most spectacular discovery, the Victoria Falls. Finding a new way to photograph such a celebrated spectacle was a problem. The least hackneyed vantage point was the tiny island in the middle of the Falls where Livingstone had landed. Getting there when the river is low is not too difficult, but now there was still enough water to create fast flowing rapids and make a landing on the island interesting. Once there, though, the view was indeed spectacular. We sat with our feet dangling over the rocky edge of the fall, just as Livingstone had recorded doing in his journal, looking over the eastern cataract with its permanent rainbow in the spray, and peering down to the brown water boiling among the boulders three hundred and fifty feet below.

Eventually we made our way down the river beyond the Kariba Dam and its lake and entered the Portuguese territory of Mozambique. There, in the little settlement of Shupanga we sought out the grave of Livingstone's wife Mary who had accompanied him on his expedition to the lower reaches of the Zambezi and there had died from malaria. The baobab under which she was buried had also blown down, though a long time ago. It had destroyed the simple wooden cross that Livingstone had erected to mark her grave. Its replacement, apparently of cast iron, was painted an undistinguished khaki colour with on one side an inscription in Portuguese and on the other in English. She was 41. I placed a few flowers on it.

* * *

Thinking about the trip now, I realise that we neglected to make the film which would have had most historical value. Instead of filming red lechwe, hippopotamus and elephant, we should have concentrated on the extraordinary range of Europeans we encountered as the country trembled on the brink of independence. Among them were some subspecies of *Homo sapiens* which surely must now be almost extinct.

In Angola, we stayed with one of the last survivors of a time long past, a Scots missionary who was faithfully keeping alight the flame of

Livingstone's zeal. Eighty-one years old, wheezing and trembling, and juggling alarmingly with his false teeth, he explained his simple straightforward faith to me. 'I'm not preaching religion,' he told me, in a voice that still retained its Glaswegian accent. 'There are hundreds, mebbe thousands of religions in the world. But we'll hae nane of them. We teach the *truth* as it is written in the Scriptures.' He was sure that political independence anywhere in Africa would be a disaster.

In a bar we met an obscenely bloated Afrikaner who referred to Africans as if they were barely human and, almost as contemptible, a European tourist guide whose claim to be a big game hunter was as false as the plastic leopard skin band around his bush hat. But there were some who still represented the twentieth century's vision of the altruistic side of empire – the British Resident Commissioner in Barotseland, a huge and modest man, well over six feet tall, still a prodigious athlete, an ex-bomber pilot and reputedly the youngest DFC of the war, who was held in great respect and affection by the people he administered. A generation earlier, he would undoubtedly have become Governor of some territory within the Empire. As it was, he welcomed the coming independence of Zambia and was preparing to vacate the grand Residency in which he lived to make way for an African, twenty years his junior, while he continued arranging for the coming of radical political change from a much smaller and unostentatious house.

The films we did make were reasonably successful but undistinguished. They gave no real insights into either history, natural history or ethnography. I was not at all sure what I should do next.

15

A New Network

Things were changing fast in British television. Technical standards were very antiquated. That was the penalty of being a pioneer. The BBC, back in 1936, had started the first publicly available television service in the world. It had done so using the most advanced technology of the time – grainy black and white pictures made up of 405 horizontal lines transmitted on the Very High Frequency waveband. When war broke out in 1939, the BBC promised its viewers that it would start up again as soon as hostilities were over. So when peace ultimately came in 1945, a few thousand television receivers were brought down from lofts, dusted off and switched on again to produce much the same black and white coarse-grained pictures. Indeed, it is said that one of the very first programmes to be shown was the one that had been so abruptly brought to an end six years earlier.

But by the 1960s, the 405-line technology was long outdated. Pictures could now be broadcast with a much higher definition – that is to say with a considerably greater number of lines making up the picture. Colour too was possible. Indeed, the BBC had been experimenting with colour systems for over ten years. And the engineers had found space in the air-waves to accommodate another channel. In 1960 the Government appointed a committee chaired by Sir Harry Pilkington to decide how these new possibilities should be handled. It decided that the BBC should start a new network to be called BBC2 and that it should be transmitted on 625 lines in the Ultra High Frequency band. At first it would do so in black and white. Colour, when it came, would have to be on 625 lines. BBC2, therefore, would be the first network to show it. BBC1 and independent television would eventually duplicate their black and white programmes on 625 lines and in colour, and ultimately 405-line broadcasting would be abandoned altogether.

That decision brought jubilation in the BBC but also turmoil. UHF transmitters required very different siting from the old VHF ones so a completely new nationwide network of transmitters would have to be constructed. New studios had to be built, engineered for 625 lines, in order to produce the extra programmes. And what is more, the Government had stipulated that all this had to be done very quickly.

Three of my ex-colleagues in Talks Department were clear candidates for promotion in the coming expansion. There was Michael Peacock who had joined the BBC on the same day as I had and beside whom I had sat on the training course thirteen years earlier. After having produced *Panorama* for several years he had moved to the Outside Broadcast Department and then became the Head of Television News. He was a young Turk, still only 34 and clearly phenomenally able.

Rivalling him as a rising star was Donald Baverstock who after *Tonight* had produced *That Was the Week That Was*, a sensational, extremely funny and occasionally scurrilous late night show labelled by the press as satire. Donald was a wild ideas man, a revolutionary. He was also unquestionably brilliant.

And there was Huw Wheldon, the ex-paratroop officer. He had made a major impact with his arts programme, *Monitor*. He edited it and also introduced it from the studio. It was transmitted fortnightly, live, and was of such quality and so exciting that the alternate weeks without *Monitor* seemed dull ones. Important sculptors, painters, architects, novelists, composers and film directors all appeared. Huw led his team of virtuoso directors with great style and much the same manner, one imagines, as he led his company during the invasion of France. On one occasion, forty-eight hours before the programme's transmission, he assembled those working on it and told them that the film stories due to be shown in the next edition were not up to standard. They would be scrapped. He and everyone else involved would start again and work all that day and throughout the following night to make a new programme. I feel sure he thought that such crises and such solutions were good for *esprit de corps*.

From these candidates and doubtless others, Michael Peacock, in spite of his comparative youth, was chosen to be head of the new network, BBC2. Donald Baverstock was put in charge of BBC1 to the great delight of the *That Was The Week That Was* empire. Huw Wheldon remained in charge of documentary output. But then things began to go wrong.

BBC2 did not have an auspicious start. After a huge press build-up, with those people who had invested in 625-line sets sitting on the edge of their seats ready to be thrilled and delighted, there was a major power failure in west London and the new network was blacked out before it even started. To make things more infuriating, BBC1 and ITV were unaffected. Even when BBC2 programmes eventually got on the air, viewers and the press took a jaundiced view of them and after a few weeks, it was generally agreed that the network's editorial policies were a disaster.

In sober truth, Mike Peacock had worked miracles to get the network on the air at all. He had insufficient 625-line studios or Outside Broadcast Units; and he had had to recruit an entire production staff to try and create new programme ideas. He inaugurated some superb new programmes – a twenty-six-part film series on the history of the Great War was one – but still he could not originate enough to fill his schedule. He did have a stock of educational programmes and he could repeat productions from BBC1. He could also buy foreign language films. To make the best of the situation he hit on the idea of giving each day of the week its own character and calling the policy Seven Faces of the Week. Tuesday Term was an evening of educational programmes, Wednesday consisted entirely of repeats, and Thursday was avowedly only for those with minority interests.

The press, who were expecting a completely new network with scintillating star-studded productions of a kind that no one had ever seen before, were scornful. The television retailers, who wanted hugely popular programmes to sell the expensive new dual standard sets, were enraged.

Nor were things much better on BBC1. Managerial tact and leadership on a large scale were not among Donald Baverstock's talents. The Governors too became uneasy about some of the more free-spoken, politically critical, and sexually explicit programmes. Hugh Greene, the Director General, stepped in. He decided that Huw Wheldon should be promoted over the head of both Mike and Donald to take command of the situation with the new title of Controller of Programmes.

The newspapers followed all these goings on with extraordinary attention. There were headlines about stabs in the back and palace revolutions and news photographers took up their positions outside the front doors of several of the protagonists. I read of all these machinations with interest. I knew everyone concerned. Many were old friends. But I did not feel closely involved except perhaps that BBC2 under Mike Peacock might want some new kind of natural history programmes which I, as a freelance broadcaster, might devise.

Then Huw Wheldon telephoned me at home. Would I call round to see him at his house in Kew, a mile or so away from where I lived.

'Let's be straightforward about this, Dai.' Huw was at his most brusque and regimental. 'I've been appointed Controller of Programmes. I accepted – provided that Donald Baverstock does not remain involved with BBC1. I've given Mike Peacock that job and I've offered Donald the job of running BBC2. It's just his cup of tea. He's a bright chap – lots of ideas – full of energy – just the fellow for Two. But Donald won't have it and is going to resign. Since he won't do it, will you?'

It was typical of Huw to be open and straightforward, making it clear that I had not been his first choice for the job. I said I wanted to think about it overnight. It would mean abandoning my anthropological project with the LSE. But BBC2 was a very exciting place to be for anyone interested in television. Its programme policy was widely regarded as a failure. Its audience was minuscule. So there was nowhere for it to go but up. Its Controller would have a reasonable programme budget and since most of the existing programmes would have to be scrapped, a virtually empty schedule. And it would mean working closely with Huw which would certainly be an invigorating experience. On the other hand, I didn't relish the thought of totally abandoning film-making in the wilder parts of the world.

The following evening I went round again to Huw. I said I would take the job, provided I was able to make a programme of some kind every eighteen months or so – just to keep in touch with the latest technologies and techniques. I would guarantee to stay for three years if I was wanted, but I thought it unlikely that I would stay longer than five. Huw said that suited him well. So, at the beginning of March 1965 I became a BBC administrator.

* * *

Administrators have their perks. The Sixth Floor of the Television Centre in White City was the heartland of BBC administrators and thus, very properly, rich with perks. I was offered my first on my first day. I could choose my own desk. The one in my office had been Donald Baverstock's, but it seemed all right to me so I declined. Nonetheless, a catalogue of suitable desks was left with me. It was not until I had been asked three times whether I had come to a decision that I realised that the system relied on a newly appointed Controller acquiring a new desk, so that his old one could move down a rung. Someone else was expecting it. So I fell into line and chose something.

The first letter I dictated came back to me typed on what looked like a child's notepaper. The top left corner carried a cartoon, in bright yellow outlined in black, of a kangaroo with a small joey peering out of its pouch. I asked what on earth this was and was told that these were Hullabaloo and Custard. Some demented public relations guru had decreed that these characters were to be the network's symbol and used on all its publicity and promotions – Hullabaloo so-called because that was what the network would create; and Custard because – well, it was just fun. My first executive decision was to banish Hullabaloo and Custard for ever.

Mrs Spicer was one of the first of my new colleagues to come and see me. I had met Joanna once or twice when I was a producer. She was one of the most powerful of all administrators at the top of the Television Service for she controlled all the Corporation's facilities and kept its accounts. She supervised the allocation of studios and outside broadcast units. She ensured that the Service did not overspend – at least, only in so far as was tactically judicious. Her empire – Planning Department– included several large offices filled with people poring over immense charts, annotating them with finely pointed pencils so that when inevitable changes were made, their inscriptions could be rubbed out. She marshalled camera crews, make-up artists, scenery workshops, recording studios, telecine machines and all the other multifarious facilities which the Corporation then maintained in-house. Over my first week she gave me a series of personal seminars in which she explained how many of these facilities BBC2 was entitled to and how much money there was for programme budgets. It was like being given a Rolls Royce, polished and perfect, with a full tank of petrol and being asked where you would like to go.

But where was that? What was BBC2's editorial policy? Everyone was agreed that the Seven Faces of the Week had to disappear. The BBC in its application to the Pilkington Committee had said it would use a second network to provide alternative programming. In a strictly logical sense, the only rational alternative to, say, a current affairs programme, was another current affairs programme. Plainly, that was not what was meant. But what was the alternative to current affairs? Football, or drama or a feature film? After struggling with such knotty problems for some time, I concluded that there was no answer. I proposed instead that BBC2's policy was simply to produce programmes of a kind that neither BBC1 nor ITV were showing and that these would be scheduled so that each would contrast as extremely as possible with what was being shown at that time on BBC1. I would ensure that every evening there were what I termed 'common junctions' when both networks started programmes simultaneously and the continuity announcer on each network would trail what was being offered on the other, encouraging viewers to switch. The two networks, after all, were not supposed to be competing but collaborating.

To bring this about, I drew a chart showing BBC1's programmes for each evening of the week with blanks alongside them for me to pencil in what BBC2 might offer as a contrast. Placing a BBC2 programme, I felt, should be like putting a letter in a crossword. It had to make sense not only when read horizontally against BBC1 but also vertically in BBC2's evening, so that viewers who wanted something very different from that being offered on BBC1 would always find it on BBC2, while those who chose to watch us from start to finish would be offered a varied and satisfying evening. This chart became one of my key documents, not only in placing programmes but in deciding what programmes to commission.

As for the programmes themselves, I declared that we were not in the business of producing carbon copies of programmes that were already being shown on other networks. Nor would we accept mindless programmes. But otherwise we would take programmes from every category one could think of, and find new approaches and neglected subjects. One measure of our success, I said, would be the width of the spectrum that our programmes covered.

Sport? The standard ones were well represented, so we would invent our own. We instituted a rugby league competition under floodlights and a series of one-day cricket matches that were played on Sunday afternoons between county sides and a touring team of distinguished old professionals.

Documentaries? There were none longer than half an hour at that time on British television, so BBC2 would introduce several strands of fifty-minuters. BBC1's documentaries strove to be balanced and objective. So BBC2 would offer, among other documentaries, *One Pair of Eyes* in which contributors would take deliberately personal and biased points of view.

We would present single gigantic productions that occupied an entire evening on subjects of particular importance that needed examination in depth. The first was an enquiry conducted in front of British judges with proper courtroom procedures into who shot President Kennedy.

Drama? We would serialise classic novels. Michael had already commissioned one of unheard-of length, *The Forsyte Saga* in twenty-six fifty-minute parts, but this would not be completed for many months to come – so far into the future indeed, that when it did appear it was I, not he, who got the credit. Following this came stylish serials based on novels by such as Henry James, Sartre, Tolstoy, George Eliot and Dostoievsky.

Archaeology had disappeared altogether from the BBC since the glory days of *Animal, Vegetable, Mineral?* and its more intellectually ambitious spin-off *Buried Treasure*. BBC2 would bring it back with a new series called *Chronicle* which would deal not only with archaeology worldwide but recent history.

Science would have a regular and continuing weekly fifty-minute slot called *Horizon*, and there would be another series, to be called *Life*, which would deal with biological subjects such as reports from the battle fronts of conservation, the latest discoveries and interviews with prominent zoologists.

Music? Michael Peacock had established that BBC2 would treat music seriously right from the network's beginnings. He had instituted a series of master classes, of which one given by the French cellist Paul Tortelier was particularly memorable. There had been analyses of difficult modern works using scores with notes that animated as the music sounded so that even viewers who were not accustomed to reading musical scores could follow the structure of the music. He had also commissioned an unforgettable hour and a half long programme directed by Humphrey Burton of Solti recording Wagner's *Götterdämmerung*.

I was only too glad to continue and expand this policy. I also discovered that some of the great American jazz and swing bands were still in existence, though struggling. British television totally ignored such music but I was told that if such bands were booked by television, it would be economically feasible for them to come over to Britain and subsequently give concerts round the country. BBC2 booked them. So Duke Ellington, Louis Armstrong, Woody Herman and Ella Fitzgerald all came to Britain and it was one of the pleasures of the new Controller, after a day at his desk on the sixth floor of the Television Centre, to go down and listen to these superbly professional bands rehearsing in one of the studios on the ground floor.

But classical music remained paramount in the network's schedules. BBC2's second Christmas – the first under my charge – was coming up. I asked Music Department to arrange a performance of Berlioz's *L'Enfance du Christ*. At that time, Berlioz's music was rarely performed, either at Christmas or any other time. This charming short oratorio would certainly be unfamiliar to most of the television audience but its appropriateness was obvious. Colin Davis, who was already renowned as a champion of Berlioz's music, would conduct and we would stage the performance in Ely Cathedral. It would bring real distinction and innovation to BBC2's Christmas. In addition I scheduled before it a ninety-minute compilation of film sequences featuring great Hollywood clowns – Charlie Chaplin, Buster Keaton, Laurel and Hardy, and Fatty Arbuckle. After the Berlioz there was to be a beautifully photographed Swedish natural history film, *Island Yearbook*, and the evening would end with the second episode of the network's current classic serial, Balzac's *Eugénie Grandet*. So it was with some pride and confidence that I released details of our Christmas programming to the press.

Going home that night on the Underground, I opened my evening paper and found an article by Milton Shulman, the paper's television critic who regularly lambasted broadcasters for their low artistic ambitions. In the text he dismissed my carefully assembled plans as unwatchable and wholly unsuited to the holiday season. He thought it a catastrophic way to run the new network and not only recommended but predicted that I would go back to the jungle. Cast down by this, I opened the memoirs of Lord Reith, the BBC's first Director General, which I had felt I ought to know about and was rather belatedly reading. Almost the first sentence to meet my eyes was a typical magisterial Reithian utterance. 'It is royal to do right and receive abuse.' It consoled me then and was to do so frequently in the next few years.

In 1965, Igor Stravinsky, 83 years old, was due to come to London to conduct the London Symphony Orchestra in a performance of some of his works. The farewell tour of one of the great composers of the century was clearly an event of real importance and I determined that BBC2 viewers should be there to share the occasion, live. There were, however, problems. Other television networks were said to be interested. The fees proposed seemed exorbitant. There were hard negotiations, but in the end BBC2 got the television rights. When the time came I went to my seat in the Festival Hall feeling like the Duke de Medici attending one of his masques incognito.

The first half of the concert was not to be televised. It was a performance of Stravinsky's most recent work – *Eight Instrumental Miniatures* – and was going to be conducted by the composer's amanuensis, Robert Craft. Stravinsky himself was saving his energies for the *Firebird Suite* in the second televised half.

The lights dimmed. Craft, a tall, elegant and grave figure, entered to appropriate applause and silence fell. The *Instrumental Miniatures* proved to be a work of extreme austerity. A few isolated notes from the piccolo; a tap on a triangle; a dissonant chord from three fiddle players. Those of us in the audience concentrated hard to follow the line of musical thought. Suddenly a stentorian bass voice from backstage bellowed 'Turn 'em on, Fred' and the entire auditorium was flooded with light blazing from banks of huge lamps that had been specially installed for the television relay. Craft continued to conduct in the dazzling brightness as though nothing had happened. The audience whispered to one another and shifted uneasily in their seats. I broke out in a cascade of sweat and wished I could hide under mine. After a few seconds and equally suddenly, the lights turned off – only to turn on again. Craft and *Instrumental Miniatures* battled on. At last the lights cut and the stage returned to the reverential half-light appropriate to the occasion. But the intellectual grip of the *Instrumental Miniatures* on the audience, if they had any, was gone.

I knew only too well what had happened. The electricians who had been contracted from an outside firm to install the extra lighting needed for television had arrived well in advance of the beginning of the scheduled transmission. They had found the hall in semi-darkness and totally quiet except for a few tinkles from the percussion and tootles from the woodwind. Presumably, they had assumed that such noises were being made by one or two musicians rehearsing a few of their trickier passages in the darkened hall, and had decided to do a little rehearsing on their own account.

Craft, in his memoirs, understandably, writes bitterly of the occasion and lambasts the BBC. However, in the end there was a great and permanent reward. Stravinsky's conducting of *Firebird* in the second half was electrifying. His craggy aged face was immobile but somehow radiated vigour and ferocity. At one point he gave a savage stab with his baton to cue a blast from the French horn. But it was two bars too soon. Alan Civil, the instrumentalist concerned, was too experienced to be fazed by this and remained silent. Stravinsky turned his eyes away from him. Two bars later, the notes came, correctly timed, from Civil's horn and a transitory but grateful beam flitted across the composer's face. It was, I think, his only change of facial expression throughout the piece. But the whole performance was of riveting intensity and as far as I know is the only visual record of Stravinsky as a conductor.

The network also paid proper attention to Britain's pre-eminent composer of the time, Benjamin Britten. We put on a production of his church parable, *The Burning Fiery Furnace* in Orford Church, the place for which it was written. We mounted a full-scale studio production of *Billy Budd* with Peter Pears as Captain Vere. This was so lavish that even TC1, the Television Centre's biggest studio, which was the largest in Europe, could not properly accommodate it all and Charles Mackerras had to conduct the London Symphony Orchestra from a second adjoining studio and communicate with his singers – and they with him – by way of studio monitors. It was a technical *tour de force* and was seen by opera critics and, I was told, by Britten himself as rivalling if not excelling its original production for Covent Garden.

So the question arose as to whether Britten would write something specially for BBC2. The invitation had been open for some time but John Culshaw, who had produced the sound recordings of many of Britten's works, including the *War Requiem,* and was by then Head of Television Music, brought matters to a head. I was to give Britten and Peter Pears lunch at the Television Centre and discuss the whole proposal. Britten already had a subject in mind – a ghost story by Henry James entitled *Owen Wingrave.* The supernatural happenings would give us a chance to use television wizardry. Britten would conduct this himself but he was not happy with the two-studio arrangement we had been compelled to adopt for *Billy Budd.* He needed to be able to see his singers in the flesh. Eventually we agreed that we would build a huge temporary studio using Outside Broadcast camera units in and around the Concert Hall at the Snape Maltings near Aldeburgh. So far, I had been able to keep up with the discussion of technicalities but then we moved onto questions of orchestration. 'How many horns can you give me?' Britten asked. I was now out of my depth. 'How many would you like Mr Britten?' I replied. 'I would far rather be told,' he said, rather formidably.

I went down to Aldeburgh during the recording of the production. There was a starry cast – Peter Pears as General Wingrave, Benjamin Luxon as Owen and Janet Baker as Kate. Britten took the English Chamber Orchestra through the intricacies of the score while the whole assembly – not only singers and orchestra but television cameramen, microphone boom operators, video engineers, property men and scene-shifters – hung on his words. We watched progress on the countless video monitors that were everywhere and listened to new music sounding for the very first time. Everyone was well aware that an important new work by a major composer was being hammered out and polished bar by bar. The tension was such that I found myself holding my breath for much of the time. In the end, I was quite relieved when the time came to leave.

* * *

John Read, who had first introduced me to a television studio at Alexandra Palace, had since then used his wide contacts in the arts world to produce a whole series of very distinguished films about British artists. He had known Henry Moore, the sculptor, since he was a boy, for Henry had been a great friend of John's art critic father. He had already made two excellent films of Henry at work. He suggested that to celebrate Moore's 70th birthday he should make a new film, which would include material from the earlier ones and survey Moore's whole career. I could think of no example of a major artist whose life had been so richly documented on film over such a long period. It was just the thing for BBC2. The finished film was as good as I knew it would be. Moore came to view it and afterwards he, John and I lunched together. After talking about the film, with which Henry was very pleased, the conversation became more general.

Henry had just come back from Florence.

'A marvellous place,' he said. 'But it was so crowded.'

He sighed.

'I tried to have a look again at Ghiberti's wonderful Gates of Paradise to the Baptistry but I could hardly see them because of the crowds. There was an English lady in front of me who leaned over and said 'Ooh look. They don't have any keyholes. Why would that would be?' 'Madam,' I said, 'they were washed away a few years ago by the floods.'

It was a delight to hear such a joke from the one artist who, perhaps more than any other, had introduced the British public to the notion that spaces within sculptures could be regarded as objects.

Brian Branston, who had succeeded me as Head of the Travel and Exploration Unit, came to see me. He had discovered that although the death of Captain Scott and his companions on their way back from the South Pole was one of the most famous events in the whole of British exploration, one important witness to it had never publicly told his story – Tryggve Gran, the Norwegian who had joined the expedition to teach the British members how to ski. He had been on the search party sent out from base camp to look for the returning party; and it was he who had found the bodies of Scott and his companions. Brian had suggested that Gran's story would make an important documentary. As I viewed the finished film, I realised that Peter Scott should be given a special showing of this first-hand account of his father's death before it was seen publicly, so I arranged a dinner for him, Gran and the only other surviving member of the expedition, Frank Debenham. After the meal, we all settled down to watch the film. In it Gran described very movingly how he had spotted the tent on the ice and pulled aside the stiff door flap and found the frozen bodies within. I glanced at Peter. He was weeping.

The archaeological series, *Chronicle*, under Paul Johnstone, quickly proved itself to be both authoritative and popular. I was keen to build on that. We decided that BBC2 should be the first network anywhere in the world to initiate and finance its own archaeological dig. We would try to solve one of the enduring problems of British archaeology – the function and origin of Silbury Hill, the immense cone-shaped mound standing alongside the A4 road in Wiltshire. It was certainly man-made and pre-Roman but few other facts about it were certain. Professor Richard Atkinson, from the University of Wales at Cardiff, agreed to tackle the project and decided to do so by excavating a near-horizontal tunnel that would pierce the hill to its very centre. The dig was to take place each summer for at least three years. Each season we would station an Outside Broadcast Unit there throughout the work, ready to transmit live any particularly exciting moments on BBC2 as well as to provide regular, maybe daily reports. The pantechnicons and marquees that would have to be put up to house the operations would also coincidentally advertise BBC2 to the busy traffic passing up and down the A4.

The press immediately decided that this was to be a treasure hunt. They discovered a local story that a statue of a horseman made of solid gold was buried in the heart of the hill. This of course was anathema to the archaeological establishment. I did my best, when announcing the project, to stamp on such a notion. Our primary aim, I said, was to find out more about the hill – its date, how it was made and what it was for.

The excavation duly took place and provided many hours of television. At the heart of the hill, Atkinson discovered a layer-cake structure created by a series of cones of different-coloured earths nesting inside one another that he described as one of the most astonishing sights he had ever seen. Eventually, he was able to deduce how the hill was constructed, even (from buried plant and insect remains) at what season of the year, and to date it fairly accurately to 2500 BC. But he could still only speculate about its builders' motives – and there was no golden horseman. It was, nonetheless, a broadcasting success, and some archaeologists were positively delighted that the only treasure it produced was knowledge.

* * *

In 1967, two years after I had taken over BBC2, the Government suddenly and abruptly told the BBC that it could now introduce colour television. This would be done on 625 lines, of course, and since BBC2 was still the only network transmitting on that standard it would be my job to organise the first colour programmes. The BBC had been agitating to do this for some time but, as with the inauguration of BBC2, the Government, having taken the decision, after so much procrastination, was now impatient for it to take effect. We had to start colour as quickly as possible.

Although we had several prototype colour cameras at our disposal, none was in commercial production and the decision had not yet been made by BBC engineers as to which to select. Work started immediately on converting two studios at the Television Centre. They had to be completely rewired. They also had to have new lighting installed. Jimmy Redmond, then Chief Television Engineer, explained to me that this was not as simple a problem as it might have seemed. If he provided the huge amount of lighting demanded by the most advanced cameras that existed at the time, the whole of the studio roof would have to be reinforced to carry the weight. That would be immensely expensive and disruptive. Alternatively, he could gamble that by the time we were ready to start studio production, new cameras would have been produced that were much more sensitive and the existing roof would be strong enough to carry the smaller number of lights needed. We decided to gamble.

Meanwhile, the best cameras we had were installed in one of the tiny Presentation Studios on the fourth floor below my office. There were two of them, neither much bigger than an average domestic living room. They had been built to accommodate announcers, weather forecasters, test cards, the model of Big Ben with which we had been ending transmissions each night since the days of Alexandra Palace and other similar bits of visual continuity. One of these studios had been commandeered in the very first days of BBC2, when 625-line studios were at a premium, to accommodate a nightly interview programme called *Late Night Line-Up* in which the evening's output was reviewed by critics, and contributors talked about programmes to come. Colour cameras there would give us experience in lighting for colour and test the whole transmission system.

One of the requirements of our colour policy was that the pictures could also be viewed equally well in black-and-white. Viewers to *Late Night Line-Up* therefore, if things went as they should, would be unaware of the change. Meanwhile senior engineers and the Controller of BBC2 would have the monster prototype receivers in their homes and would be seeing the pictures in glorious colour.

Except that the colour, quite deliberately, would not be glorious. Flamboyant colours – scarlet, vivid green, burning yellows – were easy, the engineers said. The tricky things were flesh tones. So the standard *Late Night Line-Up* interviews would provide a perfect test-bed. But even the engineers found this austere self-denying regime rather restricting. One morning, the Director of Engineering, Sir Francis McLean, rang me to discuss the previous night's transmission. We agreed that it had gone well.

'I tell you what,' Sir Francis said. 'Do you think that in tomorrow night's programme we might introduce' – he paused to emphasise the daringness of his suggestion – 'a bowl of fruit?'

BBC2 needed a special symbol to indicate to those viewing in monochrome when we were transmitting in colour. I asked Presentation Department for suggestions. One of the brightest of its producers, Richard Drewett, came to see me. 'I have an idea,' he said. 'We could get Picasso to do the visual, perhaps an image being drawn on glass or something like that, and Stravinsky to provide the musical sting.' I thought he was joking but he insisted that he knew a way to reach Picasso and he thought that the idea might appeal to him. I told him to investigate. BBC2 would be nailing its colours to the cultural mast, all right. So much for Hullabaloo and Custard!

A week or so later, Richard was back. Picasso was interested.

'What about his fee?' I asked cautiously.

'Apparently he suggested a colour television set.'

'But France does not have colour television yet,' I said.

Richard laughed. 'That,' he said 'is apparently what appeals to him about it. But he will only do it if Stravinsky also agrees.'

Two weeks later, Stravinsky died. So in the end we had to settle, rather feebly, for the existing BBC2 numeral with a dot in its loop and the word 'Colour' beneath.

We started test transmissions in colour during the day to enable the retail trade to begin selling sets to the public. We did after all want to have an audience when the service was eventually launched. Had we shown our own programmes at these extra times, it would have cost us money, but we managed to find various films, mostly promotional documentaries, that were of sufficiently high colour quality to show off the system and that could also be transmitted an unlimited number of times without incurring any fees. It was said that one of these, a film publicising tourism in southern Ireland, had to be withdrawn because the pillar boxes were green and some viewers thought this a deficiency of the new television system. Jokes like that were encouraged to get colour television talked about.

Meanwhile, we had to prepare a schedule of colour programmes sufficiently exciting to induce viewers to buy sets – and pay the additional colour licence of £5, which had just been introduced. The problem here was that there was still no way of recording colour on tape. So all electronic colour programmes, whether from the two studios that were now undergoing their tests, or the two Outside Broadcast Units that were about to be delivered, would have to be live. We did, however, have colour telecine, the apparatus which showed films. I knew that most of the films produced by the Travel and Exploration Unit had originally been shot in colour. So one of the first programme strands I decided upon for the new colour service was a fifty-minute documentary early on a Sunday evening to be called *The World About Us*. It would alternate natural history films from Bristol with anthropological and more general travel ones from London. Problems eventually arose when one unit blamed any drop in its viewing figures on the influence of the previous week's programme supplied by the other. Eventually, the London input dried up, but the strand, entirely from Bristol and under varying names, continued to run in the same place and at roughly the same time for the next thirty years.

It had always been apparent that there was no way in which we could offer a complete schedule in colour from the very beginning of the launch. I did not think this was intolerable and began to talk freely about starting with a 'piebald' service. If we could manage to get about half our transmissions in colour, this would be a fair beginning. German television was also making its preparations. They too planned to start in 1967. Rather childishly, perhaps, I determined that we should start before them. The BBC had produced the first public television service in the world in 1936. The United States and Japan had preceded us into colour but at least the BBC could be the first in Europe to transmit in colour. But how?

Wimbledon provided the answer. With both our colour outside broadcast units installed at the tennis championships, we could produce several hours of colour transmissions every day. And we would be ahead of our rivals. It was that which determined the precise day of the launch of colour in Britain – the first day of the All England Tennis Championships, 1st July 1967.

* * *

Colour had a bad name among those who had never seen our programmes. In part this was due to the staggeringly garish quality of the first colour programmes shown in the United States. I was sure, from watching the test transmissions, that ours would be in a different class, full of tonal subtleties and wholly comparable, from the point of view of colour reproduction, with any printed colour pictures. But something was needed to show off its quality and demonstrate to its sceptics, whether in the press or the public at large, that our new system was not so much colour television as high fidelity television, a medium which relayed a more complete, more informative picture of what was in front of the camera than ever before. Joanna Spicer, at one of our routine weekly meetings, murmured in her tactful way that she had set aside some of the network's programme allowance just in case I had it in mind to celebrate the arrival of the colour service with a special spectacular series of some kind.

I brooded on this. I recalled a part-work issued when I was a boy called *An Outline of History* written, or at least edited, by H.G. Wells. I remembered how excited I had been that a great writer had decided to tackle a major slab of human knowledge about which I knew little and was going to give me an outline of it. Once I had absorbed what he had to say, I believed I would be able to slot historical events into position. I remembered how eagerly I had waited, after reading one part, for the next to drop through the letterbox the following week. Why not commission a television equivalent? Television was a visual medium. Why not survey the most beautiful and influential works of art created by European artists in the last two thousand years and examine them accompanied by the loveliest music composed at the time they were created? It was an obvious idea. As to the person who might select the pictures and provide the spoken narrative, that seemed to me to be obvious as well. Sir Kenneth Clark.

Clark had been a celebrated Director of the National Gallery during the war. He had written much praised books about landscape painting and the nude as well as scholarly works on Leonardo da Vinci. And he had shown that he thought television was of some importance by becoming, rather to the BBC's chagrin, the first Chairman of the Independent Television Authority. I invited him to lunch to discuss the idea. Stephen Hearst, then Head of Arts Department, came too. Huw Wheldon said he would look in briefly for coffee – just to see how things were going. Clark wrote in his memoirs that it was my use, during our meal-time conversation, of the word 'civilisation' that set his imagination soaring. He spent the rest of the meal, he said, working out the series in his mind, barely conscious of what was being said at the table.

So the *Civilisation* series was born. Stephen selected the two directors, Peter Montagnon and Michael Gill, who would take charge of the series. We decided early on that since one of the main objectives of the series was to show off the quality of televised colour, we must not take any technical risks. We should shoot it, not on 16mm film, which would be our standard colour stock when the service was fully launched, but on 35mm, the gauge used in the cinema. That was very expensive, and Joanna's sequestered fund would have to be bolstered. But she and I worked out how we might do that and the series went into production.

Peter and Michael proved to be inspired directors. They developed such rapport with Clark that they were able to criticise his scripts from a television point of view with considerable severity. The first that he submitted they rejected totally. Clark recognised their skills and took their advice. As the production progressed I was shown roughly edited sequences. That, of course, was good tactics on Peter and Michael's part, for after I had declared several times how delighted I was and how splendid the series was promising to be, they explained that if they were to maintain such quality, a major overspend was inevitable. This led to what I think was the only successful solution to a financial problem that I managed to devise entirely by myself during my career as a managerial mogul. It was breathtakingly simple. I would declare to the press and to viewers that the series was so good that every licence payer should have at least two chances to see it. It was so full of pleasure and information, that even those that had seen an episode once would want to see it a second time. That way, I was able to give it the money I had allocated to pay for another documentary series which would otherwise have filled the time taken by the repeat.

The series was an astonishing success. The press lavished praise on it. Viewers with colour sets were equally enthusiastic and found themselves besieged by friends without colour sets who wanted to come and see this highly praised sensation. News of its quality spread to America and when the National Gallery in Washington arranged showings of each programme on film, immense numbers of people queued to see them.

Civilisation brought lustre not only to BBC2's reputation but to the Corporation as a whole. And it also started a trend. No sooner had the series established itself as a smash hit than Aubrey Singer, the ebullient Head of Science Programmes, came to see me. He was outraged. How could I, as a man with a scientific education, have selected an arts subject as the first candidate for such prestigious treatment? By now I knew that we should be preparing for a successor and I promised Aubrey immediately that the next such series would be a scientific one. Aubrey knew exactly who should be selected to do it. Dr Jacob Bronowski. So *The Ascent of Man* went into production.

Such series – twelve or thirteen fifty-minute weekly programmes about some major subject – soon acquired a name in the profession. They were 'sledge-hammers'. Others followed. Alastair Cooke on the American Bicentennial; J.K. Galbraith on economics. But there was one subject that seemed to me to be an obvious candidate with even more visual potential than its distinguished progenitor. Natural history. I would have loved to write and present it myself, but I certainly could not do so as long as I remained an administrator. Doubtless someone else would emerge who would take it on. But I could not bring myself to urge the Natural History Unit in Bristol to try and find that person immediately.

* * *

Soon after colour began, Mike Peacock who had been running BBC1 announced that he was resigning. The time had come for the commercial television networks to renew their franchises. The Government had made it clear that this was not going to be an automatic process. New production companies would be allowed to bid for them. Mike had decided to join a consortium, London Weekend Television, that was applying for one of them.

Huw and I discussed the situation. I was happily immersed in the problems of colour. We agreed that running BBC1 was probably not the job for me, anyway. I was pretty sure that I would not enjoy it as much as BBC2. The obvious man was Paul Fox, the powerful and wise Head of Current Affairs and a strong contender for Controller BBC1 when Mike Peacock had got the job in 1965. Huw put the proposal that he should be asked to run BBC1 to the Director General, Hugh Greene. Paul was appointed.

Then began what were for me undoubtedly the best and happiest years in my life as an administrator. Paul and I had adjoining offices. A file with copies of all his correspondence was put routinely on my desk for me to see, as mine was on his. We rarely had disagreements. If he wanted any programme that BBC2 had developed or any contributor that we had established, then he was welcome. The number of viewers with sets that could receive BBC2 was still comparatively small. It was not in the public interest that a network that could not be seen by a substantial section of the public should deny its successes to a wider audience. BBC2's job, as I saw it, was to innovate. If BBC1 took over some of our successes, then BBC2 had schedule space and money released to continue to innovate. When the Corporation as a whole gained access to some great and long running event – the Wimbledon Tennis Championships, World Cup football, space walks on the moon – the two networks could co-ordinate their coverage and give interested people a continuity of viewing that the commercial networks could not match.

We both worked to Huw. Every morning we met in his office and there the three of us discussed the previous night's programmes and the problems that were approaching. Huw was an inspiring leader who energised the entire Service. His tall, slightly hunched figure was seen, just as his raucous laugh was heard, almost everywhere. He declined to dine in the select senior dining room but ate at midday in the general canteen. In the evenings, if we could, he and I would stroll round the circular corridor at first floor level in the Television Centre to stand in the observation gallery of the control rooms of all eight studios to watch – and be seen watching – what was going on. And then, since he had the only personal car in the Television Service and he now lived in Richmond, he would give me a lift home.

He stuck his nose into everything. His typed memoranda were worth reading if only for their entertaining and sometimes pungent style. He was certainly loquacious and knew it, but excused himself by explaining 'How can I know what I think until I hear what I say?' He had favourite phrases which many of us adopted, I dare say, unconsciously. Producers he thought weak he would dismiss as 'babies'. Those he thought well of 'carried guns'. A programme maker, he would explain to directors and producers, had three obligations: to his subject, his public and himself. If he betrayed any one of these, selling them short, then his programme would be flawed. He backed talent and fought for freedom of expression in plays and comedy with great courage against a very reactionary Board of Governors to whom he was continually being called to account. Often enough, those writers whom he defended with such vigour behind closed doors in Broadcasting House abused him in public for being a censor. He bore their insults with equanimity. 'You have to back talent,' he would say, 'no matter if the talented are not grateful.'

He believed it imperative that the organisation he headed should actually and visibly be controlled, not by managers nor accountants, but by programme makers. Being a regimental man, he thought it important that this should be apparent in the titles of its officials. Thus Joanna Spicer, clearly one of the most intellectually able and experienced people in the whole of the Television Service, had the title of Assistant Controller, Planning, while someone running a network, even one as inexperienced as I, was called a Controller. I believe her salary was considerably greater than mine. It certainly should have been.

Working with us were a group of programme makers of great experience and unparalleled accomplishments. Bill Cotton handled Light Entertainment which in those years produced such enduring successes as *Dad's Army, Till Death Us Do Part, Monty Python's Flying Circus, The Morecambe and Wise Show* and *Porridge*. Sydney Newman, a flamboyant hard-swearing Canadian, was responsible for Drama. Aubrey Singer, by now Head of Features Group, supervised a whole swathe of non-fiction output. Norman Swallow was in charge of Music and Arts. Current Affairs was looked after by Bryan Wenham. Brian Cowgill, a highly competitive not to say combative Northerner, was in charge of sport and secured such an array of important fixtures, covering them with such polish and expertise that our television competitor came in second in almost every event.

Colour began to get into its stride, but still we were short of cameras. The engineers were reluctant to buy a great number, even if they had been available, since the technology was advancing almost monthly and they did not want to be saddled with obsolete gear. It was camera shortage that attracted me to the suggestion from Sports Department that, now colour was here, we might televise snooker. They pointed out that with all the action concentrated on a small table, a whole game could be fully covered with only three cameras. If we were to create our own knock-out competition, with a small number of frames in each match, we could record the entire event in only a few days with a minimal number of cameras and then transmit the recordings, round by round, as a weekly event in a series that could run for months. It would break our principle that all colour programmes should be as comprehensible in black and white as in colour but we agreed that the commentators could deal with that problem. So arose the probably apocryphal story that one of them said, 'Steve is going for the pink ball – and for those of you who are watching in black and white, the pink is next to the green.'

Pot Black, as the competition was called, became an instant success and established the sport as a television spectacle that captured a devoted following of many millions. I was glad of it, and proudly put it alongside *Civilisation* when defining BBC2's character. It was a network that tried to appeal to all levels of brow.

16

Exotic Interludes

Life, the Natural History Unit's magazine programme for BBC2, dealing with the current affairs of natural history, conservation and ecology, was shaping up well. It was edited and presented by Desmond Morris, then Curator of Mammals at the London Zoo. But then Desmond wrote a book entitled *The Naked Ape*. It had an extraordinary reception. Pre-publication sales were stupendous. Suddenly, Desmond saw a chance of breaking away from his scientific work to give time to his other career as a surrealist painter. But taxation was then at its most punitive. If he was not to lose nearly all of his royalties, he would have to leave the country almost immediately. So he and his wife Ramona had decided to go and live in Malta.

The Bristol Unit had to consider who – if anyone – might take over the series in the long run. But there was an immediate problem. A trip had already been fixed to visit a zoologist, Iain Douglas Hamilton, who was doing extraordinary work with a group of elephants in Tanzania. The producer responsible, Richard Brock, came to see me in my office. Everything was arranged. The whole trip, including an interview, would take little more than a week. Since Desmond was about to disappear, would I do it? I had asked, when I first accepted Huw's invitation to become an administrator, if I might be allowed to forsake my desk for a few weeks to make a programme every year or so. Here was my first chance. The thought of a few days of African sunshine, let alone of getting close to a group of wild elephants in the company of someone who really understood them, was altogether too tempting to be passed by. I agreed.

A few days later, feeling unhealthily pallid and very unprepared, I walked into the bar at Manyara safari lodge, on the edge of the Rift Valley in Tanzania. As I ordered a beer, a tall glamorous bronzed woman in fashionable khaki safari gear materialised at my elbow.

'Hello,' she said leaning towards me and lowering her voice confidentially. 'Weren't you once David Attenborough?'

While I was wondering how to respond to this, a snigger came from behind a nearby pillar. It was Alan Root and he who had put her up to this. He was undoubtedly at the time, the greatest natural history film-maker in the whole of Africa and we were old friends. I might have guessed that he would not only know about any other film-maker who ventured into his territory but that he would prepare some kind of welcome for them.

Later that evening Richard Brock appeared. He had come out earlier to make the final arrangements for the interview with Iain Douglas Hamilton. He told me what he had in mind for the following day.

'You will go up towards the elephants with Iain in his Land Rover. The cameraman and I will be in our own car on the other side of the herd, filming with a long lens. You and Iain wearing radio-mikes will then get out and walk towards the elephants – and us – and we will film you talking together about the elephants with the herd in the foreground.'

I protested. One doesn't walk up on foot towards elephants. It would be a lunatic thing to do. Richard gave me a tolerant smile. Iain knew every one of these elephants personally. He knew which was amiable, which might be a little more tetchy. He understood every nuance of their body language. He walked up to them daily. There was no danger of any kind. Very reluctantly, I allowed myself to be persuaded.

The following day, I met Iain. He wore very short shorts, sandals and was bare chested except for a curious kind of loose flapping jacket without sleeves. His shoulder-length hair was bleached by the sun. He was clearly a man who spent all his days in the bush. I was reassured.

The following day, everything went to plan. Iain found a family of elephants he knew well, led by a matriarch he called Queen Victoria. They were feeding a hundred yards away, nonchalantly ripping branches from trees and tearing up great tussocks of grass with their trunks. Beyond them, through my binoculars, I could see Richard and the cameraman, standing in the open-roofed camera car.

Iain got out and I followed him. He was holding a small bag made from one of his wife's silk stockings, full of talcum powder. He gave it a little shake. A fine haze of the powder escaped through the mesh of the stocking and drifted slightly towards me.

'Excellent,' said Iain. 'What wind there is, is coming this way. They won't be able to smell us.'

Walking cautiously forward, putting each foot down as gently as possible so as to make the minimum noise, we advanced towards the herd. Every yard or so, Iain gave the bag another little shake, just to make sure the wind hadn't shifted. It occurred to me to wonder why, if he and the elephants were such pals, it was so important that they didn't smell us. But

Iain was already talking about the elephants ahead of him. He had names for them all. He described each of their temperaments. This one was a bit of a joker. That was a young female just coming into oestrus for the first time. Over there was a young male, now getting a bit frisky and approaching the time when he might leave his mother and the family group. I didn't have to say much, which was as well for my mouth was so dry my voice would have sounded curiously croaky.

Eventually we were within twenty yards of the group. I could clearly see Richard standing in the camera car beyond the elephants. He raised his hand and gave the thumbs up. He had got what he wanted. We were dismissed.

Quickly I slipped behind a baobab tree where the elephants couldn't see me and then, as fast as I could, I made my way back to Iain's car. I was still panting with relief and excitement when Iain, in a more leisurely way, strolled back. As we drove off, I reproached myself for being so nervous. I was, after all, with one of the world's leading elephant experts. He knew what he was doing. I should have had more faith and confidence in the scientific method. While I was pondering this, there was a distant drumming sound.

'Did you hear that?' said Iain. 'We were charged by a rhino.'

I expressed alarm.

'Oh, it was only a dummy charge,' said Iain. 'I'll show you.' And in spite of my protests that he needn't bother, he put his car into reverse, drove back down the track and stopped. Once again there was a drumming noise but this time, instead of fading it got louder and a huge rhinoceros charging at full speed burst out of the bush.

It hit the rear of the Land Rover with a crash. Then it thrust its horn under the back of the chassis, and lifted the whole car so that the back wheels left the ground. It shook it as though the car were a toy. Next, it set about one of the rear tyres, ripping a great slit in it with its horn. Then it had another go at the chassis. We both sat tight, holding on for dear life. I noticed that Iain was gripping the steering wheel so tightly that his knuckles were white. The rhino backed off again and then trotted away into the bush.

'Some dummy charge,' I said. 'Thank goodness he didn't give us the real thing.'

We were now stuck. The car was undrivable. Neither of us fancied the idea of walking back through the bush towards the lodge. All we could do was wait. Eventually another truck came by. It was, inevitably, Alan Root. He gave us a lift back to the lodge and hardly stopped laughing all the way.

A few days later, I was back sitting behind my desk in the Television Centre. Several weeks later still, a postcard with a Kenyan stamp appeared in my in-tray. It showed Gertie, a famous rhinoceros which had one of the longest horns ever recorded. Scrawled on the back was a message from Iain. 'Am in hospital with a few broken ribs and torn muscles having been trampled by a rhino. Persistent little devils aren't they?'

* * *

The following year, I decided to take properly scheduled leave of absence to make a film in Bali. I had never forgotten the idyllic time Charles and I spent on the island on our way to find the Komodo Dragon. Back in Britain, I had continued to listen to Balinese music and through this interest met John Coast. John, serving in the British Army in the 1940s had been captured by the Japanese and had laboured on the murderous Burma Road, which had been made famous by the feature film, *The Bridge on the River Kwai*. When at last peace came, he had gone to Indonesia, then ruled by the Dutch, and spent some time helping the Indonesians in their fight for independence. He had married a Javanese, had become particularly interested in Bali and its music and had eventually brought a group of Balinese dancers and musicians to Europe on a concert tour. He was now a musical impresario but his passion for Bali and its music had never abated. We decided together that BBC2 viewers might like to see something of the island's spectacular traditions and hear its beguiling music. So I took three weeks off to go to Bali with a camera crew headed by one of the most imaginative of the staff film cameramen, Nat Crosby.

The island had changed considerably since I had been there in 1956, eleven years earlier. Now there was an airfield big enough to accept jumbo jets, and holiday-makers were flocking to it in great numbers from all over the world, especially from Australia. Nonetheless, John's wide range of influential contacts got us to places where the artistic traditions of the island were still strong and in our short time we were able to make three fifty-minute films. The first was entirely about the gamelan and its music. The second concentrated on Balinese Hinduism and even more ancient rituals such as one in which young girls fell into a trance and were then able to stand unsupported and with eyes tightly shut on the shoulders of men as they gyrated in front of devotees. The third in the series was a straightforward recital of music and dancing in which I hoped some viewers might be interested, as a result of seeing the first two programmes.

It happened that Benjamin Britten saw the series and he invited me to give a lecture on Balinese music at his next Festival at Aldeburgh. I knew he had visited the island in 1955 and had been captivated by the gamelan. John Coast said that Britten had told him that he could drive through the Balinese night and know exactly where he was entirely by the harmonies of the gamelans at their nightly rehearsals, for each village tuned their gamelan to a slightly different scale. I was, of course, greatly flattered by the invitation and took a lot of trouble in thinking out what I would say and how I would illustrate it. In the process of doing so, I realised that Britten himself, in his ballet score for *The Prince of the Pagodas* had included an exact transcription of sections, note for note (or rather chord for chord) of a famous traditional gamelan piece. I thought it might be interesting to illustrate this by comparing recordings of the Balinese original with Britten's masterly orchestral recreation of it.

I walked out on to the stage of the Jubilee Hall at Aldeburgh confidently enough. But then I was hugely disconcerted to see sitting in the very front row Britten himself and his partner Peter Pears. My embarrassment was not simply because both certainly knew very much more than I could possibly know about the subject that I was about to address, but also lest Britten might think that in making my *Pagodas* comparison, to which I was now unavoidably committed because of my sequence of recorded illustrations, I might be suggesting that he had been guilty of plagiarism. In the event, at the party afterwards he seemed only too delighted that someone had recognised the particular piece he had used as a model.

* * *

After a further eighteen months sitting behind a desk, I decided that it was time I had another break. It would be most valuable, I told everyone, perhaps a little unconvincingly, to have first-hand experience of the latest generations of film stocks, cameras, and editing equipment.

I hankered to return to New Guinea. I still had not adequately filmed any birds of paradise; the people, particularly those in the Sepik valley, produced some of the most dramatic sculptures in the world which I longed to see; and, perhaps most tempting of all, there were still quite large patches in the heart of the island where it could be truthfully said that no white foot had yet trod. There, thick rain forest blanketed steep mountains. The rivers were so narrow and twisting, so broken by rapids, that travelling up them by boat was impossible for any distance. Planes or helicopters could not be used because there were no open grassy areas on which to land. There was only one way to explore such country and that was by using the classic method that all explorers had used, perforce, from Marco Polo to David Livingstone and right up to the last few years – by walking, accompanied by a long line of porters carrying supplies. Maybe it was the last place in the world where that was still the case. That in itself made such an expedition worth filming.

I had asked friends in the Australian Broadcasting Commission who had contacts with Papua New Guinea, to let me know if and when the Australian administration of the island was planning an exploratory patrol. And in 1971 word came through that someone was.

There was quite a large patch of unexplored mountainous country south of the huge Sepik River that runs roughly parallel to the northern coast of the island. No one had paid much attention to it, partly because travel there was so difficult, and partly because as far as anyone knew, there were no native people living there either. But now a mining company had applied for permission to send in a party of geologists to prospect for minerals. An aerial photographic survey had been made and that, to everyone's surprise, had revealed a number of tiny clearings in the rain forest. There were people living there after all.

The Australian administration had a policy that Government officers should make first contact with such people before anyone else was allowed in. A patrol to try and do so was going to leave from Ambunti on the middle stretches of the Sepik in a couple of months' time. My friends in the ABC had made enquiries and got permission for me and a film team to join it if I wished. What a chance! I started to rearrange my diary immediately. I bought a new pair of boots and began methodical tramps around Richmond Park in the evenings and at weekends to try and get them – and my flaccid muscles – into proper shape.

The Head of Film Department suggested that the staff cameraman for the trip should be Hugh Miles, a keen ornithologist who had been trying for some time to specialise in making natural history films and that Ian Sansam should be the sound recordist. So in April 1971, the three of us, together with Keith Adam, a film director from the ABC who had made all the arrangements on our behalf, flew in to Ambunti. We were met by the officer in charge, Assistant District Commissioner, Laurie Bragge. He was, I guessed, in his early thirties. He looked alarmingly fit.

We were in good time. It would be a week or so before preparations for the patrol would be complete. Supplies were being flown in and Laurie was just starting the process of recruiting carriers. He explained the basic arithmetic that he used to calculate the number he required. If one person who is not carrying food is accompanied by two others carrying full loads of provisions, the three of them will have enough food to last a fortnight.

If a march lasts any longer than that, the number of food carriers needed starts to escalate very rapidly indeed and eventually becomes impossible to meet. A march can, of course, be extended if a carrier line can find food locally, but the New Guinea forest has little edible vegetation and very few animals that can provide meat worth eating. Nor could he rely on getting any sustenance from local people. We might not, after all, meet any. So he had planned that after two weeks there would be an air-drop of food. The patrol could therefore be away for a month.

He, together with us and the men needed to carry equipment and camping necessities such as tents, trade goods, portable radio and similar things, would number thirty. We would therefore need sixty porters in addition to carry food – which would be mostly rice. It was going to take a week or so to collect sufficient men from nearby villages up and down the river.

That gave us a little time to acclimatise to the muggy heat of the river and to get used to the unremitting assault of biting insects. We spent it travelling up the Sepik by canoe to visit some of the villages. We filmed in several of the spectacular cult houses which were filled with extraordinary sculptures. I had also hoped that we might find and even film some of the species of birds of paradise, in particular the twelve-wired bird which lived in the swamps around the river, but in this sadly we had no success.

At last the day of departure arrived. One hundred and two of us sailed away from Ambunti down the Sepik in two large launches. We turned up the Karowari, one of the great tributaries that flowed into it from the south. Eventually the river narrowed so much that our launches could go no further and we transferred into small canoes driven by outboard motors. On the evening of the third day, we reached a small riverside hamlet called Inaro. From there on, we would have to walk. Laurie had been to Inaro before – but only once. As far as he knew, no European had been beyond it.

One of the major problems facing travellers in New Guinea is making oneself understood. The island, almost unbelievably, has over a thousand mutually incomprehensible languages. The people of Inaro, together with those in two smaller villages inland – some two hundred souls in all – spoke a language that only one or two of our porters could understand. They however could serve as interpreters – 'turnim-talks' as they are called in pidgin. The mountains further to the west where we hoped to go were inhabited by a nomadic people called the Bisorio. Laurie was hoping that one of these people would be able to speak the language of the next group we might meet, who he believed were called the Bikaru.

Then we would have a chain of interpreters so that a message could be translated from pidgin to Inaro to Bisorio to Bikaru – and then perhaps even beyond that. The possibilities of Chinese-whisper style of misunderstanding seemed alarming but there was no alternative.

The turnim-talk chain was in danger of breaking before we had even used it. No one in Inaro had seen anything of the nomadic Bisorio for many days. Laurie dispatched Kaius, one of his policemen who spoke Inaro, to try and find tracks in the forest that might lead him to the Bisorio. We with all our loads would be travelling slowly. If Kaius did find a Bisorio turnim-talk the pair of them would be able to catch us up without much difficulty.

The following morning we started our march. Before we set off, Laurie issued our remaining two policemen with bullets for their rifles. As he handed them out he formally gave them strict instructions. No one must fire without his direct order. If he, Laurie, was not there, then they were only permitted to do so in self-defence. Self-defence meant that they had already been struck by an arrow. An arrow that missed them didn't count.

I suppose my tramps in Richmond Park had made some difference, but the first days of walking were nonetheless very painful and exhausting. There can be few kinds of country that are worse to walk through. Thick undergrowth clogged the ground between the tall trees. Nowhere was flat. We were either clambering up steep mud or slithering down it. We needed hand-holds to steady ourselves but when we reached out to grab a stem or a creeper, it as likely as not proved to be studded with spines. It was remarkable how quickly we acquired the botanical expertise needed to identify which plants we could grab in safety and which were better avoided. Leeches lurked everywhere. If you stopped walking, you could see them looping across the ground towards you. As you wiped your brow, small black sweat bees would settle on you in droves. They didn't bite or sting. Instead they infuriated you by crawling over any patch of exposed flesh and drinking the sweat which oozed from you.

Every morning at half past five we got up and struck camp. The march started promptly at seven. We plodded away in single file and continued until midday when we stopped for a half-hour rest and allowed the rest of the line to catch up. As the last of the carriers staggered in and was counted, to make sure that none had dumped his load in the forest and gone home, those who marched at the head of the column started off again. Keith had provided us with rations that were used by the Australian Army. Each contained a few boiled sweets but, even better, sweetened condensed milk in tubes like toothpaste. I kept my daily allocation in my pocket and consumed it on the move. An inch and a half of condensed milk squirted on to the tongue at moments of exhaustion was bliss.

At about four o'clock, Laurie would call a halt and we would make camp. The porters cut saplings in the forest and lashed them together to form simple frames over which they stretched tarpaulins. Then at about five o'clock, it would rain. Torrents would continue to fall late into the night, drumming on the canvas tarpaulins above us as we tried to sleep. By dawn it would have stopped and once more we would struggle into damp clothes and moist boots to start again.

After a week, Kaius, the policeman who had gone in search of the Bisorio, caught up with us. He had found no sign of them. A crucial link in our turnim-talk chain was still missing. We would be unable to talk to the unknown people we sought, even if we found them.

Every three or four days, we had a rest day. Laurie decided on our route by taking compass bearings and trying to relate them to an aerial photograph that had been taken during the survey. After we had been walking for two weeks, we found a trail. I cannot say that I did so. Indeed, I saw hardly any sign of it, but the carriers noticed a turned leaf, a broken twig, a smudged toe mark. There were three men ahead of us somewhere, they said. We followed their tracks all that day.

The day afterwards, after three hours' marching, we walked into a large clearing. Several huge trees had been felled. Bananas and cassava had been planted round the stumps and the prone trunks. In the middle stood a large house on stilts. A notched tree trunk led up to its entrance which was closed by a large wooden door. No smoke came from its roof. We called. There was no response. It could be that there was no one at home. Alternatively the three people whose tracks we had been following could be inside but were frightened to come out and were watching us through cracks in the bark walls.

We camped in the forest at the edge of the clearing. Laurie put out gifts in several places in the gardens. He stationed one of the porters on the stump of one of the felled trees and asked him to sing out in every language he knew that we were friends and that we brought gifts. But nothing happened.

The following morning, there was still no sign of life in the hut. Laurie decided to investigate. It was a brave thing to do. If there were people inside who didn't want visitors, he could easily get an arrow in his chest. Calling out and carrying cakes of salt, he cautiously crossed the clearing. He reached the foot of the notched pole and climbed up it. Once at the top, he managed to push aside the door and disappeared inside. We all waited. Eventually he reappeared and called out to us. The hut was deserted.

I joined him. The house was ingeniously constructed. The door opened in the middle of a corridor running transversely across the front of the house. It was so narrow one could barely squeeze along it. Anyone who entered it could easily be stabbed through the cane wall by someone within. The corridor led to a single room. A dozen or so pig jawbones were strung out on one of the side walls. A sheaf of elaborately decorated arrows stood in a corner. A dagger made from the thighbone of a cassowary was tucked behind one of the side beams. In the centre was a fireplace made of a large stone. It was still slightly warm. The people we had been following either didn't belong here or had not dared to stay. We had to go on.

Three days later, we were due to get our air-drop. To prepare for it we cleared a relatively flat patch of ground. Tarpaulins were laid out on it and green leaves put on fires to create smoke which would act as a signal. Promptly on time, the plane appeared over the mountains and dropped six bags of rice and bully beef. It came in twice more, again dropping with remarkable accuracy. It was good to get more food, and the porters did not seem to mind that their loads which had been getting steadily lighter over the previous two weeks, were now heavy again.

Two days later one of the porters had a fall and badly injured his leg. Two others were seriously sick with what Laurie thought was pneumonia. He decided that we would have to summon a helicopter. It was not a decision to be taken lightly. A dozen or so huge forest trees would have to be felled to give the helicopter enough space to descend and a platform of logs would have to be constructed on the ridge on which it could settle. That in itself would take two days of hard work. That evening, on his regular radio schedule, Laurie spoke to the administration's headquarters in Lae and soon all was arranged. We decided between ourselves that I should accompany the injured and sick men in the helicopter, to see that they were met and properly cared for. I would also take with me our exposed film and bring back some food more appetising than rice and bully beef.

The arrival of the helicopter caused a great sensation among the porters. I too thought it almost miraculous that a whirling deafening machine should suddenly appear in the sky and come down to settle beside us in this wilderness of trees. The pilot did not get out. He did not even stop the rotor. We hurled out the first load of stores that he had brought with him, and helped the sick men aboard. I clambered in beside them and within minutes we swooped up into the sky and were skimming across the forest through which we had been plodding so laboriously and painfully for the past three weeks.

Medical people met us in Lae and took charge of the injured porters. An hour or so later I was flying back over the forest. But I could see no sign of our camp or the platform we had built. The pilot flew up and down a narrow valley and then yelled to me that he was going to land on a sandbank in the middle of a river. Once down, he explained that he was so heavily laden that he was using far too much fuel flying up and down the valleys searching for our camp. So he was going to unload both the stores and me. He would come back and pick me up just as soon as he had discovered exactly where our camp was. As seems always to be the case when helicopters are involved, every word had to be shouted, every gesture hugely exaggerated, and every action taken at high speed. The baggage was pitched out, thumbs-up exchanged and within seconds the helicopter had gone.

The silence, after the din of the helicopter, was almost oppressive. I looked around and, with nothing better to do, rather pointlessly made a neat pile of the cargo. Then I sat on it. I wondered if anybody was in the surrounding forest watching me and if so what they would be making of a strange human being squatting unaccountably in the middle of their river. I then started to speculate about the pilot. Suppose he had as much difficulty in finding me again as he was having in finding the camp. I had food, it was true, but otherwise I had not the least idea where I was. It was a relief when after about ten minutes the helicopter reappeared, a tiny dot over the far mountain and thundered down to me once more.

Back in the air, I was able to see the pilot's problem. The forest trees were over a hundred feet high. Even though Laurie's men had cut down several of them, it was impossible to see the clearing unless you were almost directly overhead. Bearing in mind that there were no reliable maps of where we were, it was extraordinarily skilful of the pilot to have found us in the first place. Once again the helicopter came down to our camp, discharged its cargo and vanished. And I, having been skimming magically above the trees, was reduced once again to a stumbling, panting earth-bound biped.

The patrol now had only three days left. Thanks to the helicopter and the air-drop, we had enough stores, but in three days' time we should be back in known country and start on the descent of the April River, another big tributary of the Sepik where Laurie had arranged for boats to meet us. We went to sleep that night in low spirits. Laurie had not found the people he sought. We had no climax to our film.

I don't know what woke me the next morning, but when I opened my eyes I saw a group of strange small men standing outside the fly-sheet beneath which I had been sleeping and staring at me. They were tiny, about four and a half feet high. They carried woven bags on their backs and sprigs of leaves tucked through a bark belt, fore and aft. Their hair was twisted into spikes. Two had tucked it into a long woven snood that projected from the back of their head. Their ear lobes were pierced and hung with sections of shell. Each man's nose had a pair of deep punctures at the end in which he had inserted stout wooden pegs, except for two of the younger ones who wore, instead of pegs, long cassowary quills. I swung swiftly out of my canvas bed and went to greet them. Hugh, who habitually kept his camera beneath his bed, was filming before I even spoke.

The strangers gazed at us, open-mouthed. I smiled extravagantly and they smiled back. I was struck by how eloquent the human face is. It was clear that while we were inquisitive about one another, we equally clearly wished each other well. Laurie joined us. He was carrying sheets of newspaper which he offered to them. Almost everywhere in New Guinea, paper is a much appreciated gift, for people roll tobacco in it and make cigarettes, as I had discovered when I had been in the Jimi Valley fourteen years earlier. These people looked at it wonderingly, folded it in a tentative way and handed it to one another. Clearly they had no experience of it. Laurie tried another gift – glass beads. He put a spoonful on a leaf and offered it to them. They looked at them unemotionally and stirred them with a finger. Beads seemed to be just as unfamiliar as newspaper. Laurie tried again. This time he produced a sack of salt. They sampled it with their fingers and grinned broadly at its taste. That they certainly recognised and valued. They accepted it with every sign of gratitude. But who were they? What did they call themselves?

'Bukaru?' asked Laurie.

They grinned and pointed over the mountains from which we had come.

'Bukaru! Bukaru!' they said. It seemed that we had walked right through Bukaru country.

'Biami,' said one of the smaller, grizzled and more forthcoming of the group, and pointed at his own chest. 'Biami! Biami! Biami!' he added pointing to each of his companions in turn. This was a group Laurie had not even heard of before.

Laurie pointed down to the river at the bottom of the valley.

'Hiyami?' he asked.

They nodded and smiled delightedly and then started to list the name of other nearby rivers, pointing to the direction in which each of them lay. Laurie repeated the names as if he wanted to know how many there were – as indeed he did, though not because he wanted to collect the names but because he wished to discover which gestures these people used for counting. Neighbouring groups of people might not share a language but they often shared the same gestures to indicate numbers, so if he knew how the Biami counted he might get some idea of their tribal connections. The grizzled smiling man gladly showed us how he did so. One – and he touched his wrist; two, his forearm; three, his elbow; and so on all the way up his arm and shoulder until he got to eleven, which he indicated by tapping the side of his neck.

With this success Laurie went on to suggest in sign language that we would like to trade for food. That they seemed to understand. He went further. If they brought food, would they bring their womenfolk with them? That suggestion, however, they either failed to understand or ignored. Then, waving happily, they walked back into the forest.

Laurie explained to me that if both men and women appeared the following day, it would be a sign that they trusted us and were peacefully inclined. We should have to wait and see. The following morning they did reappear carrying taro roots and plantains with them in their net bags. But no women or children came with them. We gave them salt in return and Laurie asked once more about their families. They beckoned to us and started walking away. Hugh grabbed his camera and we followed. Laurie gestured to the police to stay behind. It would show confidence if we went by ourselves.

The Biami walked into the forest along a barely discernible track. It wound down a slope, round the trunk of a huge tree – and suddenly we realised that they had disappeared.

We stopped. Laurie called. 'Biami! Biami!'

Silence.

Why would they have beckoned to us to follow if they did not wish us to do so? Were we walking into an ambush? Should we turn back? We decided to continue.

There was no real path, but their tracks were fresh enough for us to follow without too much difficulty. Two hundred yards more and we came to a small shelter roofed with freshly cut leaves. Beneath, the embers of a small fire were still smoking. And by the side of the shelter, a grinning human skull on a stick.

'Biami! Biami!' we called.

Again no answer. Uneasily, we walked back to camp.

The Biami never reappeared. Three days later we reached the April River where launches awaited us. Five days after that, I was back behind my desk in the Television Centre trying to understand the papers we were to discuss at the next Management Methods Committee and pondering on the budget demanded by the Computer Steering Group. I suppose the reason I had asked for occasional breaks away from my office was in order that I might reconcile myself to such things. I was not sure that, as a plan, it was working.

17

The Threat of the Desk

In 1967 the Chairman of the Corporation's Board of Governors, Lord Normanbrook, retired. Harold Wilson, the Prime Minister, who had a long-standing resentment against the BBC which he regarded as recalcitrant and insufficiently responsive to his wishes, decided to bring the Corporation to heel and appointed Dr Charles Hill as the new Chairman. Hill was a Tory ex-Minister and had been, until that moment, Chairman of the Independent Television Authority. In that capacity he had frequently locked horns with the BBC's Director General Hugh Greene and it was well known that the two men had a great personal dislike for one another.

The news was greeted in the Television Centre with deep dismay. We all felt that it was a kick in the teeth. After all, Hill seemed to represent all the commercial values of ITV that we, the BBC, opposed. Soon after the announcement, I had to go overseas – to America, I think – so I was not there when the new Chairman made his first state visit to the television part of his empire. From all accounts, some senior staff in the Centre did not welcome him with any real warmth. That was not tactful, bearing in mind that he was to be our new boss and we would need, somehow or other, to get him on our side.

Soon after I returned, Hill appeared again at the Centre. He now had his own office there – something his predecessor had not thought necessary. I was sent for.

'You are the only member of the senior staff here I haven't yet met,' he said as I went in. He poured me a hospitable drink.

'People here have been far from welcoming. Why should that be?'

'Well, Chairman,' I said, 'you have to remember that we here have seen you until now as the opposition, if not, indeed the enemy. It is as though soldiers of the Eighth Army – who had, after all, already won a few battles in the desert – woke up one morning to be told that Rommel was to be their new Commander-in-Chief.'

'Oh, I see,' said Hill. 'Don't they think that I am a good general?'

'Yes, they probably think you are,' I replied, 'but that is not the issue. The question is whether you are fighting for the same things as we are.'

That exchange seemed to satisfy, maybe even to please him. At any rate, I got on very well with him thereafter.

I suspect the reason that Hill himself was so deeply offended by the fact that he had not been warmly received at the Television Centre was largely because he had always seen himself as representing the best of public service broadcasting. He had owed his first prominence to the BBC when he was Radio Doctor during the war, dispensing homely advice on the radio about healthy feeding. His coy reference to prunes as 'black-jacketed warriors' in the battle against constipation was one of the war-time jokes. That reputation, created on the BBC, was probably the reason why he was credited with knowing about mass communications and the reason that he was appointed to take charge of ITV. When Wilson transferred him to the BBC, Hill saw it much as though he was at last returning to his spiritual home, which is why the stony reception he had received had offended him so much. In the event, he proved to be a doughty defender of the BBC and its independence and any idea that Wilson might have had that Hill would 'tame' the BBC and make it more amenable to Government pressure proved to be misplaced.

One of Hill's first acts was to introduce management consultants to examine the whole structure of the BBC. The Corporation needed an increase in its licence fee and Hill had to demonstrate to the Government that the place was being properly and economically run. The consultants after they had spent many weeks and been paid many hundreds of thousands of pounds, produced an immensely long report and a list of recommendations. One of these was that there should be a change of designations of top management. Kenneth Adam, then in charge of television with the title of Director, retired. His job was to be retitled Managing Director and would be given to Huw Wheldon. Beneath him, effectively his deputy with a seat on the Board of Management, was to be a Director of Programmes. I was told that this would be me.

I viewed the prospect with some dismay. It would take me one step further away from programmes. However, the BBC was going through some rough waters. This was not the moment to resign and I did not. Paul Fox was going to continue as Controller BBC1 and Robin Scott, who had been a very effective producer of television outside broadcasts and was then Controller of Radio One, came back to television to take over from me as Controller of BBC2.

The number of meetings I was required to attend as Director of Programmes was formidable. Every Head of Department expected to have a routine meeting once a fortnight. Those I welcomed. They, it seemed to me, were what my job was about – discussing the previous weeks' output from the department, hearing about ideas that might be coming up and needed encouraging, or dismissing before they got too far. But there were other larger meetings attended by a dozen or more people which dealt with managerial matters. These I found less engaging. I did my best to concentrate on the technicalities of the subjects in hand, but the short time I had spent at the LSE learning about the techniques of observational anthropology had not been lost on me. I couldn't help noting how interesting seating behaviour at these meetings was. Very soon after any such committee was instituted, its participants settled down into regular positions around the table. Alliances were formed. People with similar views tended to sit together. If someone felt himself likely to be at odds with someone else, he or she would choose to sit opposite rather than alongside. When the chairmanship changed, there could be a great reshuffle and one could then deduce what the new alliances and oppositions were going to be. I wondered whether or not there was material here for a paper in one of the anthropological journals entitled perhaps *Some Observations on Seating Rituals in BBC Committees and Steering Groups*.

On one occasion, my observations had some practical value. I presided over the Television Service's Computer Steering Group. It not only determined the way we would use computers in our planning and financial systems, but spent a great deal of money. I therefore had encouraged the Chief Accountant to sit next to me, on my right. On this occasion, however, he was late and his chair was empty. I waited as long as I could until, at the very last minute, he rushed in, full of apologies. But instead of coming to sit in his usual position next to me, he walked right round to the far end of the table where there were several empty seats and sat down opposite me. It was clear that he was going to oppose me on something. Rapidly I cast my eye down the agenda and saw to my alarm that there was a proposal for a major piece of expenditure that I had neglected to clear with him before the meeting. When at last we reached it, I apologised to the meeting. I was not happy with the paper as it stood and had decided to defer it. As the meeting broke up, the Chief Accountant came over.

'I'm so glad you didn't take that paper,' he said, 'I was going to have to fight you about it.'

* * *

One of the few pleasures of sitting on the Board of Management was lunching with fellow Board members after our weekly Monday morning meeting in Broadcasting House. Sometimes, a guest – perhaps someone in the news, or an expert whose wisdom we might imbibe – would be invited to join us. Frank Gillard, Managing Director of Radio, had served during the war as General Montgomery's favoured press correspondent and the friendship had endured. He suggested that the Board might like to invite the Field Marshal, as he had become, to lunch with us.

In due course the Field Marshal turned up. After a somewhat awe-struck meal, Frank asked permission to ask questions and led off with one which he clearly knew would get an entertaining answer.

'Field Marshal,' he said, 'how do you think history will judge General Alexander?' Alexander had been a handsome and dashing rival of Montgomery's during the war and there was said to be no love lost between them.

The Field Marshal rose to the occasion. 'Well,' he said, squaring his shoulders and looking round the table, 'it is my belief that General Alexander will go down in history as the best...' – he paused and surveyed the expectant faces around the table – 'as the best-dwessed general of the war.'

He smiled, well pleased at having got his performance off to a good start.

'And what about Churchill?' said Frank. 'Do you think that he had a personal feud with Hitler?'

The Field Marshal was now well into his stride.

'You must wemember,' he said, 'that the Pwime Minister didn't like the Fuhrer. Didn't like 'im. I well wecall the occasion, soon after D-Day, when the Pwime Minister came to my cawavan in Fwance to visit the fwont. We set off in my field car. After we had been dwiving for an hour or so, I said to him, Pwime Minister, would you like a sanitawy stop? – because to tell the truth, I wanted a sanitawy stop myself. "How long will it take to get to the Siegfwied Line?" he said. "About forty minutes," I said. "In that case ," said the Pwime Minister, "I'll wait."'

The Field Marshal beamed round the table. The naturalist within me stirred.

'Some animals,' I ventured, 'mark their territory with sanitary stops. Perhaps he was doing the same.'

'Vewwy interwesting,' said Montgomery, as though the thought had not occurred to him before. 'I can give you another example. Later on in the campaign he paid me another visit. My twoops had just cwossed the Whine.'

It took a second or so for us, hanging on his words, to make allowances for the Field Marshal's articulation.

'The Pwime Minister wanted to see it for himself. The Pwess Corps was following us and just before we got to the wivver, he asked me to stop. He got out and ordered the pwess to be kept back. And then the Pwime Minister, by himself, walked down to the wivver's edge and *urewinated* in the Whine.'

* * *

A battle was going on at the very top of the BBC. Everyone knew that the Chairman was at odds with the Director General. So it was no surprise when suddenly, to everyone's dismay, Hugh Greene announced that he was retiring. He had been a great defender of editorial freedom. We were more surprised when the statement continued by saying that he had accepted a place on the Board of Governors. Perhaps a Governor's seat was part of the conditions he demanded for his resignation but if so it was hardly a wise one for it confused still further the respective role of the Governors and the Management. The Governors should be there to represent the public interest and keep an eye on professional broadcasters, to hear their arguments and justifications for their actions and ultimately either approve or forbid. To introduce a professional broadcaster on to the Board was to increase the likelihood, already apparent under Hill's leadership, of governors assuming that they were broadcasting professionals and encourage them to take over the tasks and responsibilities of the Board of Management.

Charles Hill was now able to appoint a Director General of his choice and thus increase still further his personal influence over the Corporation's policies. Unlike later occasions, only internal candidates were considered. The outstanding one was Huw Wheldon. But Hill mistrusted Huw who he thought was too clever. There was also Charles Curran. He had been Secretary to the Board of Management and then Managing Director of the Overseas Service. He was not a man to rival Huw in terms of charisma but nonetheless a gifted administrator of great intellect and a practising Catholic with high moral principles. Three less obvious candidates were also summoned – Kenneth Lamb who had also been Corporation Secretary as well as Head of Current Affairs; Paul Fox, who as Controller BBC1 had already demonstrated his great strength; and to my surprise, me. Hill didn't ask me whether I wanted to apply. He simply told me that the Governors wished to interview me.

I turned up as I was instructed to do. I suspect I was a little surly. I didn't want the Director General's job and I didn't, in truth, think I was the right man for it. It was therefore hard for me to expatiate to the Governors on what I would do if I were given it. In the end, it went to Charles Curran.

* * *

The BBC's addiction to titles and the habit of its staff of reducing those titles to acronyms was often mocked. Assistant Head of this, Chief Assistant to that, Deputy Controller of the other. They became even more mockable when they were reduced to initials, as they always were on internal memoranda. Many notes circulating between offices and desks carried rubrics at the top consisting of three or four lines of capitals, interrupted only by full stops and semi-colons. To an outsider they must have seemed as impenetrable as Babylonian cuneiform. Nonetheless, once you got used to the system – and that did not, in truth, take very long – they made life simpler. You did not have to remember the surname of the engineer in charge of telecine, nor even whether there had been a recent change in departmental staffing. One simply put H.E.Tel.Rec., which stood for Head of Engineering Television Recording. And when someone was promoted to a different job, all the files stayed exactly where they were and the new occupant took over with scarcely a hiccup. With twenty thousand people in the organisation, that was a sensible way to run things.

The practice even extended into everyday conversation. There too it had its value. Referring to the Director General, even to his face, as 'DG' is a convenient way of expressing respect, but was also a very useful way of underlining the mode in which one is speaking. Is this a casual conversation, not to be taken seriously? Is it a formal instruction that is being received and perhaps contested? The custom becomes a little more difficult to justify when titles become hugely extended. Who except a Corporation aficionado would know that A.C.Tel.Ops. is the Assistant Controller Television Operations? The practice becomes even more absurd when it leaks into general social life. Even I thought it funny when Jane arrived at the Television Centre for a social evening and was welcomed in the entrance hall with 'Good evening, Mrs D.P.Tel.'

My title of Director of Programmes was, however, somewhat misleading. In my new job I had far less to do with programmes than I had as Controller BBC2. Then Huw had given me great freedom of editorial decision. I felt I had to do the same to my successor, Robin Scott. So apart from general discussions about overall plans for both networks, programme issues only landed on my desk when things were going wrong

– when vast overspends loomed on a production, when judgments had to be given on questions of taste and language, when there were accusations of distortions of facts, risks of libel and other serious departures from the Corporation's editorial principles. Otherwise my time was taken up with changes in management methods, staffing problems including firing people, disputes with the unions, the introduction of computerised systems and, overwhelmingly, financial matters.

The Television Service did not overspend. But inevitably every producer wanted to improve the quality of his productions and only too often that meant a request for more money. We could increase our income by marketing our programmes overseas and great efforts were being made to do so. The only other way to meet rising costs was to get an increase in the licence fee. Governments, naturally and properly, resist such an idea and the easiest way for them to do so is always to state, without any particular evidence, that the BBC is grossly wasteful and inefficient. It also suits commercial television to support that opinion.

Around this time, I went to a cocktail party at which people from all sections of the television world were present. Across the room, I saw Sir Lew Grade, then the most powerful mogul in all commercial television. He came across, beaming, to clap me on the back and congratulate me on my promotion. What did I see as my main problems, he asked, conversationally? I told him that there was the perennial question of an increase in the licence fee.

'Nonsense,' he said, amiably. 'The BBC is rolling in money. Cut out waste. Improve efficiency. I only wish we in independent television had the sort of income you have.'

'Now that's very interesting,' I said. 'Independent television costs the nation, overall, twice as much as the BBC. Its annual income from advertising is 93 million pounds. The BBC's income from the licence is almost exactly the same. But for that money we provide, not one network, like ITV, but two, and four radio networks as well.'

'Are you sure of your figures?' he said.

'They come from your official handbook and ours,' I replied. 'How do you account for the difference?'

'Well, the whole of the regional structure of independent television that was imposed on us by Government is grossly wasteful.'

'Okay. Let's allow ten million for that. An hour of your programmes still costs the nation around twice as much as ours. If we are so wasteful, how do you explain that?'

Lew's eyes narrowed and he leaned towards me.

'It's the iniquitously high salaries we have to pay our senior executives,' he said.

* * *

Establishing exactly what a particular programme did cost was not of course easy and I was learning very fast about the subtleties and complexities that accountants can use to transform the appearance of their balance sheets. What proportion of the cost of running our studios should be allocated to a particular production? How much did it cost to run our own make-up department? Would it be cheaper to hire freelance make-up assistants? Should those programmes that were made on film and never went near a studio be charged a share of studio costs? And what would happen if the costing structure caused a stampede into film, leaving our expensive permanently-manned studios half-empty?

It was not only ourselves who needed to know the answers to questions about programme costings. The Government, anxious to create a flourishing and independent television industry, had instructed the BBC to reduce its own output and buy a specified proportion of its programmes from independent production companies. When the BBC said that a budget demanded by an independent was far too high compared with the Corporation's own costs, the independent producer complained, understandably, that many of our real costs never appeared in a BBC producer's budget. What about telephone calls? Were they charged to individual programmes? Who was paying for all those libraries that staff producers used so freely? And the subsidised canteens where they ate? And, come to that, what about the Director General's salary? So a more comprehensive costing system had to be introduced and one that, while giving BBC managers better financial information on which to base their own decisions, would also allow individual producers the greatest possible freedom in spending their money, buying what they wanted in the most economical way. The complexities of doing this, and of persuading producers and engineers, scenic designers and canteen supervisors that these changes in administration and decision-making had to be made and in the long run would be to their advantage, took an increasing amount of my time.

My title of Director of Programmes became less and less apt and I became more and more restive. It was not what I had come into television to do. Nor did I think I was very good at doing it. I decided after four years, that it was time for me to return to programme making.

* * *

It took some time to find the right moment to resign. There were still battles in which I was engaged that could not be abandoned until they were won, but at last in December 1972 the right time came. I had continued my practice of every two years or so taking several weeks away from my desk to make a programme. I had scheduled the next for February 1973 – a trip to Indonesia with a team from the Natural History Unit. It was an ideal break point. I could clear my desk and appoint various people to look after particular issues in my place without starting awkward rumours. Then three weeks or so before I was due to depart, I announced that I would not be returning to my office. The Press tried to invent a story that I was being pushed out but I managed to scotch it.

Internally, I made it clear that this was not retirement. I was, quite straightforwardly after eight years administering – four years with BBC2 and four as Director of Programmes – resigning from the staff in order to return to programme making. There was no cause for farewell parties of any kind. I hoped that, when I got back from Indonesia, I would be reappearing in the bars and offices of the Television Centre as a freelance broadcaster. My replacement would be announced in due course.

The time came for my last major administrative commitment – presiding over the weekly Programme Review Board. This took place every Wednesday morning and was a key meeting in which any programme transmitted during the previous seven days could be discussed. It was a forum attended by the Head of every production department in the organisation together with other key producers who were asked to attend if there was likely to be an important discussion about their programmes. Regular membership was regarded as a privilege and there were usually around forty people there. It was the sounding board, perhaps even the heart of the Television Service.

The *Radio Times* provided us with our agenda. Using its programme pages we reviewed the output, day by day, on both networks. Running the meeting, one had to ensure that there was a fair share of debating time for each evening, especially if there were some important or controversial programme that had been transmitted towards the end of the week under review. On this occasion, it went quite well. Seven minutes before the meeting was due to end, at eleven o'clock, I said, as usual, 'And now, last night's programmes.'

Everyone turned their page of *Radio Times*. As one, the whole room burst into a roar of laughter. Instead of the billings for the previous evening's programmes, there was a list of wholly fictitious programmes every one of which involved my name – either in the title, the sub-text, the list of contributors or technicians. Not only that, but the illustrations to these imaginary programmes were all photographs, more or less embarrassing, of me.

The whole enterprise had been devised and plotted by Paul. Every departmental head had been asked to devise a list of joke billings for his kind of output. Paul had edited them and persuaded the *Radio Times* to print fifty or so copies of a new double page insert. Everyone's secretary – including, in secret, my own – had been given the job of unstapling their copy of *Radio Times* and inserting the new pages. I could not have asked for a more heart-warming and entertaining farewell gift.

I walked out of the meeting with very mixed feelings. But I knew I was right to go.

18

Back to the Jungle

By the 1970s, Broadcasting House in Bristol had, arguably, established itself as the pre-eminent centre for natural history broadcasting in the world. The fact that the Natural History Unit developed there was not because the west of England was particularly rich in wildlife, but because of the Corporation's long-standing regional policy. BBC centres outside London produced programmes about local subjects which were only shown locally. But regions also, understandably, yearned to make programmes for national consumption. To support this ambition, each region was encouraged to develop a speciality. The 'national' regions – Scotland, Wales and Northern Ireland – had a lot to offer. Scotland was best at this. They scored major successes with Scottish dramas written by Scottish writers and performed by actors who made their Scottishness fully apparent by their accents. It also contributed programmes about Scottish dancing and, since the English were deemed not to understand the significance of New Year's Eve, about Hogmanay.

The English regions had a more difficult job when it came to contributing to the national network. The North, based in Manchester, had one of the last surviving music halls, the Palace of Varieties, and transmitted regular programmes from it in a series called *The Good Old Days*. The unpaid audience dressed up for it in Edwardian costume and even competed to do so. Birmingham tried hard to make programmes about industry. It, for reasons of regional *amour propre*, had been given a palatial regional centre at Pebble Mill. Programme people there recognised that its imposing entrance hall was as big as many a studio and suggested using it as such for a network lunch-time chat show. This ran for a long time under the title, *Pebble Mill at One*. When I was with BBC2 we had established a drama unit there to produce a regular series of half-hour plays which flourished for some years but ultimately withered on the vine.

The one spectacular and lasting success, however, was West Region, based in Bristol. Desmond Hawkins had established natural history as West's particular interest immediately after the war on radio. He skilfully and amiably kept me off his patch when television began to develop in the

1950s. He had formally established the Natural History Unit in 1957 and the *Look* series presided over by Peter Scott had been a great success. The Unit also provided special programmes about animals for children and regular fifty-minute natural history films for BBC2. No other such unit in the broadcasting world had two television networks and four radio channels to broadcast its output. If it was not the biggest and the best unit of its kind, it certainly should have been – and Desmond, first as Head of the Unit and later as Controller West Region himself, made absolutely sure that it was.

When Desmond had suggested that I might make a series in Indonesia, I had put it in my diary as a perfectly genuine sabbatical from administration. Now it became my first assignment as a freelance.

The producer was to be Richard Brock who, two years before, had persuaded me to substitute for Desmond Morris and walk recklessly close to elephants in Africa. He had already planned the trip and outlined the programmes. My task would be to appear in picture, describing what I saw, adding a bit of information here and there, allowing viewers to identify with me and thus, from the way I looked and the things I said, give them some feel of what it was like to be where I was – whether it was hot, cold, dangerous, exhausting or exciting. In short, I would be what is called these days a 'presenter'. It was a bit of a come-down, after having devised, directed, scripted and researched programmes as I had done in the early *Zoo Quest* days, but that was the arrangement made while I was a full-time administrator and the series a mere sideline for me. It could not be changed now.

The filming team was somewhat larger than it had been for the early *Zoo Quests*. Instead of having just one companion, I would be part of a team of five. Maurice Fisher, the cameraman, had an assistant, Hugh Maynard. There would be a sound recordist, Dickie Bird, whose first name is Lyndon though as far as I know, no one apart from his wife ever uses it. And there would be Richard, as director, to tell me what to do. All were to become good friends with whom I would work on many projects to come. In particular, Dickie would, off and on, be patiently recording my words, gently correcting my grammar, criticising my enunciation and fumbling down my shirt front to try and conceal his tiny radio microphone, for the next thirty years.

Richard had decided that one programme would feature the famous volcano of Krakatau. Its eruption in the nineteenth century was one of the most dramatic ever described. A small group of islets, lying in the strait between Java and Sumatra, were pulverised in a series of titanic explosions. One created the loudest noise in the whole of recorded history. The sun was entirely blotted out for days on end. A tidal wave was produced that crashed on to the nearby shores of Java and Sumatra, carrying fishing boats bodily up into the hills and drowning thousands of people.

Twenty years after this titanic explosion, a small cone of ash appeared in the sea where the island had once been. Local people called it Anak Krakatau, the child of Krakatau. It has continued to erupt with varying intensity ever since and has now formed a reasonably substantial island. It was this that we were to film.

It seemed to me that telling the dramatic story of the eruption really did give a presenter a chance to justify his presence. I could provide the facts while a scarlet fountain of fire played into the night sky behind me; or maybe, if Anak was not being too explosive, I could stand beside a river of molten rock, my face illuminated by its sullen glow. Richard agreed. Either way, however, we were both sure that we should get the best effect at night.

We arrived at the little fishing village of Labuan on the western tip of Java in the middle of the afternoon. We chartered a fishing boat and in the early evening set out towards the small group of irregular cones on the far horizon, one of which, we assumed, was Anak Krakatau.

The sea was oily and dead flat. Above, the moon was sailing through a cloudless sky with such brilliance that I could have read a book by its light. But ahead of us it was dramatically different. A huge arch of clouds framed the islands ahead, stretching down on either side to the distant horizon. We seemed to be sailing towards an immense cavern. Discharges of lightning played within and across it. As we got closer and entered it, the ceiling of clouds above blotted out the moon and it became very dark. The scene was positively Wagnerian. Now we could see in the gloom ahead a red glow. This must be coming from Anak Krakatau. Suddenly it disappeared and I thought for a moment that the eruption had stopped, but it was just that one of the surrounding islets, invisible in the blackness, was screening it from us. We rounded the islet and there in front of us was exactly what we had hoped for – a fountain of fire.

It jetted high into the sky from behind the silhouette of a low dune of ash. It was difficult to tell how far away the eruption was, but the scarlet bombs of lava that shot into the air seemed to be falling so slowly that I guessed we must still be at least half a mile from it. We were close enough, however, for ash to drift down on us, stinging our eyes and matting our hair. The noise it made was not great – a steady blast like the clearing of titanic lungs rather than a series of deafening explosions.

We landed and walked along the beach of black volcanic sand to find the best view and then trudged a hundred yards or so inland until we could see where huge lumps of lava were falling heavily into the ash directly in front of us. Hugh set up the tripod, Maurice took out his camera and I started to think of the words that I might say. Maurice looked apprehensively at the sky. It was starting to drizzle. The raindrops would certainly be loaded with glass-sharp particles of ash. If just one of those got on the lens or in the magazines, we could be in trouble. Maurice very sensibly took the camera off the tripod, put it away in its metal box and waited for the drizzle to stop.

But it didn't. It turned to rain, falling gently at first and then, within minutes, drenchingly. There was nowhere to shelter – no building, not even a tree or a bush. All we could do was to stand out on the open ash plain and submit to being soaked to the skin.

After ten minutes or so, the rain eased. Then it stopped. Maurice got out his camera again. He framed up a picture and started to check focus and exposure. And once again it started to drizzle. The drizzle turned to rain, and Maurice was forced to put his camera back in its box.

By the time this had happened four times I began to suspect that Anak Krakatau was creating its own microclimate with its own repetitive cycle. Maybe the immense electrical discharges with which it was generating the lightning in the sky above, were a factor. Whether that was so or not, the time between the ending of the heavy rain and the next onset of drizzle was almost exactly the same as that which Maurice needed to set up his camera and prepare to film. By three o'clock in the morning, we were still standing on the ash plain, sopping wet, and there was no sign of the cycle changing or stopping. Richard decided we should give up. Miserably we returned to our boat.

There things were hardly any better. The only dry corners on board had already been claimed by the crew, who were curled up in them. All we could do was to lie on the open deck while the rain pelted down on us and swilled across the planks beneath us. The three hours to dawn seemed very long indeed.

As the sun came up, the rain at last stopped. Anak Krakatau was still squirting lava, but in the cold light of dawn it was no longer the spectacular display it had been in the darkness of night. We too had lost our fizz. Wretchedly cold, with our wet clothes clinging to our chilled bodies, we sat miserably on the deck. Richard tried to rally us. The sequence might not be quite the dazzling pyrotechnic spectacle he had imagined, but the story of the eruption was nonetheless a tremendously exciting one, and he was sure I would be able to produce words that would compensate for the lack of dramatic pictures.

I did my best to work out a piece in my head. The trouble with the Krakatau story, from a presenter's point of view, was that it was full of numerals: the year in which the eruption happened (1883), the size of the island that was destroyed (7 kilometres long and 5 kilometres wide), the amount of rock that was ejected (15 cubic kilometres) and the area over which it was spread (4 million square kilometres), the distance that boats were carried inland (2 kilometres), the number of casualties (36,000), the distances away that the detonation was heard (3224 kilometres away in South Australia where it woke sleeping people; 4811 kilometres away in Rodriguez Island where the noise was thought to be gunfire). All these were valuable elements in the story. But my short-term memory for numbers is poor. Nor was I feeling at my most mentally acute. Get one figure wrong, and the spoken piece would be ruined, and would have to be done again. I decided that the only solution was to chalk the numerals concerned on various parts of the ship in places where I, but not the camera, could see them. So with Dickie recording my words, and Maurice filming while walking slowly backwards in front of me across the deck guided by Hugh with his arms round Maurice's waist, I meandered around the tiny boat glancing apparently abstractedly at the mast, the side of the dinghy, the boom and finally a bucket on the deck. The story took nearly two minutes to tell. That is a very long time in television terms for someone to talk in close-up unrelieved by any other picture, but it seemed to hold the attention and I didn't get one date or dimension wrong.

I reckoned I had won my spurs as a presenter.

* * *

Richard had other plans for making the most of a presenter in his series. As well as conveying basic facts and figures about a place or its history, he wanted me to describe in words my subjective reactions to some of the exciting places he planned to take me. None were more dramatic than the caves of Borneo.

I had some idea of what was in store for me because, when I had been running the Travel and Exploration Unit, I had overseen the production of a series of films made by the eccentric anthropologist and war hero, Tom Harrisson, who then was Curator of the Sarawak Museum. One of his episodes, which subsequently won several prizes, dealt with the extraordinary harvest of birds' nests from the great Bornean cave of Niah.

Cave swiftlets, tiny birds somewhat smaller than sparrows, build their nests in this cave. Clinging to the slightest roughness in the limestone walls and ceilings, they produce a sticky saliva which they plaster on to the rock. Over a period of several days they build this into a tiny cradle-like ledge on which they deposit two eggs. Chinese gourmets consider that these nests make the most delectable of soups and are prepared to pay a great deal of money for them. So every season men climb rickety bamboo poles and risk their lives, two hundred feet above the cave floor, collecting these nests.

Since Niah had already b een filmed, Richard had decided to go to another nest-producing cave, Gomantong, on the other s ide of Borneo, in Sabah. Here the technique used f or h arvesting the nests is slightly differ-ent. Instead of climbing up poles, the nest-collectors walk up the lime-stone escarpment above the cave entrance and get into the main chamber through a h ole in its ceiling. From this, many years ago, someone had hung a long flexible ladder made of rattan cut in the surrounding forest. A rigid ladder swivelling from it, like the jib arm of a crane, enables the col-lectors to get to most parts of the roof.

Watching and filming this was dramatic enough, but Richard had a surprise for his presenter. He told me that I was not to go deeper into the cave until he was ready. Then with Maurice filming me as I walked, I had to advance beyond the chamber occupied by the swiftlets, farther into the cave where something else would await me to which I should respond in words. Being primed to react with eloquence to a surprise of some kind is enough to freeze the words on anyone's lips, but as I rounded a corner in a passage that led from the swiftlets' cave deeper into the mountain, I could not have asked for a greater astonishment.

At the far end of the chamber ahead of me, lit theatrically by a shaft of brilliant sunlight shining down from a hole in the ceiling, I saw a golden dune a hundred and fifty feet high. It rose from a broad base on the cave floor to within a few feet of the ceiling. Its surface, unaccountably, seemed to be shimmering. It was not until I was within a few yards of it that I realised what I was looking at. The substance of the dune was the accumulated droppings of an immense colony of bats that covered the entire ceiling two hundred feet above, and the shimmer of its surface was created by a moving, glinting, slithering carpet of cockroaches. They had the rich handsome colour of newly opened horse chestnuts. They were stolidly eating their way upwards through the bat guano. But they also consumed anything else that lay in their path. Here and there lay the bodies of bats that had fallen from the roof. The cockroaches were swarming all over them. One I saw fall. It seemed to have an injured wing and was still alive. The cockroaches immediately scurried over that too and within a few seconds it had disappeared beneath the chestnut blanket. But cockroaches are themselves mortal. Black sexton beetles were clambering among them seeking their dead bodies and munching their way through them. On one side, the dune flowed over a scree of tumbled boulders. These were shrouded with silver webs. Huge spiders crouched beside each, ready to pounce on any cockroaches or beetles that might stray away from the surface of the dune.

The bat droppings were only slightly moist and did not cohere, so the top layers of the dune were slipping slowly downwards. I wondered where it might be going and saw that the whole mass was slowly sliding through a hole in the floor several yards across, like sand flowing through the waist of an hourglass, into a chamber beneath. Presumably, the droppings of centuries must be accumulating in some lower cave system. I kept well away from the hole.

I did my best to describe the elements of this astonishing scene to camera. I marvelled at the interconnectedness of nature – the way each evening the bats would fly out across the canopy of the surrounding forest, and return to the cave with cumulatively several tons of insects in their stomachs, and how their droppings still contained sufficient nutriment to sustain the carpet of cockroaches. I pointed out that the cockroaches in turn were consumed by beetles whose own droppings and ultimately bodies also contributed to the dune; and I described how the resultant substance, known politely as guano, was then carried away in sacks by the local people and used as fertiliser for their cultivated fields. That in turn produced plants growing outside the cave on which caterpillars could feed before they turned into flying insects and were caught, once more, by bats and their substance returned to the cave.

But Richard wanted more. At his behest, Maurice and Hugh, Dickie and I started to climb the dune, kicking our way up it, as one kicks steps in a steep snow-field. My boots sank in a good twelve inches, but neither they nor my trousers seemed to have any particular appeal to the cockroaches, and having them slither around my legs caused me no particular problem. The smell from the droppings in the lower part of the dune, where it had been processed by the cockroaches, was somewhat sweet – a little sickening, but not intolerable. As we climbed higher, however, and the droppings were fresher, the fumes became somewhat ammoniacal and I began to wonder if I would manage to talk at any length once we got to the top.

The summit of the dune was within twenty feet of the cave roof and from it I had a clear view of the bats. There were two kinds. Large ones hung in dense ranks from the ceiling like stalked black fruits, and twisted round agitatedly, to look at us as we shone our torches on them. The second kind were as small as mice. They occupied another part of the limestone ceiling that was deeply pockmarked and honeycombed. As our torch beams played over them they scuttled around in their tiny vertical shafts in a way I found quite unpleasant.

We assembled ourselves. I had to say something to camera which would evoke the atmosphere of this bizarre place. Our torches were far too weak for us to film by, but Hugh was carrying a powerful battery light. Maurice asked him to switch it on. Immediately, the whole population of bats peeled off the ceiling and took to flight. At first there was great confusion with animals rocketing in all directions but within a minute they were all flying in the same direction round and round in a circle forming a great vortex just above our heads. The air was filled with the noise of their leathery wings and their squeaking calls. We could feel the heat from their bodies and I was almost choking from the ammoniacal fumes that came from their all too fresh droppings around my feet.

'Running,' said Maurice, with his camera to his eye. 'Say something.'

'Some people,' I shouted, 'are said to dislike bats because they fear they might get entangled in their hair. Of course, there is no danger of that, because bats have an amazing navigational system, a kind of echo-location based on supersonic calls, far beyond the range of the human ear. It is one of nature's marvels. As you can see, none of these tens of thousands of bats are even colliding with one another, let alone me.'

That was about as much as I could manage before I choked.

Maurice cut the camera.

Hugh turned off the battery light.

And a huge bat crashed straight into my face.

* * *

Our main work was in the rain forest. We decided to devote one programme to the animals that live in the airy world of dangling lianas and pillar-like tree trunks that lies between the leafy sunlit canopy and the dark forest floor. Here flight is obviously an invaluable asset and a number of reptiles and mammals have evolved ways of getting around by air.

Most squirrels, of course, are capable of impressive leaps, but in Borneo, as in North America, some manage to extend their range by gliding. The North American species are only a few inches long, but the Borneo flying squirrels are giants, nearly three feet long from nose to the tip of the long tail and a rich brown colour. They have a flap of furry skin that stretches from wrist to ankle. When they launch themselves into the air, they stretch out all four legs and, using their tail as a rudder, they sail through the air for as much as a hundred yards. We found a pair that lived in a hole seventy feet up in the trunk of a giant *Koompassia* tree. Each evening, they emerged to feed but before they left, they apparently played games with their neighbours, chasing one another around a rectangular aerial circuit, scampering in their billowing cloaks along horizontal branches, then leaping off to glide down to a distant trunk, galloping vertically up it to regain their original height and then racing out along another branch to glide back to the tree in which they started. One evening, after giving us a particularly delightful display, they provided a spectacular climax to their performance. Three of them followed one another up a particularly tall tree and launched themselves in succession off its topmost branch. The last did so at the run so that all three were in the air at the same time until each executed a neat wheel in a different direction to land on its own tree. It was like the gala ending of a display at an air show.

Lizards too glided between the branches. They are small creatures barely six inches long with greatly elongated ribs. Sitting on a branch with their long ribs lying alongside their flanks, they look unremarkable. But every now and then in order to pursue an insect or to challenge a rival, they leap off the branch and pull their ribs forwards and outwards to reveal a flap of brightly coloured skin on either side. There is also a species of frog that can glide. It has membranes between its long toes so that it is able to sail through the air as though standing on four small parachutes.

We found and filmed examples of both these remarkable flyers, but the most surprising of our cast of aeronauts was a snake. The Paradise Flying Snake certainly spends most of its life in trees, climbing swiftly and skilfully, using the transverse scales of its underside to give it purchase on the bark of a tree. But naturalists we spoke to, both in Borneo and Britain, doubted whether it could truly fly. It does not have elongated ribs like the flying lizard nor any other visible feature that might serve as an aerofoil. The best a flying snake could do, sceptical naturalists said, was simply to drop out of a tree and then to wriggle across the ground to climb up another.

Clearly we were unlikely to find the truth by wandering through the forest hoping that a small snake might suddenly and unexpectedly descend upon us from the branches above. Even if one did so, there was no way in which we could have a camera, framed and focused and ready to record the exact manner in which it travelled. We should have to find one of these reputed aeronauts and somehow persuade it to take to the air.

The first part of this was easily arranged. We were staying in a government rest house on an agricultural station. The local men knew the snake very well and soon caught one for us. It was a beautiful little creature, about a foot long and scarcely thicker than my little finger, with green-blue scales flecked with gold and red. But how could we encourage it to fly?

Richard had an idea. The station had a water tower, about fifty feet high standing in the middle of a neatly clipped lawn. We climbed up to the platform at the top and tied a leafy branch to the safety rail that ran round it. Maurice, standing on the lawn below with his camera on a tripod, carefully focused one of his longest lenses on the tip of the branch. I took the snake in a cloth bag up to the top and put it on the branch. It moved cautiously through the twigs and curled up at the far end among the leaves. I shook the branch gently. The snake curled up a little tighter. I gave it a delicate prod with a stick and it fell off the end, dropping like a stone. Being so small and light, it landed on the grass without trouble and started to slither away, but it was immediately retrieved by one of the crowd of local workers who had gathered around the tower to watch the goings on. No one could have described its descent as a flight. Maybe the sceptics were right.

Richard's view was that even if the snake were able to fly it would probably prefer to do so from – so to speak – a running take-off. He had to devise a way to enable it to get that. He found a coil of transparent plastic tubing in the station's stores. Maybe that would help. We took a yard and a half of this up the tower and replaced the branch with it. Holding the snake by its tail, I dangled it above the opening of one end and eventually managed to persuade it to put its head into the tube. It wriggled down inside with some enthusiasm and since the tubing was transparent I could see just how it was getting on. As the snake wriggled swiftly towards the other end, I called down to Maurice telling him to start filming. The snake arrived at the far end, stuck its head out for a couple of inches – and hesitated. I put my lips to my end of the tube and blew what I fancy was a low E flat. The snake shot out, fell steeply downwards for about fifteen feet and then started to glide forwards, skimming through the air like a frisbee.

Watching Maurice's film later, we were able to see exactly how it did so. It pulled its ribs forwards and outwards so that its circular body became a flat ribbon, with a slightly cupped underside. At the same time it drew itself into a number of S-shaped coils with each bend nearly touching the flank of the other, so that its silhouette changed from being long and thin to short and wide. As a consequence, it managed to catch the air beneath it very effectively indeed. If a glider can be said to fly, then that snake was a flying snake.

It landed almost daintily on the clipped lawn to the accompaniment of cheers from the assembled crowd. Before it could wriggle away, someone picked it up by the back of its neck, droppedit in a bag and sent it back up to me. Maurice wanted another shot on a different lens.

Once again I dangled it over the top of the tube and dropped it in. I blew my trombone note and once again the snake came out of the farther end at speed to sail gracefully through the air. We tried a third time. By now, however, the snake had apparently made an accurate assessment of its surroundings, for this time instead of gliding down towards its cheering admirers, it executed a sharp left-hand turn and headed straight for a thick clump of bamboo. It could not only glide, it could steer. It landed within a few feet of the bamboo and within seconds was lost within it. Even if we had been able to recapture it, we would not have done so. We felt it had earned its freedom.

* * *

The flying snake was not our only coup. We managed to film several other animals that, as far as Richard and I knew, had never been filmed before.

In a coconut plantation, we found another accomplished glider, a colugo. This is a mystery animal. Zoologists don't know how to categorise it. It is sometimes called a flying lemur – but it is certainly not a member of the lemur family. Some suggest that it may be very distantly related to the fruit bats, others that it is a remote cousin of the tree shrews. It is about the size of a large cat and has the most extensive cloak of furry skin possessed by any mammal. This stretches not only to the tips of its fingers and toes but up to the sides of its neck and back to the end of its long tail. Colugos are said to be able to glide a hundred and fifty yards but we were not able to put this to the test for the coconut plantation had all its trees neatly and uniformly planted within a short distance of one another. The colugo's flight, however, was undoubtedly impressive.

One night, searching in the forest with our torches, we caught the red eye-shine of a tarsier, a tiny primitive primate the size of a small bush baby. Its eyes are so gigantic that it cannot move them in their sockets so that if it wishes to glance to one side, it has to swivel its whole head. We caught it and filmed it grabbing insects and ferociously ripping them to pieces with its needle teeth.

Best of all, we filmed a troop of proboscis monkeys feeding in the mangroves and with the help of the latest lenses managed to do so much more successfully than Charles Lagus and I had done when we had worked in Borneo on our way to the Komodo Dragon seven years earlier.

* * *

At the end of April, after three months in the field, the rest of the team went back to Britain. I, however, stayed a few days longer. In my new life as a freelance, I had accepted a commission to write an essay about a short nature walk of my own choice. I chose to climb Mount Kinabalu.

The mountain stands thirty miles inland from the north-west coast of Borneo, not far from the Malaysian city of Kota Kinabalu. Seen from there, it looks like a huge rectangular battlement crested by minarets and turrets. I did not select it because of the animals I might find, even though it has some specialities that live almost nowhere else. I had already seen a great deal of Borneo's mammals. In fact, some of Kinabalu's mammalian inhabitants, I hoped not to see at all. It is home to six species of rats including, unhappily as far as I am concerned, one that was described in an official guide as being particularly bold and the biggest of its tribe, with a body, discounting its tail, nearly a foot in length. Nor did I go hoping to see some spectacular bird that I might have missed in the lowland forests, for the mountain's birds are, for the most part, small and undramatic. Had I been a more knowledgeable botanist, I might have made the climb to see its plants, for it has several hundred species of orchids and seven of that extraordinary group of carnivorous plants, the pitcher plants. One of these is *Nepenthes rajah*, a giant which is said to produce a pitcher so monstrous that occasionally it succeeds in drowning one of the mountain's rats. My interest in the mountain was neither botanical nor zoological. I wanted to climb up it because of its geology.

Kinabalu is a young mountain. It rose a mere seven million years ago as a gigantic hump of molten granite, welling up from deep in the earth's crust, lifting the mudstones and shales that constitute this part of Borneo. That rocky overburden has now been eroded away exposing the bare solidified granite. Nonetheless the mountain is still nearly thirteen and a half thousand feet high, the tallest peak between the Himalayas away to the west and the snow-covered mountains of New Guinea to the east. According to geologists it is still rising by a few millimetres each year.

I plodded up through the thick rain-drenched bush that covers its lower flanks for a day and spent the night in a hut that had been erected some years earlier by a team of British scientists making a biological survey of the mountain. The next morning, I left the hut at three o'clock and clambered in the darkness through low bush and up steep rock gullies. An hour or so before dawn I reached the summit plateau.

I had never seen anything remotely like it. The rock beneath my feet was bare of any vegetation, except for a few dwarfed plants that here and there had managed to insert their roots into tiny cracks. Ahead, black against the paling night sky, rose two extraordinary pinnacles that have been given the undignified but nonetheless accurately descriptive name of Donkey's Ears. The tropical crescent moon, lying on its side, was hammocked between them. On the other side, the plateau rose gently into a peak, its highest point. It was as though some titanic rasp had filed and chamfered the rock into these extraordinary shapes, smoothing off any awkward corners. And indeed that was so. The mountain is so high that not so long ago an ice cap had formed on its summit. Even now a thin crust of ice occasionally glazes the rock pools. Looking more closely at the granite, I could see here and there long parallel grooves scored across it. They had been made by small rocks frozen fast to the underside of the ice cap and dragged over the granite as the ice slipped slowly downwards under its own weight. The granite itself was beautiful, bespangled with small crystals of black hornblende and studded with much larger white crystals of felspar. In places differential erosion had left these huge crystals standing slightly proud of the surface, giving me a reassuringly firm foothold as I walked over them. In one place I found a dark brown patch a foot across. This was evidence that not much of the original rind of the granite mass had been eroded away for it was the metamorphosed fragment of the shales that had once overlain it.

I sat down, ate a bar of chocolate and drank deeply from the clear pool of rainwater that had formed in a dimple in the granite. Far away, across the dark carpet of rain forest flecked with tatters of mist, I could see the distant lights of Kota Kinabalu, outlining the coast. Clouds began to race up the vast rock flanks of the mountain. The glancing rays of the rising sun struck the tips of the pinnacles and moved slowly downwards until they streaked across the wet flanks of the plateau making them glisten like mercury.

If I had ever had any doubts about abandoning a career that would keep me sitting behind a desk in London, I lost them there.

19

The Tribal Eye

During the years that I had been making the *Zoo Quest* series, I had inevitably accumulated souvenirs. The helmet mask I had been given by the chief in Sierra Leone had started the habit. In Guyana I had made a small collection of the bead aprons traditionally worn by Amerindian women. In Borneo, on my way to the Komodo Dragon, I had acquired an alarming wooden mask with hornbill feathers stuck in its top and small circular trade mirrors as eyes. In Australia I had assembled a dozen or so Aboriginal bark paintings as well as Yurlunggur, the spectacular gigantic didgeridoo. As more and more such things accumulated on my shelves in Richmond, I began to realise that they were not merely mementoes. They had other less trivial qualities and I began to look for similar objects in galleries and auction rooms. I had become a collector of tribal art.

London did not have as many dealers as Paris, Brussels or Amsterdam, but there were one or two if you knew where to look. Herbert Rieser was one of the least pretentious. He had a small gallery in a back street not far from Marble Arch. In his youth he had been tall and lean, but age had crumpled him and he walked with a kind of loping stoop. His primary affection lay with African objects, which was hardly surprising for he was born in South Africa of German parents, but his knowledge was comprehensive and his eye for the qualities of an object was as keen and as perceptive as anyone's. There was no telling what you might find on his shelves.

His customers tended to treat his gallery as though it were a small and exclusive club. Herbert would be sitting hunched at the back, wreathed in Gauloise smoke, handing out glasses of whisky from an apparently inexhaustible bottle he kept beneath his chair, while two or three of his customers stood around debating the quality, origin or function of some new piece he had just acquired. Innocent strangers who might occasionally poke their noses in merely to ask the price of an object in the window were likely to retreat in bafflement.

I dropped in one day to find Herbert looking rather more lugubrious than usual. 'Big trouble,' he said, and he pulled aside a small curtain that screened an alcove behind his chair. There I saw a spectacular sculpture of two snouted animal-headed figures, male and female, seated on a bench with their arms on one another's shoulders. I recognised its style as that of the Senufo people of Mali. Why should such a thing be trouble? Was it a fake? I thought it was magnificent.

'Yes, of course it's magnificent, my dear,' said Herbert. 'It's just been brought in by my best African runner. I had to buy it. If I hadn't he would start going to other people first. So I've given him a cheque. But it's going to bounce. I don't have the money.'

I asked him how much it had cost. It was far beyond anything that I had ever paid for a tribal object. On the other hand, I did have the money, for being a freelance, I had put some aside to meet a tax bill at the end of the financial year. And the Senufo figure was a really superlative piece. So we came to an agreement. I would give him a cheque which would cover the one he had given his runner and also give him a reasonable profit. Then I would take the figure, provided he gave me a note promising to take the figure back and return my money if at the end of the financial year I found I couldn't pay the taxman.

The end of the year came. I managed to pay the taxman. I kept the figure. I had, I realised, become a *serious* collector.

Immediately after I resigned my job as Director of Programmes, I suggested to Stephen Hearst, the Head of the Arts Department, that there really ought to be room on television screens for totem poles and masks as well as old master drawings and post-impressionist paintings. Perhaps I might make a series about tribal art to set alongside *Civilisation*. It would be shorter and it certainly would not have the mandarin authority of a Kenneth Clark series. But it would take viewers to exciting places and show them intriguing things. Now that I was a freelance, I was able to suggest that I might both write and present it. Stephen was equally free to turn it down, but he didn't. *Tribal Eye* was commissioned.

He allocated two experienced directors to the series. One, David Collison, I knew well for he had been one of the main producers of archaeological programmes on BBC2. He was a thorough-going television professional, bonhomous, practical and an excellent leader of a team. The other, Mike McIntyre, I hadn't met before but Stephen said he had great visual imagination and an unquenchable appetite for the exotic. We were joined by a brilliant young film editor, Anna Benson Gyles, who was wanting to leave the cutting room and tackle other aspects of film-making. She would help with research and also try her hand at directing. All of us, I suspect, for differing reasons, were making major shifts in our careers and had a sense of release and excitement at moving into new fields.

David, nominally in charge, decided that the four of us should travel together round the capitals of Europe – Paris, Brussels, Berlin and Vienna – to survey the range of material that the series might cover. For two weeks, we spent the days looking at wonderful objects in the galleries and storage rooms of great museums, and the evenings sitting in restaurants debating how we might compile a series that visited all the inhabited continents of the world and surveyed a full range of the techniques that tribal people use to create such things.

I had the idea that, early in the series, we might include a sequence investigating whether the qualities that European connoisseurs admired and wrote about with such erudition and confidence, bore any relation to those that tribal people perceived or valued in their objects. We decided to do so in the first programme that was to be devoted to the Dogon. These people live in desert country just south of the western Sahara in Mali and carve in an austere sometimes almost abstract style. Few African sculptures have received as much aesthetic praise or have fetched such phenomenally high prices at auction as theirs. The obvious thing to do was to assemble a selection of Dogon pieces that, by European estimates, were good, bad and indifferent and then show them to a group of Dogon people to see if they rated them in the same sort of way. But there was a problem. No European museum or collector would be likely to allow us to take their Dogon masterpieces back to Mali; and while it would be easy enough to find rubbish there, it would be next to impossible to find superb objects. They fetched such enormous prices in Europe that virtually all of them had been bought up by traders long since and were now sitting artfully lit and elegantly mounted in museums or on collectors' shelves. I decided to solve that problem by showing the Dogon not the objects themselves but photographs of some that had appeared over the past few years in London's sale-rooms. I would find out how they rated them and compare their evaluation with European opinion as expressed by the prices that they had fetched at auction.

However, when we arrived in Bamako, Mali's capital, we heard of a wealthy dealer there who was said to have a superb private collection of Dogon sculpture. We went to see him and he offered to lend us a small selection including one of his most prized masterpieces, a monumental carving of a horse and rider. He insisted, naturally enough, that they should be properly insured, though I was astounded at the high valuations he put on them, and in particular on his horse and rider. It also occurred to me that it would not harm his business to have some of his objects given a little publicity.

The road to Dogonland, in those days, was as rough a one as you could find in Africa. Our truck rattled over boulders, bucketed into deep dust-filled pits and shook over corrugations until our teeth were loosened in their sockets. We had done our best to cushion the horse and rider by wrapping it in blankets. All the same, as we hit yet another unseen boulder or careered out of a deep rut, I was fearful for its safety.

After a couple of days we reached the little Dogon village of Sangha which we planned to make our base. David Collison and I unwrapped the horse and rider with great care and some trepidation. I had been right to be worried. There was a crack around the base of its neck.

The crack was not immediately obvious for the figure was covered with a congealed crust of the millet gruel and chicken blood that the Dogon pour over sacred figures during their ceremonials. I had learned from Herbert Rieser what to do in such circumstances. You must separate the two pieces so that their edges don't grind together and chip. That would make it harder to achieve an invisible repair. So I tried to ease them gently apart. The neck wobbled but remained firmly attached. That was mystifying. David held the body, I held the head and we pulled. The crack widened. Inside, I could see nails – bright, shiny, unrusted nails. We pulled more uninhibitedly. With a prolonged squeak, the crack widened to half an inch.

Inside we could see that the wood was not the hard well-seasoned timber that such sculptures should be made from, but new, roughly sawn softwood. The neck had been crudely nailed to the base and the join concealed beneath a crust of libations, doubtless newly applied and given some appearance of age by being gently baked in an oven. Our horse and rider was a fake.

Nonetheless, I went ahead with my investigation. I showed an assembled group of Sangha villagers my photographs and asked them to put them in order of quality with the best – whatever that might mean – at one end, and the worst at the other. My halting French was translated into Dogon by one of the younger men who had spent some time working in Bamako. Whether my sample group of villagers understood what I was trying to ask, I do not know. In their answers, however, they spoke not of beauty or expressiveness but of correctness. This mask had an ear that should not be carved that way. That figure looked odd but it came from another part of Dogonland so they would rather not comment. An old and impressive mask, of a kind known as a *kanaga*, which had fetched a very good price in Paris, they thought little of. Every young man, they explained, carved his own *kanaga* which he would wear during his initiation ceremony. Why would anyone else want it? One old man was particularly taken by a photograph of a small figure. He stared at it fixedly. I asked why he liked it so much. He had never seen it before, he said, but he recognised its style. It had been made by a sculptor he had known well but who was long since dead. Looking at it was like hearing a voice from the past.

I showed them the objects we had brought from the Bamako dealer. They dealt with them in the same sort of way. None had come from Sangha. One or two were from distant villages and were 'correct' as far as they could see. The horse and rider with its wobbling head attached by only four nails was simply discounted. None was singled out because it was deemed to be particularly beautiful. Beauty, it seemed, was only one of a whole cluster of qualities they looked for and it was by no means necessarily the most important.

It occurred to me that although the formal aesthetic qualities of tribal objects are those most spoken and written about by European critics and collectors, in practice we too value objects for a whole cluster of disparate qualities. In addition to beauty, we seek rarity and antiquity. We value a rich subtle patina that comes from long use and are particularly pleased if a figure has an interesting history or was once owned by a famous collector. Such qualities contribute to the pleasure that European collectors derive from a tribal sculpture but have no relevance whatever in the societies from which the objects came.

Back in Bamako, we returned the horse and rider to its owner. I naturally apologised for the damage. He looked at the raw wood exposed in the neck, and the glittering, shiny nails. 'Don't worry,' he said, 'it is easily fixed,' and he hurriedly took it away. But he didn't suggest that he would claim the huge sum that theoretically was due to him from the insurance.

* * *

European taste for works of art from other cultures is a comparatively recent phenomenon. The Spaniards who invaded the New World in the middle of the sixteenth century had little or none. The people they conquered, the Aztec and the Inca, worked gold with an unparalleled skill and produced a vast range of wonderful objects – pendants, gorgets, necklaces, earrings and nose rings, beakers, masks, all made of gold. The Spaniards seized every golden piece they could find and melted them all down. One jewel, an articulated silver fish inlaid with gold, was such an obvious wonder and made with such extraordinary skill that it was sent back to Europe undamaged as a gift to the Pope. Benvenuto Cellini, the greatest goldsmith of the European Renaissance who was credited by his contemporaries with skills that were virtually supernatural, saw it and declared that he had no idea how it had been made and that he could certainly not match it. Of all those wonders, virtually nothing remains – except the metal itself which must still exist within the gold ingots lying locked away in the basement of national treasuries.

But between the great empires of the Aztecs in Mexico and the Incas in Peru, there lay a whole group of small chiefdoms. The Spanish found them very hard to deal with. They couldn't be quelled simply by seizing one overall ruler, as had happened both in the north and the south. They had to be beaten into submission one by one. And there was another difference. In these smaller societies, gold was not reserved for priests and aristocrats. Anyone could own it, and most people did. Many, in fact, took it with them to the grave. So today in Panama, northern Venezuela and Colombia, magnificent golden jewels and miniature sculptures of ravishing beauty, are still being dug from the ground.

Tribal Eye, to be reasonably comprehensive, had to include something about the great accomplishments of the New World goldsmiths, so as well as showing some of the very few Aztec and Inca jewels that survive, we decided to try and film grave-robbers at work.

* * *

Grave-robbing, of course, is illegal but it has been going on in Middle and South America for centuries and is widely accepted among the local people themselves as a perfectly respectable profession. Indeed it has its own code of practice comparable, we were told, to the rules of a trade union. A tract of land that might contain an ancient cemetery may be bought and sold with that possibility in mind and digging for its gold is simply regarded by the landowner as a rather specialised and particularly rewarding way of exploiting mineral rights.

In Bogota, we had little difficulty in arranging to film the *guaceros*, as the grave-robbers are called. We met one of their representatives in a market place in Bogota and were put in the back of a battered Land Rover. Black cloth was taped over its windows. We would not be allowed to know exactly where they were working, not because of worries about the legality of what they were doing, but because they did not wish others to know in case rivals tried to take a share of the crop. We had no idea in what direction we were being taken, but it was clear that we were going up hill. After an hour or so, the truck stopped. The back doors were opened and we clambered out.

We were on an open hillside. It was cold and misty. We were taken to a thatched hut from which blared the sound of a transistor radio playing South American pop music at full volume. Inside we were introduced to a small black-haired sallow-faced man, lounging in a chair with a revolver on his hip and a rifle across his knees. He poured us glasses of neat Scotch malt whisky and we drank his health – effusively. We were welcome to look around, he said.

The open moorland behind the hut was cratered with pits, like shell-holes on a battlefield. Digging was still going on in a dozen or so. To anyone who respected the painstaking procedures of academic archaeology, the sight was certainly a very dismal one. The diggers were interested in only one thing – gold – and in finding it as quickly as possible. We filmed one man working in a pit that was already five feet deep. He had just discovered pottery – fragments of a burial urn. Whether it had been broken by the weight of earth above or whether his spade had smashed it, we could not tell but we watched him pick out fragments of it and pile them beside the pit. He uncovered a skull. As he lifted it, its jaw dropped loose. He cradled it with his hand and set it beside the broken pottery. A few minutes later, he uncovered a stone axe blade. That too was put beside the hole. It was worth little and was certainly not what he was seeking.

He then laid aside his shovel and started to flick away grains of earth with a twig. He had glimpsed a glint of gold. Careful probing established the dimensions of the buried object. He carefully loosened the earth and after a minute or so exposed a small lip plug in the shape of a miniature parrot. He spat on it and rubbed it with his sleeve, cleaning off the last particles of soil. The gold, uncorroded, still shone brightly even after five hundred years underground. He handed it out to one of the men and it was taken back to the hut.

We moved from pit to pit, watching and filming. By now it was early evening. The whisky-drinking man from the hut came down to tell us that the truck was waiting to take us back to Bogota. We went back to where we had piled our gear. Mike picked up his jacket. It was plain, from the way he was searching through his pockets, that he had lost something. The whisky drinker asked if anything was wrong. Mike confessed that he had carelessly left some money in his jacket pocket and now it was gone.

'How much?' the whisky drinker asked

'Two hundred dollars,' said Mike, abashed.

The whisky drinker took out a wad of grubby dollar bills from his back pocket, peeled off two hundred dollars and handed them to Mike.

'No, no,' said Mike, 'it was my fault.'

'Don' worry, señor,' said the whisky drinker grimly, 'I will get it back.'

* * *

The most accomplished of all the Colombian goldsmiths were the Tairona. They could cast semi-circular noserings in filigree gold so fine that it looks like lace. They knew how to treat gold alloyed with copper in such a way that the copper on its surface was selectively etched away leaving a microscopic layer of pure gold that could then be burnished so that the casting looked like solid gold but was, in fact, merely gilded. One of their favourite subjects for a pendant was the image of a warrior chief no more than two inches high wearing a wealth of personal jewellery, all modelled in exquisite detail, and from such figures we have an accurate idea of how they wore their splendid regalia.

Once the Tairona had lived in large towns with paved roads and complex buildings. They had resisted the Spaniards for longer and with more bravery and determination than any of the other local people. The Spaniards ruthlessly destroyed their towns and executed their leaders. Even though their lands were occupied, the Tairona rebelled again and again, but eventually all organised resistance was crushed. Only a few families managed to escape to the mountains farther north, the Sierra Nevada de Santa Marta. There their descendants still survive. They are known today as the Ika.

The Ika, we thought, might still give us some idea of how their gold-working ancestors had lived. They do not, however, welcome outsiders. Who can blame them, knowing their history, but once again we managed to find the right contacts. We drove to one of the small towns in the southern foothills of the Sierra Nevada, but thereafter we had to travel on horseback. After two days we reached a deserted monastery. Christian monks from Spain had settled here in the late nineteenth century, preaching peace and forgiveness to the people their predecessors had treated with such brutality. They allocated an extensive plot to be used as a cemetery so that their converts could, after death, rest in holy ground. We went to visit it. It was empty but for half a dozen graves clustered in one corner. The headstones revealed that they contained the bodies of monks who had died here. Not a single Ika lay among them. After several decades of fruitless preaching, the Christian Fathers had decided that the Ika were unconvertible and had left.

The Ika lived in a village nearby. They were a quiet undemonstrative people. Both men and women wore their hair down to their shoulders. All were barefoot. Their dress was still strictly and uniformly traditional – white long-sleeved cloaks over baggy trousers with dome-shaped hats. Every man carried a *mochila*, a striped woven shoulder bag in which he kept dried coca leaves and a *poporo*, a small gourd full of powdered lime. Coca leaves and lime, when chewed together, produce a juice which has a mildly anaesthetic effect, numbing the pains of cold or hunger. Most of the men chewed incessantly and even if they were not doing so actively, they kept a wad of coca in their cheek.

The village had a sacred enclosure planted, they said, with an example of all the plants that grow in Ika country. It was a symbol of the whole world. In the centre stood two dome-shaped window-less thatched houses, one for men and one for women. A path in the shape of a phallus, paved with pebbles, led to the door of the men's house. Within it was dark, for the only light came from a small circular hole in the roof through which escaped the smoke from a small fire smouldering in the centre of the earth floor. In the gloom just beyond its glow, several men sat staring at the embers, slowly and meditatively chewing coca. Beyond, another man sat at a loom weaving cactus fibre thread into the white cloth the Ika use for their costumes. They spoke quietly in low voices. The only other noise was the tapping of the sticks on the sides of the *poporos* as the men added another small quantity of lime to the coca leaves they chewed. I wondered if, when their *poporos* had been of solid gold as they once certainly were, the atmosphere would have been so subdued. There are tales that every now and then, the Ika suddenly vanish from their villages. They have gone to a secluded valley away in the mountains. Sentinels are posted to make sure that no strangers approach undetected. Then, it is said, the priests appear in front of their people dancing in magnificent masks of solid gold. If that is true, no outsider has ever seen it.

* * *

Africa too has its virtuoso metal casters. The most famous are those in the city of Benin. Bronzes cast in that city arrived in Europe during the nineteenth century. They at least, unlike the golden objects from the New World, were not melted down. Perhaps the mere fact that they were not made of a precious metal, saved them. At any rate, they were recognised as works of art and treated as such. Indeed, so accomplished were they that some blinkered Europeans could only account for them by claiming that their makers must have been taught their craft by Europeans, perhaps by the Portuguese. *Tribal Eye* could not ignore them either.

The bronzes also came to Europe as the result of military conquest. At the end of the nineteenth century, Britain had established trading posts along the coast of West Africa. One of them, at Sapele in the delta of the Niger River, had a trading agreement with the Oba, the ruler of Benin, to supply palm nuts, palm oil, ivory and incense. One clause of the agreement stipulated that the Oba would stop human sacrifice, which he had been practising on an appalling scale. The Oba, however, regularly disregarded all agreements including that concerning human sacrifice. In 1897 a young newly-appointed British Vice-Consul decided to travel to the city and demand that the agreements should be kept. When he and his party were a few miles from the city, they walked into an ambush set by warriors from Benin. Eight Europeans and two hundred and forty African porters were slaughtered. Only two Europeans escaped to tell what had happened.

The British Empire reacted swiftly. Twenty-nine days later, a Punitive Expedition armed with rifles and Maxim guns was marching on Benin. The city was soon taken. Inside the palace, among the putrefying bodies of the men the Oba had sacrificed in a last desperate attempt to stave off disaster, the expedition found thousands of bronze plaques and figures. They seized them all as the spoils of war. Back in Europe, some were kept by members of the Expedition as mementoes of their West African adventures. The rest were put up for auction by the British Government to help pay for the cost of the expedition.

The technical accomplishment and beauty of the bronzes created a sensation. European museums competed to buy them and, either directly or through dealers, representative collections were acquired by Berlin, Vienna and Leiden. Others went to museums in Philadelphia and Chicago. The biggest group of all, however, was bought by the British Museum and has been on display there almost continuously ever since. In recent years, many pieces that members of the Expedition kept for themselves have also appeared in the sale-rooms and Nigeria has been able to reclaim some of its heritage.

David Collison decided to direct the Benin programme. Anna Benson Gyles would come too. And we would have as our guide and general contact man, Prince Humphrey, one of the sons of the present Oba, who had studied archaeology and anthropology at university.

Bronze casters still work in a special quarter close to the Palace walls. The metal they and their predecessors used is not in fact bronze, which is an alloy of copper with a little lead and tin, but brass, which contains zinc instead of tin. West Africa is rich in copper and the skill of smelting it from ore has a long antiquity. The metal we saw being melted down in earthenware crucibles, however, came from fragments of engine cylinder heads and empty bullet cartridges left after the Biafran war. We watched Osiasefe, one of the master casters, make a small wax head, modelled around a clay core, representing the Oba. This he wrapped with clay. The whole was then baked. In the process the wax melted and was poured off through a channel in the base – which is why the process is called 'lost wax'. Molten metal was then poured in through the same channel to replace the wax. After it had cooled, the baked earthen mould was broken away to reveal the bronze. The process is capable of great refinement, but the castings Osiasefe produced were comparatively crude and, unlike earlier bronze-smiths, he relied on filing their surface to remove roughnesses and give them some detail.

An Oba still rules the city of Benin. His palace is a huge compound, surrounded by tall mud brick walls. Every morning, courtiers arrive at its gate. They are bare-chested and wear voluminous white kilts, with around their necks strings of the precious red coral beads that the Oba gives to his closest and most favoured advisors. Some come on foot, some in taxis, a few riding side-saddle on motor bicycles. Every morning a small crowd assembles to watch them, just as people wait outside Buckingham Palace in the hope that they may see an Ambassador wearing his gold-laced diplomatic dress with a white feathered cockade hat on his head, arriving in a horse-drawn carriage.

Thanks to Humphrey's influence, we were allowed to film within the Palace itself. Much of what goes on there is aimed at maintaining the Oba's health and well-being for he is, in his people's eyes, semi-sacred and the well-being of the whole Benin people is intimately bound up with his own person. No one is allowed to see him eat. When he speaks, he does so with only the slightest movement of his lips so that the minimum of his life force escapes.

We were allowed to film the courtiers reaffirming their loyalty to the Oba by dancing in front of him to the sound of drums, as they do every morning. We lurked unobtrusively in a side court to watch him handing out judgements to resolve disputes among his subjects. But there was one place that even Humphrey had no access to – the Oba's harem. Only women were allowed there. Anna volunteered to go into it. She is not tall, but she looked particularly small as she disappeared, a slight, pony-tailed figure clutching a small cine-camera, through the mud brick gateway that led to the forbidden quarters. Her enterprise did not, however, bring as big a reward as it deserved. The harem, far from being a luxurious pleasure garden, consisted only of mud-brick rooms around a dusty courtyard where the Oba's wives cooked for their royal husband and sat playing with his children.

Humphrey also took us, with the Oba's special permission, to see the royal shrines. These were built around the walls of a large open enclosure, one for each of the Oba's predecessors. Each contained a number of long elephant tusks, standing vertically with their tips resting against the wall and their bases in spectacular bronze supports cast in the shape of an Oba's head. Wooden staffs, with rattles at the top, were propped alongside them. Many of the bronzes taken by the Punitive Expedition came from shrines such as these.

Humphrey took us round, telling us which of his ancestors was commemorated by each shrine. I noticed that on one of them, among the bronze figures, the bells and the elephant tusks, lay four stone axe blades. I asked him about them.

'They are thunderbolts,' he said. 'They are very, very powerful. People find them in the fields. They are not made by human beings.'

'They look very like the stone axe blades used by the people many centuries ago, don't they?'

'They do,' said Humphrey, 'but they are not. They are hurled to earth by the gods. They are thunderbolts.'

I had no more wish to pursue the matter than to press a Christian priest to say whether or not the wine, after the sacrament, contains haemoglobin.

* * *

The importance people place on objects and their symbolic significance was also a key element in the film we made among the people of the north-west coast of America. The tribal lands along this coast are extraordinarily rich. The sea is full of halibut, herring and cod; the rivers at the right time of year swarm with salmon; the woods contain elk and beavers and yield great crops of berries. In consequence, their communities are very rich and the people place great importance on an individual's personal wealth. A successful man, to show just how great he is, will once or twice in his life give a huge feast called a potlatch. There will be magnificent theatrical dances in which masked figures enact traditional stories. Food and drink of all kinds are lavished on all the guests. And at the climax, impressive quantities of gifts will be distributed to everyone.

There is a competitive edge to these events. One man will feel obliged to give a potlatch in order to show a rival that he is the wealthier. He may even destroy some of his possessions to demonstrate that he has so much that one or two treasures, here and there, are of no consequence. Saving for such events may take many years. So central are potlatches to the native traditions and beliefs that when the Canadian Government, urged on by Christian missionaries, was doing its best to force native peoples into European-style moulds, the potlatch was for many years officially prohibited. Only in 1951 was that repressive law repealed.

Mike McIntyre, who was to direct the programme about this culture, had learned that these traditions were particularly strong in the settlement of Alert Bay. The Kwakiutl people who live there have among them some of the most gifted carvers on the coast who make spectacular masks for their rituals and still carve immense totem poles. The Kwakiutl never wholly abandoned their traditions, in spite of great pressure from Government and even when potlatches were made illegal, they continued them in secret. In 1921, they planned to hold one in a remote village. Precious ceremonial objects were brought out from secret hiding places and assembled in preparation for the festivities. The police got wind of what was happening and raided the village. They arrested thirty people and confiscated over eight hundred masks, feast bowls, ceremonial ladles and other carvings. The episode was still remembered with fierce resentment in Alert Bay.

Mike contacted Arthur Dick, head of one of the most senior families in the settlement, whom he had discovered was preparing to hold a potlatch within the next few weeks. It would be an ideal opportunity for us. Mike's long persuasive letters got replies that were far from welcoming. Arthur Dick was agreeable but approval did not rest entirely with him. The Alert Bay Tribal Council would also have to give permission. Some members objected to television cameras intruding on such an important, intimate and sacred event. Others argued that potlatches were intended to be public displays. Would not television enhance their effectiveness rather than detract from it? The Council was divided. After each meeting Mike was telephoned in London and given a lengthy report of the debate. Eventually a vote was taken. The Council had decided, narrowly, that we could come.

The strength of Alert Bay's respect for tribal custom is immediately plain to the visitor. Memorial poles, carved and painted in distinctive Kwakiutl style, stand in its burial ground. There are more carved poles in front of people's houses on the hillside above the harbour. Just outside the town there is a huge ceremonial hall with a boldly painted façade of traditional design and an immense pole that, we were told, was the tallest anywhere along the coast – and therefore surely, they said, in the world. It was in this house that the potlatch was to be held.

Our first night in the town was an uncomfortable one. Part, at least, of the reason why the Council had trouble in deciding whether or not to allow us to film was that an ultra-traditionalist group on the Council wanted to commission their own film which would cover not only Arthur Dick's potlatch but also resurrect the whole story of the suppression of the ceremony in 1921 and the confiscation of the sacred paraphernalia. They and their supporters did not want us to be there. In the end the council had decided that both films could go ahead. But the rival team was already in town. As we sat in our hotel that evening, a noisy crowd assembled outside in the street. One of our supporters came in to tell us that there had been rumours – he didn't know where they came from – that our real purpose was not, in fact, to make a documentary featuring the potlatch but to film Kwakiutl people getting drunk in the local bars. Our film shots would then be used to support a movement forbidding native Americans to drink alcohol and so humiliate them. We had to go out into the street and assure people that we were doing no such thing. Kwakiutl men are large, heavy and powerful. Some, it must be said, had certainly been drinking. The atmosphere was not pleasant.

The next day we went up to the ceremonial house. It was immense, an enlarged version of the building in which, traditionally, ceremonies were held during the winter months. In accordance with custom it had no windows. The only natural light came from a small hole in the roof which served as a chimney for the big log fire that burned in the centre of the floor. Each end was blocked off by immense cloth screens hung between the massive carved house posts and newly painted with highly stylised representations of totemic animals – whales, shark, beaver and bear. Behind these screens, dancers would prepare for the theatrical performances.

The building was lit by a few low-power electric bulbs which gave just enough light for people to see their way around. Such dimness was sufficient for the human eye but not nearly bright enough for filming. Mike, right at the beginning of the negotiations, had explained that we would need to install our own lights and we had been given permission to do so. But now objections were being raised. Another meeting with the organisers was hastily assembled. The traditionalists argued that additional lights would wreck the event. I could see their point of view. On the other hand, we had come all this way with the assurance that we had been given such permission. At last, we were told that we could proceed and the electrician we had brought with us from Vancouver could start installing extra lighting.

Now the rival film team appeared. They had decided that we would be a part of their story. I felt sure that they would take every opportunity they could find to suggest that we, by making our film, were being patronising and disruptive. Every time we had a problem, such as how best to position a particular lamp, they filmed us in close-up, with the result that negotiations with tribal elders were encouraged to develop into confrontations. This was ironic, for we knew that our rivals would themselves take advantage of our lighting for their film. It took all day before our lighting set-up was finished.

The ceremonies started in the early evening. Every single inhabitant of Alert Bay seemed to be there, squatting on the ground, sitting on benches, standing around the walls. Two dozen old men from distant villages walked out from behind the screen, sat down on both sides of a long wooden drum and began to beat out a driving rhythm with wooden mallets. A group of old ladies, their long grey hair combed down over their shoulders, wearing scarlet capes, decorated with pearl buttons, slowly filed in and sat down in front of the drummers. The audience was now totally quiet. The men began to sing. One of our guides whispered to me that they were chanting the names of the members of the family who had died since the last potlatch was held. The spirits of the dead were now present within the house. Several of the old ladies were in tears.

And all the lights went out.

For a moment there was silence. It was pitch black. Not only our lights but all those that were normally used in the house had gone out as well. Then there were angry shouts. People bumped into one another as they stumbled about in the blackness trying to get out of the building. Some struck matches. A few switched on hand torches. In the confusion, our electrician managed to reach the entrance door. He knew what had happened. The transformer he had attached to the mains supply to feed our lights had blown. It would take him at least an hour to fix it. As an understanding of what had happened, and who was responsible for it, spread through the audience, I could feel hostility rising. Real trouble was prevented, perhaps, because of the blackness. No one could see who was who. By the time some people had left and returned with candles and torches, the angry mood had simmered down a little. It was too dark for fights and anyway, people wanted to get on with the feast.

It was two hours before the lights went on again. The audience settled down once more. Strange whistles came from behind the screen. They were the songs of spirits. A line of dancers, wearing traditional blankets over their shoulders, emerged shaking rattles. A terrifying masked figure of a huge bird pranced out with its great beak clacking. Slowly the magic of the occasion and the drama of the dances took hold and we finished our film without further difficulty. The masks we saw that night would look impressive in any museum, but no matter how imaginatively displayed they are in such a place, they could only convey a small proportion of the energy, ferocity and other-worldliness that they had that evening.

* * *

Mike McIntyre was also to direct a programme about the objects produced by peoples of the Pacific. I had already shot material when I was in the cult-houses of New Guinea three years earlier, which was in the BBC vaults and had not yet been shown. We went to Vanuatu and on the island of Malekula walked up into the mountains and the country of the Big Nambas people to film their extraordinary masks and cult houses. And we went to the Solomon Islands primarily to film the last of the *tomakos*, the long, slim, magnificently decorated canoes that were once used in head-hunting raids.

We had also heard of a new cult that had started in the village of Makaruka on the southern coast of the main island, Guadalcanal. It seemed similar in some ways to the John Frum movement, the cargo cult I had filmed in Tanna fifteen years earlier. The people had rejected most things European and reverted to ancient traditional ways. The cult had been started in 1957 when a local man, a baptised Catholic named Moro, had fallen into a trance and seen a vision explaining the origin of the world. He declared that the material wealth that had come to the islands had been wrongly appropriated by Europeans and was in fact intended for the native peoples. It could only be reclaimed if the people returned to their ancient traditions and practices. The cult was, by definition, anti-European. Nonetheless we sent messages ahead asking if we could visit the settlement and film there. And we had been given permission to do so.

Landing at Makaruka is not always easy for it lies on the southern 'weather' coast and sometimes the surf is so heavy that it is impossible to get ashore. We were lucky. The sea was almost flat calm as we left our yacht and paddled in a canoe towards the shore. As we approached people emerged from the dark shadows of the palm trees at the head of the beach. They were all dressed traditionally, the men in tapa loin-cloths, the women in thick fibre skirts. There was not a stitch of European clothing to be seen. Many of the men carried long spears.

As our dinghy breasted the surf, the men waded out towards us and, slightly to my consternation, suddenly picked up the canoe bodily with me in it and carried it on their shoulders out of the sea and up the beach. I was met by a group of men playing pan pipes. I climbed out of the canoe and walked up an avenue of warriors holding spears who had painted their faces with white stripes and dots. Ahead of me a small man was standing beneath a ceremonial arch. He wore a waistcoat of shell beads and a broad-brimmed hat with a bead veil all around its rim. This must be Moro. Everyone was shouting. I felt I was taking part in a 1930s movie of a white man arriving in an unknown South Sea island paradise. I half expected Dorothy Lamour to materialise from behind a palm tree. When I was a few yards from Moro and about to shake his hand a choir broke out with God Save the Queen. I knew, at least, what *British* ritual demanded at this point. I stood to attention. So did Moro. The anthem came to the end. I put my hand forward towards him but had to withdraw it. The massed choir had started on a second verse. They knew more words of the British national anthem than I did.

Now Moro put forward *his* hand. I shook it, and he started on a long speech. It was translated paragraph by paragraph into excellent English by one of his senior men who had been a teacher in a mission school. There were many references to how far I had come from across the sea and to my wife and family. A suspicion formed in my mind. Maybe this elaborate and respectful reception was a case of mistaken identity. 'Attenborough' sounded only too like 'Edinburgh'. Certainly my name had been used when we sent messages asking if we might visit Makaruka. However that might be, it was too late for explanations. I should just have to do my best to play the part.

I was still wondering how best to impersonate royalty when it was explained to me that if I, together with Mike and the camera team, were to come any farther ashore, we should have to discard all our European clothing and adopt the same style as everyone else there – that is to say, stark naked except for a narrow tapa loin-cloth. I was not sure how the Duke of Edinburgh would have behaved under these circumstances. However, small cubicles had been provided for us. Tapa loin-cloths are put on wet. That way, they stretch or contract to hug the anatomy. Feeling unhealthily pallid, like creatures that had crawled from beneath a stone, we stepped out. Moro reclaimed us and led us round a series of special displays. I clasped my hands behind my back and leaned forward from the waist to display interest in what I was being shown and attempted every now and then to ask an intelligent question.

One group of women were breaking pink shells into fragments and shaping them into the beads that traditionally were used as money. Another group were splitting and preparing breadfruit. Yet another were plaiting mats. Then we were shown into a hut. This, it seemed, was the community's treasury for it was hung inside with hundreds of strings of shell money, like bead curtains. Bare-breasted women sat cross-legged and silent around the walls. They were Moro's wives. We were also shown into another house which our interpreter told us was called The House of Memories. Tables of matting stood around the walls and in the middle. On them were exhibits – small sculptures of blackened wood, stone axe blades, woven baskets, tobacco pipes made from cone shells, shrivelled yams, oddly shaped water-worn stones. Each had a label 'Memory of yam', 'Memory of axe', 'Memory of how a possum brought wealth'.

The school teacher explained that the purpose of this house and its contents was to demonstrate that the island belonged to Moro and his people. They were staking their claim to it not only by reviving their old customs but also by imitating one of the stranger customs introduced by the western invaders themselves often to serve a similar purpose. They had created a museum.

* * *

Responsible museums – and responsible collectors – record whatever is known about the provenance of any object they acquire. Where did it come from? Who made it and why? Who collected it and when? Without such details any object is shorn of much of its scientific value. Only too often exciting, lovely and intriguing pieces appear on the market without any provenance whatever. If a collector acquires one such he will set about establishing what its provenance might be. That is part of the fun.

In 1985 I bought a small wooden figure of a man. It was eighteen inches long but scarcely thicker than my thumb. Its body was set in a graceful curve, doubtless following the shape of the thin branch from which it had been carved. Its fingers, lying over its belly, were enormously long and almost threadlike. There were six on each hand. It had spectacularly large, elegantly carved male genitals. Its face was grotesque. The mouth ran from ear to ear in a thin, toothless semi-circle. The head carried a ridged crest running from the centre of the forehead to the nape of the neck. Its eyes were not set in oval sockets, but surrounded by two raised circular rings so that they goggled.

This strange image had appeared in a New York auction. The catalogue said that it came from Easter Island but estimated its value as being far lower than that of a genuine Easter Island carving of any antiquity. That suggested that the auctioneers thought it was either very late in date or even that they had some doubt about its authenticity. Those sitting in the auction room must have agreed that it was of little consequence for it was knocked down at a price even lower than the estimate. The dealer who bought it normally dealt with pre-Columbian antiquities. He said that it was going so cheaply that he couldn't resist it. And when he offered it to me for only slightly more than he paid for it – neither could I.

While working on *The Tribal Eye* I had accumulated a range of books about Pacific tribal sculpture. I also had a long run of illustrated auction catalogues. I leafed through them all to see if I could find an illustration of anything that resembled my purchase. The standard wooden male figure coming from Easter Island is a naturalistic portrayal of an emaciated man with a normally proportioned body, eyes in lenticular sockets, bared teeth, a goatee beard and prominent ribs. Such figures have been carved in considerable numbers from the early nineteenth century until today. The islanders say they represent human ancestors, not gods. Recent ones can be bought quite cheaply. Earlier ones can fetch very big prices indeed.

Mine, however, was not in the least like these. I could find only one figure that resembled it at all closely. This belonged to the Museum of Anthropology in St Petersburg and it was not male but female. Nonetheless, its likeness to mine was remarkable. It was about the same size, with the same extreme elongation of the body. It had the thin-lipped grin, the odd ridged head-crest, and the circular double-ringed goggling eyes. Whether it had six fingers or five was impossible to tell for its midriff was worn, presumably from repeated handling, and had lost its details. The two seemed to be a pair, male and female. It was a clue.

I wrote to the St Petersburg Museum asking about the provenance of their figure. Their reply was a disappointment. Its acquisition number painted on its base – 736 – showed that it was one of a group of Pacific objects that had been transferred to them when the Museum of the Imperial Russian Admiralty was closed in 1828. Another Easter Island piece had come with it that represented a human figure with the head of a bird and wings instead of arms. Both pieces, said the Museum, must have been collected from the island by official Russian expeditions to the Pacific. But they did not know which or indeed when – except that obviously it must have been some time before 1828.

I did some more reading. Only two Russian expeditions, I discovered, had visited Easter Island before 1828. One spent only a day there, the other only a few hours. Both retreated in a hurry after having been attacked by the islanders. Neither recorded doing any trading or collecting. That seemed odd. Explorers' journals are usually meticulous in recording such details. But the trail seemed to have gone cold as far as the St Petersburg figure was concerned.

I had even less success with tracing the provenance of my own. All the auctioneers would tell me was that it had come from the estate of a deceased New England dealer who kept few records. They themselves, they said, had tried their best to find out where it had come from but had failed.

The Easter Island man sat on my desk surrounded by mystery. The more I looked at him, the more impressed I became by his qualities as a sculpture. His six-fingered hands gave me a further clue, though only a vague one. Unnatural numbers of digits are used elsewhere in Polynesia to indicate divinity.

Then in 1985 a catalogue was published which reproduced all the known drawings and paintings made on Captain Cook's second expedition, the one during which, in March 1774, he had called at Easter Island. And there I saw drawings of the two St Petersburg figures, the female figure and the birdman, side by side on a single sheet of paper. They had a certain amateurishness about them, suggesting that they had not been done by a professionally trained artist, but the portrayal of the figures was correct in every detail, even down to the number of grooves in the wings of the bird-headed figure. The sheet had come from an album that had belonged to Captain Cook himself. How could Cook have acquired a drawing of two figures that are now in Russia?

Cook was not the first European to visit Easter Island – a Dutchman and a Spaniard had preceded him – but he was the first to stay there for any length of time and the first to make a proper survey of the island. I consulted his official account of his voyage. In it he recorded a great deal about Easter Island. The people told him that the gigantic stone figures that ultimately would make the island so famous, had been carved by their ancestors to commemorate great men. Cook noted however that they themselves were now poor and destitute and seemed to lack the skills that such work must have demanded. But he made no mention of wooden figures. I looked in the text of his daily journal which sometimes contains details that were left out of the official polished-up account. But there was no reference to any such things there either.

However, Cook was not the only one to keep a journal on the voyage.

Two Germans, Johann Forster and his son Georg, who were the expedition's official naturalists, both did so and both published them. I consulted them too. Johann's added nothing much to Cook's account. But Georg's, as I read with mounting excitement, gave a more detailed account of what happened on Sunday March 14th 1774, the second day of the expedition's visit.

Cook himself had been feeling ill and did not accompany the main party that went to explore the island. In the afternoon, however, he felt a little better and went ashore taking with him young Georg Forster and a Tahitian lad named Mahine who had joined the expedition in Tahiti as an interpreter. They began to trade. The islanders badly wanted European cloth. This is how Forster described what they offered in exchange – 'several human figures made of narrow pieces of wood about eighteen inches to two feet long and wrought in a much neater and more proportionate manner than we would have expected after seeing the rude sculpture of the stones. They were made to represent persons of both sexes and the features were not very pleasing and the whole figure much too long to be natural. The wood of which they were made was finely polished, close grained and of a dark brown.' I could hardly have hoped to find a more accurate description of the St Petersburg female figure. Or, come to that, mine.

Forster went on to record that neither he nor Cook acquired any of these figures. But Mahine did. He said they were better than the ones carved by his own people back in Tahiti and that they would be greatly appreciated by his friends and relatives there. So he bought 'several'. So such wooden figures were certainly taken back to Cook's ship. Perhaps, as the expedition sailed away, the Forsters regretted that they themselves had not acquired any. After all their job was to make collections of animals, plants and what were called 'artificial curiosities' – that is to say, objects produced by the people they met. Since Mahine would not be parted from those he had bought, Cook could well have instructed one of the ship's draughtsmen, whose usual job was to draw profiles of coastlines, to make a recording of two of them before Mahine took them all away.

What happened next? I turned the pages of Forster's journal with increasing excitement. Young Georg did not let me down. When the party got back to Tahiti, he says, Mahine was immediately surrounded by his relatives 'who were extremely numerous and expected presents as their due. As long as the generous youth had some of those riches left, which he had collected at the peril of his life on our dangerous and dismal cruise, he was perpetually importuned to share them out. And though he freely distributed all he had, some of his acquaintances complained he was niggardly.' So the Russian voyage I had to identify was one that collected figures not from Easter Island as the St Petersburg Museum had suggested, but from Tahiti.

And there was one. Forty-six years after Cook left, the Russian Admiral Bellingshausen arrived there. By this time, Tahiti had been converted to Christianity. The King, Pomare, had a great passion for all things European and in particular for European linen. He particularly wanted the sheets from the Admiral's bunk and offered Bellingshausen all kinds of objects from his kingdom to get them. The Easter Island figures must have been obvious candidates. They never had any religious significance in Tahiti anyway. When the Russian expedition got back to St Petersburg, all the objects they collected in Tahiti, as Bellingshausen records in his journal, were deposited in the Museum of the Imperial Russian Admiralty.

So now I knew that the strange sculptural style of the St Petersburg female figure was not a late aberration of little consequence. On the contrary, it was an ancient one. And she, together with the bird-headed figure, had a particular importance, for it was now clear that they were the first sculptures to have left Easter Island in a European ship.

But what about mine? Could it too have been one of those collected by Mahine? The bizarre features it shared with the St Petersburg piece were so similar that it was inconceivable that whoever carved one was ignorant of the other. Could mine have been carved in recent years as an imaginative re-creation of a male partner for the female figure? Was it, to put it bluntly, a forgery? I tried a new line of enquiry.

The laboratories at Royal Botanic Gardens at Kew are able to identify many woods by the microscopic examination of their structure. They looked at a tiny sliver taken from the male figure and pronounced it to be the wood of the toromiro tree. This species grew *only* on Easter Island. Its wood was hard and close-grained and was used by the islanders for all their most important carvings. That confirmed that my piece was carved on the island. But it did more. It also told me something about the figure's date. The toromiro tree became extinct on the island in 1956. So the figure must have been carved before then. But pictures of the St Petersburg female were not published until 1973. So by the time that a picture of it could have reached the island there was no toromiro wood from which to carve a mate for her.

Had the male been carved in the years immediately after Cook's departure while the visual memory of this ancient style still lingered? That was possible, though it seemed extremely unlikely. After all, Cook suggested that the island was in cultural decline at the time of his visit and archaeologists in recent years have confirmed that this was so. The fact that the islanders had readily offered the St Petersburg figures to strangers for barter suggested that the style no longer had any importance for them. Why then should any carver wish to resurrect an old abandoned cult? Nor was the style used by the islanders when during the nineteenth century they began to carve male figures to sell as souvenirs to the ships that started to call at their island.

The remaining explanation was that my male figure had indeed been among the 'several' figures 'representing both sexes' that Mahine had taken away with him to Tahiti. If it was, there was little difficulty in imagining how it might have crossed the Pacific to America for American whalers were regular visitors to Tahiti throughout the nineteenth century. It seemed that my figure, after all, was an early and highly important one. I dearly wished I could go to the island, simply to walk around the place about which by now I had read so much. However there seemed no chance of doing so as part of a natural history project for the island is barren indeed and has no wild fauna except for a very few species of seabirds.

A year or so after I had worked all this out, the Natural History Unit asked me to write a short ecological series to be called *The State of the Planet*. I discussed possible ideas with Kate Broome, who was to direct the last programme in the series. One evening she telephoned. She had had an idea for the final closing sequence. Recent archaeological work on Easter Island had established that the environmental and cultural impoverishment that was already apparent in Cook's time was the result of the destruction of the island's once rich cover of trees – including the now vanished toromiro. She knew I was very busy, she said. She hesitated to add to my travelling commitments, she said. But could she possibly persuade me to go to Easter Island?

I could hardly believe my luck. I showed her the figure. If I agreed to go, I said, could we add a couple of days to our stay and shoot material for a programme telling its story? We drew up a very cheap budget and I sketched an outline for a film that would be a kind of ethnological detective story. We got the money.

We made a quick trip to St Petersburg. I took the male figure with me in a little velvet bag and filmed it and the female figure, lying alongside one another in the Museum. I also asked to see some of the other pieces that shared the same acquisition number as the female figure, just to make sure of the correctness of my Bellingshausen attribution. The curators produced a superb drum, a god figure, a coconut splitter and some sacred mats. They all had the same painted acquisition number, 736, as the female Easter Island figure. They were all Tahitian, as I knew they would be.

When we got to Easter Island we had no difficulty in identifying the exact beach on which Cook landed and where Mahine had bartered for the figures. Cook marked it clearly on his chart and anyway the people still know today. It is the nearest landing place to Orongo, a sacred village of low stone houses that stands a mile or so away on the top of a thousand foot high cliff. It was the ancient and once the most important ritual centre on the island. As I walked along Cook's beach, I imagined islanders coming down to greet the strangers who had come from the tall ship that lay at anchor off-shore. I thought of how men must have rushed back to their villages to find objects to offer for trade, of how one might have run up to Orongo and gathered up what figures still lay in the abandoned ritual houses and brought them back to Mahine, young Forster and Cook.

We went up to Orongo. The rocks around it are carved with reliefs of bird-headed human figures very like the one in St Petersburg which Mahine had collected. And there among them was a mask-like face with circular staring eyes. There were other similar masks carved on rocks elsewhere in the island as well. What could they represent? The people told us. Circular eyes were a sign of divinity. The masks represented Make Make, the ancient and most important god, the creator of all the world. Could my figure, with its circular eyes, six-fingered hands and spectacular genitals also represent him? If it does, then it is the most complete surviving image of the supreme god of Easter Island.

20

Life on Earth

The Natural History Unit had been exploring the idea of a big series in the mould of *Civilisation* and *The Ascent of Man* while I was still sitting behind a desk. Presenting it would be a full-time job for someone for two or three years, so at that time I was out of the running. But soon after I resigned, Chris Parsons, a senior producer in the unit, came to talk to me about it. Now there was nothing to prevent me from saying outright that I would dearly like to write and present it. Chris welcomed the idea. We took it to Robin Scott, the Controller of BBC2, and he accepted it for his network, provided that co-production money could be found from one of the American networks. That would take time, but in any case it would be some months before the producers Chris wanted to work with him on it would be free. That was why I had made *The Tribal Eye* first.

Chris and Desmond Hawkins, now Head of Programmes at Bristol, also had another proposition. Peter Scott's *Look* series had come to an end and BBC1 had no regular natural history programmes in its evening schedules. The Natural History Unit wanted to offer a series of half-hour programmes about animals of all kinds but they needed to devise a style that would make its disparate components cohere as a series. They suggested that this could be done if I were to provide all the commentaries. Still nervous about the unpredictability of a freelancer's life, I was very grateful for the proposal. It might give me a small but regular income for several years to come. Narrating them would also enable me to re-establish my public connection with natural history without wearying the audience by appearing too frequently in vision. So *Wildlife on One* was started.

As soon as I got back from Indonesia with Richard Brock I started to sketch programme synopses for the big series. I was quite sure how it should go. I would start at the beginning of life, trace the long story of its evolution as both plants and animals spread from the sea and on to the land in increasingly complex forms. We would use living species to illustrate how fish hauled themselves out of the water and became amphibians, like newts, frogs and toads. Early amphibians, in turn, developed impermeable skins and so gave rise to the reptiles. Ancient reptiles, such as dinosaurs, developed warm blood and eventually gave rise to mammals. The series would end with monkeys, apes and mankind. Not everyone, however, agreed. When the idea was presented to one of the American network executives whom we hoped might be a co-producer, he was reported to have said 'You mean you are going to try to hook people for a long series with a first episode that's all about green slime? You must be crazy.' I suppose I should have foreseen that. More unexpectedly, there were objections from within the Natural History Unit. One of the senior producers maintained that telling the story in the way I was suggesting was a very outdated approach – nineteenth-century natural history at its most boringly conventional. The Unit's first big 'sledgehammer' series, he argued, should be more forward-looking and contemporary. It should include a survey of the latest behavioural theories and the findings of molecular biology. Chris Parsons, however, backed my approach and I started work on the detailed scripts.

We decided that the series should be called *Life on Earth* and that it should have thirteen episodes. I had established 'sledgehammers' at this length when they had started with *Civilisation*. A weekly series thus neatly filled a whole three months of the year, which was how schedules were planned at the time. Any longer and it would extend into a new quarter when the pattern of the schedules might change and loyal viewers would be irritated if it had to move to a new day and time. Any shorter and there would be the problem of filling a few odd weeks before another big series could begin. Happily, when I sat down to survey geological time, the schedules of creation seemed also to be conveniently divided into thirteen parts. The episodes were to be fifty minutes long. This was because American television to whom we hoped to sell the series scheduled programmes in one-hour segments. We therefore had to offer them fifty minutes of programme to give them space for their ten minutes of advertisements.

Tackling a series of such length as a single production brings a great advantage. It makes it economically possible for any one programme to include a number of quite brief sequences from several far distant places. In Programme One, I could suggest a sequence of no more than a minute or so to be shot in Australia about the earliest fossil evidence of life, because in Programme Eight, when dealing with the origin of mammals, I would certainly have to go there to film the duck-billed platypus.

As I finished each script, I sent it to Chris and the production team who listed the sequences according to their geographical location. By the time half a dozen or so were completed we were able to start planning long trips – to Kenya and on to Madagascar and the Comoro Islands; to the prairies of the United States and up to the tundra of Alaska; to the rain forests of the Amazon and down to the bare wind-swept pampas of Patagonia. On each tour I would be filmed talking to camera for sequences that would eventually go into half a dozen different programmes. While we were doing this, a dozen or so other cameramen, often by themselves, would each be concentrating on one particular animal and trying to film it doing the things that I had specified in the scripts. All in all, it would take us three years to complete the entire series. That sounded extravagant, but as I pointed out to the planners, almost any fifty-minute documentary takes about three months to make, so if we did manage to complete thirteen programmes in three years as we planned, we would be doing particularly well.

* * *

Our first trip was to the Galapagos. John Sparks was the director. With us came the crew with whom I had recently worked in Indonesia – Maurice and Hugh with cameras, Dickie with recorder.

Galapagos was a happy choice for a beginning. It was there, after all, that Charles Darwin, famously, had made the observations that led him to his theory about the mechanism of evolution the results of which we were trying to chronicle. It is commonly supposed that it was the variations in the beaks of Galapagos finches that started Darwin's speculations. In fact, as his journal makes clear, tortoises were the trigger. He noticed that those on well-watered islands, that could graze on low-growing plants, had a rounded front to their shells. In contrast, those from dry islands that could only find leaves on the branches of trees, had a sharp peak in the shell front that enabled them to crane their extremely long necks almost vertically upwards. Only the first sort survive today in any numbers, and even they can only be found in more remote parts of the islands. The biggest population lives in the huge crater of an extinct volcano called Alcedo.

Climbing up there was hard going. Each of us had to carry two gallons of drinking water, our own kit, and a share of the filming gear. The whole of the first day was spent walking over great fields of roasting black lava that lay on the volcano's lower slopes. The next day we climbed to the crater rim, 3700 feet high, and descended into the crater itself. And there were the tortoises. About two hundred of them were plodding slowly through the low bushes, slumbering with their necks stretched out on the ground, or sitting motionless in pools of rain water. The adults were about five feet long but there were also some youngsters no more than twelve inches in length. The population up here, it seemed, was thriving. They seemed not to hear any sound we made as we walked up behind them, and only became alarmed when they caught sight of us. Then they made a hiss, like the sound of a punctured tyre and immediately retracted all four of their legs which occasionally left them in embarrassing positions, teetering on a boulder.

It was the breeding season. The big males were picking quarrels with one another, emitting open-mouthed gasping snarls. Each sniffed the rear end of any other they encountered, trying to identify females. Their mating technique thereafter is nothing if not straightforward. The male, having spotted a female, plods after her. When eventually in this slow-motion chase he catches up with her, he simply keeps on going and climbs up with his forelegs on to her from behind. There is much thudding and squeaking and sounds like that of heavy leather harness under strain. He then starts copulating with a stertorous rhythmic roar and a thrust from his hind legs which rocks both their shells and sometimes lifts the female's rear end clear from the ground.

We quickly pitched camp and started filming this primordial scene. Soon after midday the clouds started to accumulate and before long it was raining. Back at camp we discovered that not only were our locally hired tents not water-proof but two of them had been flattened, presumably by optimistic male tortoises. The night was a wretched one, cramped, cold and wet inside our leaking tents, listening to the rhythmic roars of the indefatigable copulating males and wondering when they might in the darkness make amorous advances upon us. It was an unforgettable beginning to our whole project.

* * *

Since we were tracing the history of life, species that linked one great group of animals with another were very important elements in our story. Fossils of fish called coelacanths had long been known in rocks 400 million years old. Judging from their bones, they had stout fleshy bases to their fins, and some scientists believed them to be very close to the first four-legged animals to crawl out of the sea and up on to the land. They were therefore a link to the first amphibians and thus to all terrestrial vertebrates, including ourselves. When in 1938 a South African trawler brought up one alive from a depth of around 300 feet, it caused a scientific sensation. It was a monster, five feet long and much bigger than its fossil relations which were no bigger than herrings. Sadly, it had been gutted and its entrails thrown away before any scientist saw it, so details of its soft parts that could not be revealed by fossils still remained unknown. In spite of an intensive search over many years no further specimen was found.

Until, that is, another was fished up from the seas around the Comoro Islands, off the African coast north of Madagascar. It was the discovery of this second example that Julian Huxley had talked about in the first programme I had produced up in Alexandra Palace twenty-five years earlier. Since then, two or three a year had been caught in the waters around the Comores which were clearly the coelacanths' main home. But no one had succeeded in filming a living example. It would be a great coup for *Life on Earth* if we managed to do so.

Peter Scoones believed that he could. He was – and is – one of the most experienced underwater cameramen with an extraordinary skill in devising special equipment for particular jobs. All the specimens, like the first, had been caught at depths around 300 feet. That was beyond the range of a scuba diver but well within reach of an underwater camera. Peter's idea was to lower a remotely controlled video camera on a long cable in the area off the Comores where most specimens had been caught. It seemed worth a try. There were other things of interest to us on the islands as well – a particularly photogenic colony of fruit bats, and some lemurs – so whatever happened a visit to the Comores, which we could tack on to a trip to Kenya, would be worthwhile.

The Comores, we discovered, were not at that time politically stable. There had been a series of bloodless revolutions and it was difficult to find out who was actually in charge. Eventually we contacted helpful people on the islands and in due course, Peter and I arrived in Moroni, the capital. Most of the coelacanths that had been caught had been hooked at night, when they probably came up from greater depths in order to feed. So that was the time we went out to look for them. Every evening we left harbour in a launch. When we reached the area recommended by the local fishermen, Peter lowered his camera, with lights attached, and we all gathered around a monitor in the stern to watch what it might reveal.

Things did not look promising. Any idea that a group of these prehistoric monsters would appear out of the gloomy depths and swim into the beam of our lights to take bait was soon seen to be wildly optimistic. The sea was too deep for us to anchor, and there was a strong current. So all our video camera revealed was the sea floor moving steadily past at two or three miles an hour. Even if we were to catch a glimpse of a coelacanth moping around on the bottom, we would be swept past it before we could get a good look at it. Nonetheless, we persisted.

Although the sea floor was largely flat, in places there were rocky reefs. As soon as we spotted one ahead on the monitor, we had to haul in the cable so that the camera, as the current took us towards the reef, would rise above it and not get tangled. The line was so long that it took some seconds before even the most frantic hauling on the boat had any effect. The end was inevitable. On the fifth night, we saw a huge reef coming. We hauled in the cable as quickly as we could but not quickly enough. The camera crashed into a crevice in the rocks and jammed tight. The current swept the boat on, white lines zig-zagged over the television picture and then the monitor went blank. The cable had broken. And that, it seemed, was that.

We had, however, got some outstanding fruit-bat footage, as well as some sequences with lemurs, so our visit to the Comores was not a total write-off. Nonetheless we were a disappointed group when we assembled at Moroni airport to make our farewells. The young man who had been made Minister of the Interior in the latest revolution came to see us off. We thanked him for having made our visit possible. Would I do a favour for him in return? Of course, how could I help? Well, he said, he had a sister who worked in the BBC's Overseas Service. Would I take her a special present from him – and he produced a package securely tied up with string, about the size of a large dictionary. What was it? Just a present, he said. I could hardly refuse a request from a Minister, so I stowed it away in one of my bags, shook his hand warmly and walked into Customs.

There a Customs officer gestured to us to put all our bags on a long bench and stand beside them. We did as we were told. The Customs officer worked his way along them, interrogating the owner of each. I could hear the questions he asked the others.

'Is this your bag?'

'Are you carrying anything for someone else?'

Had I fallen into a trap? How should I reply when my turn came? I was still trying to decide when a door behind the bench opened and out stepped the Minister of the Interior. This time, however, he was wearing an official hat heavy with gold braid appropriate, no doubt, for the Commander in Chief of the Customs Service. He came directly to me.

'Are these your bags?' he said, rather aggressively as though we had never met.

I nodded.

'Are you carrying anything given to you by someone else?'

He stared straight at me.

Before I could answer, he said 'Very well. Carry on' and I scuttled into the departure lounge. I still do not know what his package contained but I passed it on to his sister in London just as soon as I could get rid of it.

Peter stayed behind. He wanted another few days with the fruit bats. The morning after the rest of us had left, someone knocked on the door of his hotel room. A fisherman had just come into the harbour with a coelacanth. Was Peter interested? He grabbed his camera and rushed down to the harbour. There he found a huge coelacanth lashed to the underside of its captor's canoe. It was alive – but only just. He persuaded the fisherman to untie it and he filmed it swimming feebly over the floor of the harbour. It performed well enough to demonstrate one of the important characteristics which were of importance to our evolutionary story. It did indeed use its fins like four stout legs. The film and photographs Peter took, in spite of the fish's condition, were the first and, for many years, the only ones taken of a living coelacanth.

* * *

Other survivors from distant geological times were easier to find. I was particularly keen to include a sequence about horseshoe crabs. They belong to a very ancient group of sea creatures, older even than the coelacanth or any other fish. Chris Parsons agreed that they would make a valuable sequence and did some research into practicalities. 'It will be easy,' he said. 'I've booked a flight that will get us into Philadelphia on the evening of May 30th. The following day, the first of June, we will drive down to Cape May to a particular beach and then in the early evening a million horseshoe crabs will come out of the sea.'

It seemed unbelievable but so it turned out. We arrived on Cape May in the late afternoon. Flocks of gulls and small waders had already assembled on the beach. They knew, just as we did, what was about to happen on this one special night of the year.

As the sun sank lower in the sky, a few smooth grey-green domes, the size and shape of military helmets, broke the surface of the shallows. Within a few minutes, they had appeared all along the edge of the sea. Each had two small bumps on the front at each side, like sensors on an automaton. They were eyes. At the rear of the dome, which concealed their legs, was a second roughly triangular plate at the end of which was hinged a long spike. The shells of the females were about fifteen inches across. The males were a few inches smaller and much more abundant. As one of the large females crawled up into really shallow water, a male trundled after her and clamped himself to the rear edge of her shell. Sometimes several tried to do so at the same time, and the female had to haul a small chain of three or four behind her as she advanced up the beach.

By the time the sun started to set, there were so many crabs in the frothing surf that they formed a continuous pavement. So great were the numbers that some climbed unsteadily over others. The wavelets, as they lapped gently up the shore, sometimes caught the side of one of these and flipped it over so that it lay on its back with its five pairs of legs mechanically gyrating, trying to push itself right-side up with its rigid tail-spike. The gulls were quick to spot such accidents. They swooped and started stabbing at the crab's more vulnerable underside. But they did so in a desultory investigative way. Their main feast was yet to come.

As the females crawled up the beach and reached high watermark, they dug in the sand and each extruded hundreds of pale green eggs the size of the sand grains with which they mingled. At the same time, the males clamped on to the females' shells, discharged their milt as did other unattached males that surrounded the females on all sides.

The mass spawning continued for several hours. Then the tide turned and the crabs began to retreat. The gulls ran among them, deftly and swiftly picking out the eggs from the sand. Their harvest was so great it seemed extraordinary that the horseshoe crabs should prefer to deposit their eggs on the beach rather than in the sea. Perhaps this is because they evolved at a time so early in the history of life that the land truly was a safer place. Then there were no birds in the sky nor any four-legged predators crawling about on the ground.

But there was a bigger mystery. We had been able to predict the highest nocturnal tide of the year by consulting tide-tables and calendars. But how had the crabs known, weeks earlier in the dark depths of the sea that the time had come for them to travel into shallower waters? And how did they realise, on the evening of June 1st, that the next tide would be at its peak so that they could all emerge together from the sea as their ancestors had been doing annually for a hundred and fifty million years?

* * *

John Sparks was in charge of the programme that was to deal with monkeys and apes. One sequence in my script concerned their ability to touch their thumb with a forefinger. The development of the 'opposable thumb' as primatologists call this characteristic, was invaluable initially for it enabled them to grip branches as they clambered around in trees. It also allowed them to pick up small objects and even use them as tools. And tool-using was an important step in the history of humanity. In my script I suggested that chimpanzees might illustrate it. John thought we could do better. Why not use gorillas which then had hardly been filmed? A remarkable American zoologist, Dian Fossey, was working with mountain gorillas in the Virunga Mountains of Rwanda. With her help, we might be able to get quite close to them. I doubted whether Dian would accept us for I had heard of Dian and, by all accounts, she was very protective of her gorillas and unwilling to receive visitors. Nonetheless, John wrote to her and to my surprise and delight she agreed to help us.

John, Dickie Bird, cameraman Martin Saunders and I arrived in the small town of Ruhengeri in Rwanda, in January 1978. We were met by a young bearded Englishman, Ian Redmond, who had come down from Dian's camp to meet us. He had bad news. A young gorilla, Digit, had just been killed by poachers. Dian had known Digit since he was an infant and had watched him as he grew up and developed. He had shown particular affection for her and had become her special favourite. His killing – it was difficult not to call it murder – had greatly shocked and distressed her. Not only that, she had a chest infection so severe that even before Digit's killing she had thought of sending us away. But then she had decided that we should come, for we might be of help in telling the world of Digit's death and raising funds to help protect the surviving gorillas.

We piled into Ian's truck and drove for three-quarters of an hour through small villages and cultivated fields to the edges of the forest that covered the upper slopes of the volcano. From there we walked. As we climbed higher, the nondescript bush gradually changed into a forest of *Hagenia* trees, their long branches loaded with ribbon-leaved ferns and wispy pallid Spanish moss. Karisoke, Dian's camp, proved to be a group of half a dozen green-painted corrugated iron huts connected by paths edged by low rails.

We tapped on the door of her hut and found Dian in bed. I had seen photographs of her, a tall, raw-boned, pony-tailed Californian, but now she was pale and drawn. A blood-spattered handkerchief lay on the table at her bedside. She told us, in a matter of fact way, that she had pneumonia. She had had it before and had the drugs to deal with it but she was now spitting blood and that was new. A mountain shack at ten thousand feet, drenched by daily rains and nearly always surrounded by mist, hardly seemed the right place for such a patient. Her main concern, however, was not her health, but Digit's brutal murder. His killers had mutilated his body, cutting off his head and hands to sell as macabre souvenirs in the tourist shops of Ruhengeri. Some of her African staff had chased the poachers and caught one who she was sure was Digit's killer. She spoke with white hot passion even though she had to snatch her breath and endure sudden stabs of pain in her chest. The army had come up and taken the captive poacher away. It was as well for him that they had not left him with her.

She went on to tell us of the many dangers facing her gorillas. Even though the forest had been declared a national park, farmers from the surrounding villages were continually encroaching upon it and allowing their cattle to graze there. She had warned them many times but recently had decided that threats were not enough. She had taken the law into her own hands. She had crippled the trespassing cattle with shots in the spine and left them in the forest to die. And that was not all. She believed that there was corruption among officials responsible for the Parks. Bribes were being given. People were being allowed to hunt there or to cut down its fringes to make more fields. She was sure an expatriate Belgian, working within the Department, was the source of most of the trouble. International attention must be drawn to what was going on. We had to help in publicising Digit's death. We promised we would do what we could.

The following morning, Dian was still too weak to take us out. Ian Redmond did so instead. Before we left, he gave us some instruction in gorilla etiquette. The gorillas we would see belonged to groups that were accustomed to Dian and her assistants. With Ian to introduce us, there was every chance they would accept us as well. Nonetheless, we had to behave in a proper way. When gorillas move through the dense undergrowth of these forests, they make regular burping noises, called by the scientists who study them belch-vocalisations – BV's for short. It was now realised that these have nothing to do with indigestion. They are signals by which an individual indicates its location to other members of the family. Everyone in a group feeding in thick vegetation needs to keep check on the others so that if they hear a noise nearby they know whether it was made by one of their own number or a stranger – perhaps an attacker. It was important for us, therefore, to be seen when we were a long way away. Then, as we approached, we should burp regularly so that they could keep a check on where we were. The worst thing to do was to creep up so quietly that the gorillas were not aware of our presence until we were at really close quarters. That was the way to get charged.

When we got really close, we should show respect. Keep our heads low. That is a signal used in both gorilla and human society. Respectful human beings bow to show deference. Gorillas require the same sort of acknowledgement. Don't talk too loudly. And lastly – don't stare; that would be interpreted as a challenge and it isn't a good idea to try and challenge a full-grown silver-backed gorilla.

With these precepts clearly in mind, we set off up the mountain. Soon we were moving through dense waist-high thickets of wild celery and stinging nettles and before long we found the trail of a group of gorillas. Tracking gorillas in such country was hardly more difficult than tracking a steamroller. They leave a broad path of flattened vegetation in their wake. If the plants are beginning to grow upright again, the trail is probably a day old. If the sap from broken stems is only just starting to coagulate or is still flowing, the gorillas have been there very recently. Gorilla dung is also very informative. After about two hours it acquires tiny white dots – eggs laid by a fly. And if it is still warm, it is recent – and you are very close.

Cautiously we made our way in single file in the gorillas' wake, bent double and talking only in whispers. Ian was ahead. He stopped and beckoned to us. We crept up towards him. The ground ahead sloped gently downwards and there, sitting amid the giant celery, pulling it up in great handfuls, sat three large female gorillas. They were only twenty yards away. One of them turned her shining black face and looked directly at us. Then she looked away and ripped up another clump of celery.

We sat and watched entranced for several minutes until, one after another, the gorillas heaved themselves to their feet and moved unhurriedly away. Our initial introduction was over. We walked back to camp with our pulses racing.

The next day we returned to the same group. Once again we approached and this time, we decided to begin filming. Martin, without making any sudden movement, slowly put up his tripod, silently clipped the camera into place and focused. The camera made a faint whirr. One of the gorillas looked at us unconcernedly. Apparently, pointing a single large glass eye at her did not constitute a stare.

We worked for a week, going out each day at dawn, picking up the trail of a group we might have left the previous night and following it until we caught up with the animals. Once, Dian felt well enough to come with us. As we sat watching a small group, she glowed with pride and pleasure at seeing her magnificent animals behaving in such a relaxed way. She seemed in better spirits than at any time during our visit.

Guided by Ian, we gradually felt more confident in the presence of the gorillas. And so did they. One morning their trail led into a particularly tall and dense patch of grass. As we crept, half-crouched, it rose above our heads. We could hear sounds of them eating very close by. I went forward, my head down so as not to stare, burping perhaps a little more frequently than was absolutely necessary, and found myself in a circular arena of flattened vegetation. It was like the small clearing in a meadow of tall grass that one used to make as a child and pretend it was a room of one's own. Then I realised that to one side, only a few yards away, there was an adult female. This was the closest I had ever been. She seemed gigantic. Perhaps this was an opportunity for a sensationally dramatic piece spoken to camera. Martin had set up his camera and tripod a few yards behind me. Beyond, I could see Dickie wearing his headphones and looking at the dials on his recorder. John, next to him, gave me the thumbs-up.

'There is more meaning and mutual understanding in exchanging a glance with a gorilla,' I whispered, 'than with any other animal I know. Their sight, their hearing, their sense of smell are so similar to ours that they see the world in much the same way as we do. We live in the same sort of social groups with largely permanent family relationships. They walk around on the ground as we do, though they are immensely more powerful than we are. So if there were ever a possibility of escaping the human condition and living imaginatively in another creature's world, it must be with the gorilla. The male is an enormously powerful creature but he only uses his strength when he is protecting his family and it is very rare that there is violence within the group. So it seems really very unfair that man should have chosen the gorilla to symbolise everything that is aggressive and violent, when that is one thing that the gorilla is not – and that we are.'

At that point, the vegetation on the other side of the flattened arena parted and out strutted a silverback male, the ruler of the group. The female had seemed large but he was immense, the very epitome of strength and power. While I was wondering whether or not to retreat, he suddenly sprang into action. He raced across the arena, gave the female a loud and violent thump on the back and disappeared into the grass on the other side. She got to her feet and prowled after him. In view of the clout he had given her, I wondered whether I had been over-romantic in the last sentence of what I had said. On reflection, and consideration of the violent record of humanity, I decided to let the statement stand.

Our last day came. I had still not explained, to camera, the significance of the opposable thumb. It was not the first thing that came to my mind when the gorillas were around. That morning we found them quite early. We came out of a thicket of low bush to discover a single adult female sitting by herself, pulling up handfuls of celery and chewing it. Occasionally, she picked up a single stem. With her thumb and forefinger. Clearly the time for the opposable thumb had arrived.

I crawled slowly towards her. She looked at me in such a placid way that I was encouraged to crawl a little closer. I lay on my side and turned to look at the camera. That inevitably meant that I had to turn my back to her. I didn't like doing that. Martin seemed to be ready. There was a noise from behind me. I turned and saw her heaving herself towards me. She stretched out a huge black hand, put it on the top of my head, and looked at me with her deep brown eyes. This didn't, after all, seem to be the right moment for a chat about the opposable thumb. She removed her hand from my head and pulled my lower lip to look inside my mouth. I felt a weight on my legs. Two youngsters, the size of chimpanzees, were trying to undo my shoelaces.

I felt in no way alarmed or even threatened. I had, it is true, crawled towards them, but it was the female who had made the first physical contact. She and the youngsters seemed to have accepted me as a welcome visitor, almost as a member of their group. My overwhelming feeling was one of privilege. I lay there for five minutes, almost holding my breath with delight. The female resumed her feeding. Eventually, the youngsters got bored with my boots and galloped away to look for something else to investigate. The female followed them and I crawled back to my companions.

'Well, we got a bit of it,' said John.

A bit? I had been there for five minutes or more. Was there a camera fault or something?

'No,' said John, 'but I was waiting for you to say something about the opposable thumb. Martin only had about fifty feet in the camera before he would have to reload and I didn't dare take any general shots in case you started to speak and we ran out of film. And anyway, the way you were playing with them looked so extraordinary that viewers would have thought they were tame gorillas in a zoo. Then Martin said why not take a few feet to give the boys in the editing suite a laugh. So, yes – we did get a bit.'

I could sympathise with John's dilemma. All the same I was sorry that only a fragment of one of the most exciting encounters in my life had been recorded.

The next day we set off down the mountain. A lorry met us at the bottom of the trail and we started the drive back to Ruhengeri. I was sitting in the driver's cab with Dickie. Martin and John were in the back. Suddenly a soldier sprang out into the road ahead, waving his arms wildly, signalling to us to stop. Our driver swerved to avoid him and drove straight on. I thought the soldier must be trying to hitch a lift and the driver was having none of it. But then there were two shots and the whine of bullets overhead. This was getting serious. We rounded a corner at speed and screeched to a stop. An army truck was blocking the road.

A tousled-headed European climbed out of the truck and started to harangue us in French. What were we doing up in the mountains? We had no right to be filming. We hadn't checked with the Parks office. We had no permit. We had gone up to film Digit's dead body in order to make trouble. As he spoke, I realised that this must be the Belgian Dian had described to us as her main enemy. Our film would be confiscated, he went on. And it was no good trying to escape because the airport had been alerted and we would not be allowed to leave the country. We must go immediately to Ruhengeri and surrender to the Security Police.

As we drove off, the Army truck following behind us, we debated what to do. Martin revealed that while John and I were being questioned, he had managed to change the labels on the film cans. If they did try to confiscate our film, there was a chance he could manage to hang on to the most important rolls. At the Security Police Headquarters we were lined up against a wall and given a detailed body search. Every box of equipment was opened, examined and impounded. We were put into an open air barbed-wire enclosure and locked in.

After an hour or so, a more senior officer arrived and interrogated us all over again. Dian had sent threatening letters to officials in the Parks Department and the Police, he said. We must be involved in some way. The American consul was contacted. It was past nine o'clock, when we were eventually told that we would be released. However, we were being expelled from the country and had to leave immediately. As a plane had anyway been chartered for us the following morning, we did not argue. And we kept our film, with its misleading labels unchallenged, with us.

Back in London, I got in touch with some of Dian's ex-students who had worked with her in Karisoke and together we started to make arrangements to start a fund to help protect the mountain gorillas.

* * *

Life on Earth was gratifyingly well received. Its ability to take the viewer in a fraction of a second from one continent to another, the systematic and serious way in which we had surveyed the natural world, not taking short cuts and featuring groups of animals that had hitherto been largely neglected – sea slugs, legless amphibians, naked mole rats and other creatures – made a great impression. The critics were unstinting in their praise. The audience was huge. Each programme, following the pattern initiated by *Civilisation*, was repeated within the week and many viewers watched both transmissions. And the book of the series sold in numbers far beyond anyone's expectations or hopes.

The United States, however, was resistant. A co-production deal had been signed before we started, but that was three years earlier. Since then there had been a palace revolution in the network concerned and the new monarchs did not care very much for the thirteen fifty-minute programmes that they had inherited.

Nothing happened for many months. Then I was telephoned by someone in the BBC's overseas sales office. At last the network looked likely to take up its option. I was delighted. However, he added, there was a small complication. My English accent simply would not be understood in Peoria or Chattanooga. Acceptance was dependent on the Americans' ability to cut out my in-vision appearances talking to camera, and to re-record the commentaries spoken by a Hollywood filmstar. The changes, of course, would be done with integrity. The name of Robert Redford was mentioned. Did I mind?

I minded a great deal. The series was the most ambitious thing I had done in television and I was proud of it. I did not know whether I would ever be allowed to do anything else again on such a scale. He must have realised, I said, that I would have minded, so why was he asking me? His answer was a surprise. Apparently, somewhere in the small print in my contract lay a clause stipulating that any changes to overseas versions had to have my approval. In spite of all his blandishments, and the fact that both the BBC and I would get substantial residual payments if I agreed, I stuck to my guns.

Eventually, the network surrendered its option. Happily, America's public broadcasting system, PBS, decided to risk it and bought the series for a much smaller sum than the commercial network would have paid. When at last they transmitted it, they discovered that my accent, after all, was comprehensible even in Peoria and Chattanooga. So I began to acquire sufficient of a reputation in the United States for my name, when it appeared on subsequent series, to be reckoned more of an advantage than a liability.

21

The Living Planet

Working on *Life on Earth* had occupied a couple of dozen of us for three years. Together we had taken the project through all its stages – debating the shape of the initial scripts, selecting the species to use as examples and then zig-zagging in separate groups round the world, often with one director shooting material not only for his own episode but doing bits and pieces for others as well. Sometimes I came back having been filmed speaking the first half of a sentence that fitted neatly on to a second half that we had filmed on another continent two years earlier. It was as if we were all working on an immense and complicated jigsaw with everybody having his or her own particular section yet at the same time being able to reach over and fit something useful into somebody else's. And as the picture slowly came together, it proved to be more detailed and coherent than we had imagined when we had first roughed out its design. I had enjoyed the experience so much that I wanted to repeat it. How could I?

There was, of course, another way in which a television series might survey the natural history of the earth as a whole. Instead of documenting one particular group of animals at a time, as we had done in *Life on Earth*, we could examine one particular kind of environment wherever it appeared. We could, for example, look at deserts, examine the problems of living in dry hot conditions and then survey the ways in which different kinds of animals, in different deserts worldwide, managed to deal with those problems. I sketched an outline of twelve programmes on this basis and called it *The Living Planet*. Thanks to the success of *Life on Earth* it was commissioned straightaway.

Ned Kelly, one of the three directors allocated to the series, telephoned me in September 1980, a few weeks after he had taken up the job. He is a tough, stocky man, an excellent naturalist and a dedicated mountaineer with considerable Himalayan experience. Unlike me, the colder it is, the more he likes it. He had therefore laid immediate claim to the programme that would deal with the polar regions and he had exciting news. He had discovered that a Royal Navy ship, HMS *Endurance*, was on its way to the South Atlantic and would be visiting the various scientific research stations run there by the British Antarctic Survey. As well as giving them support, the ship also had the job of showing the flag and making it plain internationally that Britain had a continuing presence in the South Atlantic. The Admiralty clearly thought that having a television unit on board would not harm that process for they had given permission for us to join her.

Unfortunately, if we did so we should have to spend Christmas there. It would be the first time that I had spent Christmas away from Jane and our two children in thirty years of marriage. But a cruise on the *Endurance* would produce material that we could get in no other way. I could hardly turn it down. I consoled myself with the thought that instead of sitting down beside a roast turkey on Christmas Day, I might be wandering among great congregations of penguins.

But it didn't turn out that way. HMS *Endurance* was in Port Stanley harbour when we arrived there, but one of the BAS supply ships that was due to take stores to the bases had damaged her propeller. *Endurance* was going to tow her to Montevideo in Uruguay for repairs. After Christmas, another BAS ship, the *Bransfield*, would arrive in the Falklands. She would take us to South Georgia. The *Endurance*, on her return from Montevideo, would then pick us up from there. Meanwhile we would have to spend Christmas in Port Stanley's only hotel, The Upland Goose.

Ned Kelly, Dickie Bird, Hugh Maynard and I were the only guests. Tinsel hung forlornly from every light and shelf but that seemed to make Christmas feel more desolate than ever. On Christmas Eve we wandered round the bleak streets harking to herald angels singing domestically from within the small wooden clap-board houses. Then a message arrived at the hotel from Iain Stewart, the manager of the telecommunications company which maintained the island's connections with the outside world. Would we like to join him and his wife for Christmas lunch? We did and it made up for a lot that we were all missing. The Stewarts' hospitable act was to remain vivid in my mind in view of what was to come.

Another invitation arrived from the Governor. There would be a big party at Government House on Christmas night. We were welcome to that too. Most of the citizenry of Port Stanley seemed to be there. The Governor, Rex Hunt, proved to be a fanatical games player. He lured a group of us down to the billiard room to compete in a knock-out competition he called Killer. He was the only one who knew its rules, its method of scoring or indeed what it was that each competitor was aiming to achieve. Not surprisingly he won. That was followed by another game he called Billiards Hockey in which the participants hurled billiard balls at one another across the baize of the table. The climax of the evening came when he produced a wind-up gramophone with a pile of 78 rpm shellac records and we all sat drinking rum, gin and whisky listening to Louis Armstrong, Sydney Bechet and Bessie Smith singing to us through a hurricane of hisses and crackles.

Three days later we left for South Georgia in the *Bransfield* and three days after that we arrived at Bird Island, a small islet lying off the south coast of South Georgia. Here four BAS scientists lived in a tiny three-roomed hut crammed with books, stores, nets, electronic tracking gear, pots of paint, tools, and great piles of cold weather clothing. They had been there for over a year seeing no other human beings and spending their days tramping over the island's desolate moors to count albatross and penguins, measure their eggs and weigh their chicks. While stores were being unloaded from the *Bransfield* and stacked in an outhouse, two of them took us to see the colony of macaroni penguins for which the station is famous.

We toiled up a steep slope of tussock grass behind the hut to the crest of a col. The penguin colony occupied the entire valley that lay on the other side and ran straight down to the sea. There were three hundred thousand pairs, they said. I could quite believe it, for the noise of their squawking was deafening. But the valley itself was filled with cloud. I could see just three birds, ghostly shapes a few yards away in the mist. How long it might take for the mist to clear no one could say. But in any case, we could not linger. The unloading of the stores would be finished within the next hour and the *Bransfield* would certainly not wait for us. So I stood on the crest of the col shouting above the deafening clamour of the invisible birds, doing my best to describe what viewers might have seen under other circumstances. As a sequence illustrating the unbelievable richness of the Antarctic, it was scarcely impressive.

That evening we left in the *Bransfield* and sailed round to the north coast of South Georgia and the main BAS Station at Grytviken. This had once been a Norwegian whaling station. The immense animals, having been harpooned far out to sea by catcher ships, were towed back here, hauled up the vast slipways by the tail and then cut up. It was only too easy to imagine the gargantuan butchery that must have taken place here – the thick blubber being peeled away from the bodies and sliced like monstrous linoleum, huge coils of intestines spilling out over the vast slipways, immense hearts shedding torrents of blood from veins big enough for a man to swim along, foetuses as big as cows flopping out of vast bellies, still wrapped in their membranes; and men wandering through this steaming horror wielding huge knives on poles like diminutive demons in one of Bosch's horrifying pictures of hell.

But then factory ships were introduced which did this work at sea. Their arrival, together with the increasing efficiency of hunting methods and the consequent reduction in the number of whales in the Antarctic Ocean, killed Grytviken. The end was very sudden. The company must have decided that it was uneconomic to remove anything whatever. Their employees boarded the relief ships and simply sailed away.

A pair of catcher ships still lay alongside the jetty. They had sprung leaks and were listing, half-sunk, in their moorings. Rusting winches stood at the head of the slipways. A twenty-foot-long saw, used doubtless for cutting the bones, hung slumped in its mount. We opened the creaky door of a small hut and found ourselves looking at the projection room of a tiny cinema with spools of 35mm film curling all over the floor. We peered inside sheds piled high with new cordage, blocks and tackle, lathes, bags of cement and harpoon heads. Stores of grain and other food lay stacked in the corrugated iron warehouses. Wherever there was anything that was remotely edible, huge fat rats with calloused tails were boldly waddling over the floors. And above it all towered the ice-hung peaks of South Georgia, many of them still unclimbed, sparkling in the sunshine.

HMS *Endurance* steamed in after a couple of days to pick us up. Her duties were essentially peaceful but nonetheless she was armed and carried helicopters which could be used for surveys, rescues or rather more war-like operations. I had mixed feelings, as I climbed on board. I had, it is true, spent two years doing my national service as an Instructor Lieutenant RN, teaching navigation and meteorology. Most of that time had been spent either swinging round a buoy on a vessel in the Reserve Fleet in Scotland or with a Naval Air Station in Pembrokeshire. I had, however, been to sea for short periods in a destroyer and a cruiser. I had even, on one brief but alarming occasion, stood on the bridge of a destroyer as Navigating Officer. So I had some idea of how a naval officer should conduct himself. I knew that when going on board, one saluted the quarter-deck, that one went 'below' rather than 'downstairs', that one spoke of 'scuttles' rather than port holes and went 'for'ard' rather than towards the front end. I also knew that the Captain of a Royal Naval ship was an awesome and remote figure, close to God. I was a little taken aback, therefore, when Captain Barker, RN of the HMS *Endurance*, with four gold rings glittering on his sleeve, offered to carry my suitcase and showed me into the cabin next to his.

A Captain in the Royal Navy is not, by naval custom, a member of the wardroom like all other officers. He dines by himself, and is looked after by his own steward. But he can ask guests to dine with him and Captain Barker invited me to do so every night. He was an accomplished raconteur and every evening, as we steamed slowly around South Georgia, I listened appreciatively to his stories. He told me about the naval operations in the Middle East in which he had been involved in the 1960s and gave me some of the background to the increasing tension that was then building between Argentina and Britain over the Falkland Islands. There were arguments going on in Parliament about the *Endurance*'s presence in these seas. Some MPs argued that the ship, for reasons of expense as well as provocation, should be withdrawn; others that to remove *Endurance* at this moment would be interpreted by Argentina as a hint that London had no real interest in or loyalty to the Falklands and would do nothing to prevent Argentina taking them over. South Georgia was a dependency of the Falklands and there were rumours that Argentina was sending armed parties posing as collectors of scrap metal to establish illegal bases there. This was denied by Argentina. One evening over dinner as we cruised along the northern coast of the island, Captain Barker told me that on the following day he was going to 'order the choppers to hop over the mountains to see if we can catch the Argies on the other side with their pants down.'

That jocular way of putting things seemed all of a piece with the rest of conventional navy-speak. It was hard to recognise that he was talking about operations that might lead to men killing one another. Surely it was all just part of an elaborate international game of cops and robbers in which the players merely postured at each other.

Two years later, Argentina invaded the Falklands. I saw television pictures of people in Port Stanley being taken from their homes at gun-point. I thought of the Stewarts and the Christmas dinner I ate with them. I remembered that unpretentious evening playing games with the Governor and I understood how it was that Britain reacted the way she did.

From South Georgia, the *Endurance* took us on to other snow-covered gale-lashed penguin-covered lumps of rock known as the South Shetlands. One of them, Zavodovsky, was blanketed by chinstrap penguins – fourteen million of them, according to one estimate. Few people have ever landed on Zavodovsky, for its shores are continuously battered by huge waves and there is no beach. But *Endurance* had helicopters and they landed us on the island for a privileged hour or so to film the penguins and then, as the weather worsened, whisked us off before a gale marooned us. We went on to other bases on the Antarctic continent and then sailed back westwards to the Falklands.

* * *

The first programme in *The Living Planet*, I thought, should deal with the creation of new land – in other words, with lava pouring from a volcano. We would then show how animals and plants managed to colonise a new totally sterile environment. Unfortunately, all the world's volcanoes, as far as we could discover, were inconveniently quiescent at that time. There were, it is true, some places in Hawaii where molten basalt was sliding like black treacle down a mountainside, but the sequence we were planning would be the first major one in a new series and we needed something more dramatic than that. I had in mind a scarlet fountain of fire spurting high into the night sky. We had tried to get such shots with Richard Brock on Anak Krakatau eight years before and had failed. Now we were to try again. Our best chance of doing so seemed to be in Iceland.

Iceland is proud of its volcanoes. They attract visitors. They provide the subject matter for wish-you-were-here picture postcards. Some are even harnessed to provide central heating for whole towns. So Icelanders keep a close eye on them and are very skilled at predicting what they are going to do. We contacted an Icelandic geologist and he promised to let us know if an eruption looked imminent.

One Saturday morning in November 1981, I was sitting in the British Museum at a meeting of its Trustees, an extremely distinguished gathering, rich in dukes, ex-ambassadors, captains of industry and emeritus professors. I had only recently been appointed, perhaps because of the *Tribal Eye* series, and I was still finding my way and unsure of myself. We were in the middle of an earnest financial discussion when one of the Museum's messengers walked quietly in, circled the table until he came to my place and handed me a note. I read it quickly and caught the Chairman's eye.

'May I have permission to leave?' I said. 'I have to film a volcano that has just started to erupt in Iceland.' It was not my first spoken contribution to a Trustees' Meeting but it created my biggest impression so far.

It was now a race to get to the volcano before the eruption stopped. No one could know when that would be. It might be a few hours. It might be a week. I took a taxi from the Museum to Richmond, and threw some appropriate clothes and boots into a bag. Jane drove me to Heathrow. There I met Ned who had raced up the motorway from Bristol. Since most volcanoes are mountains and he was a mountaineer, he had nabbed this programme as well as the polar one. We congratulated ourselves on being just in time for the Reykjavik check-in.

An announcement appeared on the airport's video monitors. The Reykjavik flight would be delayed. It was past ten o'clock in the evening before we at last took off. Two and a half hours later we landed at Reykjavik. We were met by an Icelandic film crew and together transferred into a small charter plane which flew us through the darkness across the island to the town of Akureyri on the north coast. From there two Land Rovers took us to a guesthouse in the mountains. The geologist who had alerted us was waiting there. He told us that the eruption had begun two days earlier. A fissure five miles long had opened up across the bare volcanic landscape. Molten lava had spurted up along its entire length, but much of it had fallen back into the fissure clogging and then sealing it. Now only one plume remained. We might just make it.

It was now three o'clock in the morning. We loaded our gear on to sledges and hitched them to skidoos which are, in effect, motorbikes on skis. For some reason that I do not understand, those who drive them think that they do not function properly except at speeds in excess of forty miles an hour. With some of us driving and others, including me, riding pillion, we hurtled off into the night.

Sitting on the pillion seat of a skidoo, heavily muffled, with your arms round the waist of the driver, and your head buried for most of the time in his neck to escape the searingly cold wind, is not the best way to enjoy a landscape. But as we rounded the corner of the valley, I peered round my driver's neck and was thrilled to see a quarter of a mile ahead of us a scarlet plume of lava spurting two hundred feet into the black sky. We stopped within two hundred yards of it, clambered stiffly off the skidoos and walked cautiously towards it.

It was bitterly cold – twenty degrees below zero. A gale was blowing, and there were squalls of snow. The red liquid rock was spouting vertically upwards from a circular crater making a noise like a jet engine. As it shot into the sky, it disintegrated. The smaller particles solidified in mid-air and fell as spiky ash that blew into our faces and stung our lips and eyes. Bigger lumps, still glowing scarlet, flopped down around the edges of the crater like immense pancakes, slowly lost their colour and became part of the blackness.

We were walking over solidified lava. In places it formed a jumble of blocks with such sharp edges that they could cut the leather of our boots. Elsewhere, it had remained liquid for longer and now lay in flat pools, with a surface that was sometimes smooth, sometimes ridged with coils, like rope. How recently this lava had been liquid was difficult to tell, but as we walked closer to the edge of the crater, we crossed cracks that still glowed red in their depths. Sometimes we stepped unknowingly on great blisters that had formed close to the surface and collapsed a few inches downwards with a crash – and we were relieved when the lower side of the bubble proved to be hard enough to carry our weight.

The wind was so strong that when the cameraman put down the gear to take out his camera, it overturned the heavy metal box. Its strength, however, was to our advantage for it blew from behind us and so carried away much of the heat of the lava fountain, allowing us to get quite close. Just how much it helped we quickly realised when it veered slightly, and a sudden grilling heat hit our faces and the hail of ash increased. Larger lumps began to thud on the ground alarmingly close by and we retreated in a hurry.

Iceland is just south of the Arctic Circle, but even so in November there are only a few hours of light in the middle of the day. When we started to film, the fire fountain was jetting into a pitch black sky. But slowly, as we worked, the sky began to pale and we were able to distinguish the outline of the surrounding mountains. We circled round the fountain and found on the other side a river of lava gushing from a fissure and pouring down the slope at around 60 miles an hour. Beyond lay a field of moving lights, trembling in the haze of their own heat and looking like a network of highways illuminated by the headlights of cars. We worked for several hours. I spoke the words I had prepared for this all-important sequence. By mid-afternoon we had finished and we trudged back to the skidoos. Only thirty hours had passed since I had left the meeting in the British Museum. But we had our sequence.

* * *

One of the problems we faced with the series was how to avoid repeating subjects which we had already shown during *Life on Earth*. The episode dealing with islands would certainly have to show how reptiles often grew to giant proportions when isolated for many millennia. We would show giant tortoises – but instead of the ones in Galapagos which we had filmed in the earlier series, we would go to Aldabra in the Indian Ocean which had a giant tortoise population of its own that had evolved quite separately. But what about the Komodo dragon? There was no parallel to that. It was twenty-five years since I had first filmed them; and we were going to go to Indonesia anyway. It would be nice to go back to see how the dragons were getting on. So I wrote them in to the script.

It is usually a mistake to go back. This time we approached the island from the west, taking a ferry that now operated from Bima on Sumbawa, the place where we had camped for days at the end of my first trip. The large powerful ferry, loaded with tourists, took a mere ninety minutes to get to the island. We were taken not to the village where we had landed but to a vast thatched building, standing on stilts that was being specially erected for the visit of an International Conference on National Parks. From there we were led along the beach to a large notice that said No Smoking. The path then turned inland with numbered posts placed along it, so that visitors would know how far they had come and needed to go. Half way along it, there was one of those jokes that officialdom, in its ponderous bureaucratic way, feels obliged to crack every now and then – a sign saying 'Beware Dragon Crossing'. Another hundred yards and we arrived on a platform overlooking a ravine. This too had its sign. 'Dragon Viewing' it said.

The lizards were already there, feeding on part of a goat carcass. At first we were told that on no account must we leave the platform. The dragons were far too dangerous.

It required a whole day of negotiation before we were allowed to walk into the ravine ourselves and I was permitted to stand near the lizards and talk about them.

I could hardly absolve myself from having contributed in some measure to this sorry transformation. I told myself that if the animals did not earn their keep, spinning money from tourists, they might not, in the long run, survive at all. And that in the farther recesses of the island there must surely be dragons just as wild and romantic – and disregarded – as the ones I had seen not so long ago.

* * *

Living Planet also had to include a programme about tropical rain forests. Here too, we had to find a new approach. I decided to introduce the programme sitting high up in the canopy. Then I would descend, stage by stage, describing the changing conditions, the reduction in light and the increase in humidity. Natural history sequences would show the animals that lived in each zone and explain how they dealt with those particular conditions. As we moved down from one level to another, so I would supply linking narrations to camera, hanging from a rope, until finally I closed the programme standing on the forest floor.

Although I had written with such confidence, I was not entirely clear as to how this might be done, but Adrian Warren, a young, athletic and mustard-keen assistant producer allocated to the programme, knew exactly. We would use jumars. This tree-climbing technique was then only just being developed. To start with you have to throw a thin line over a high branch. If necessary, it could be shot over with a crossbow. Then you tie a rope to the end of the line and haul it up. Once this rope is secured, a pair of jumars is clipped on to it. These are metal handles with slings hanging from them. They can be moved up the rope but not down. You put a foot in each sling. Shifting your weight on to one, you bend your other knee so that you can slide the jumar up the rope. Then you straighten that knee so that, in effect, you step up. Then you do the same with the other leg. That way you climb vertically upwards. Adrian explained it all to me as we stood at the bottom of a two hundred foot tall kapok tree in the rain forest of Ecuador. I was not unduly perturbed. I thought back to my rock-climbing days and felt that I would not be upset by the height; and as I watched Adrian move nimbly up and down the hanging rope to show how easy it was, I thought I could manage the physical effort involved without too much difficulty. I was wrong on both counts.

The first few steps were easy enough, but nonetheless extremely exhausting. I laboured away, swinging rather erratically from side to side. After a few minutes, pouring with sweat, I had to rest. I looked down. The rope, with my weight on it, had stretched a bit. I was still barely above Adrian's head. Hugh Maynard with his camera and Dickie Bird with his recorder, were standing beside Adrian, waiting patiently for me to get to a reasonable height before they even started to record my progress. I tried again. It took me a long time before I reached the height of a hundred and fifty feet, which is what Adrian felt was necessary to make the point.

Coming down was easier, though the process of changing from the jumars on to a descending clip, depended on tying a special knot. At such a height, with the knowledge that if I got it wrong I would drop to the ground like a stone, I found that remembering whether left should go over right, or vice-versa, was a little stressful. Nonetheless, I managed and by the time I got back to the ground, though somewhat ruffled and drenched with perspiration, I had recovered some of my composure.

'Okay?' asked Adrian solicitously.

'Nothing to it, really,' I said.

'Just as well,' said Hugh. 'I've just had a look at the magazine. There was a horrible scratch on the film. You will have to go up again.'

All this was done in order to get the first scene-setting shot of my disappearing upwards towards the canopy. For the shot in which I actually arrived there and turned nonchalantly to camera to introduce the programme, Hugh Maynard and his camera had to precede me. It proved more convenient to do that in the top of another tree. Adrian had constructed a tiny platform – about the size of a tea-tray – on which he would sit with Hugh and his camera alongside. I was to speak my piece sitting astride a huge branch nearby that projected horizontally from the main trunk.

This was where I was to discover that my second assumption about confidence was mistaken. Somehow, the unconcern I felt about heights as I clambered about on the rock precipices of North Wales in my youth seemed to have evaporated with the passing years. I had only been mildly alarmed when climbing up on the jumars. The rope a few inches in front of my nose made me worry a little less about the space beneath. But as I hutched out away from the trunk along the branch, with the rope lying carelessly along it, my feet dangling on either side and a sheer drop of two hundred feet below me, I felt no such reassurance. The branch itself was loaded with ferns, orchids and bromeliads. Indeed, that was the very reason it had been chosen, for I was to talk about these tiny ecosystems that were perched up here in the canopy. I inadvertently detached one of them as I moved nervously along the branch. Watching it spiralling downwards to land many seconds later on the ground, did not settle my nerves.

'Action,' said Adrian. I looked at the camera lens and tried to keep a quaver out of my voice as I delivered my lines.

'I am sitting two hundred feet up in the canopy of the jungle...'

I hope I appeared to be unconcerned, but if viewers had noticed the agitated way in which I was clasping and unclasping my left hand on my left thigh, they might have had a hint of how high in my mouth my heart was.

* * *

Another programme in the series took me much higher than the top of a kapok tree. Very much higher. There are surprising things to be found at altitude. The animals that venture farthest away from the ground are not, as one might suppose, birds. They are spiders. Young ones of many species, soon after they hatch, climb to the top of grass stems and start extruding a thread of silk from their spinnerets. Eventually this gets so long that the wind catches it and lifts the spiderlings skywards and they are swept away to colonise new territory. They have been recovered by researchers from altitudes of twenty thousand feet. My script suggested that we might watch them being collected from within one of the jet aircraft that had done so by towing a net behind it. Adrian did not think much of this. Sitting inside an aircraft would hardly give the sensation of what

△ Joy Adamson caresses Elsa the lioness. George Adamson lies alongside.

▽ I cautiously get acquainted with Elsa, who has commandeered the vehicle in which I am to sleep.

The new managerial team at BBC Television in March 1965, inspecting the first 625-line BBC2 studio in the Television Centre: Huw Wheldon, flanked by Michael Peacock, Controller BBC1, and the newly appointed Controller of BBC2.

△ Dressing down. The *Tribal Eye* film team are allowed to film in the anti-European village of Makaruka in the Solomons, but if they discarded European clothing and wore loin cloths.

▽ Moro, leader of the Makaruka people, dressed in a costume made from shell-money, with the statues he keeps in the community's cult house.

Bested by gorillas. Trying to explain the importance of the opposable thumb of the primate hand cannot survive the attentions of a young mountain gorilla keen on a romp.

△ Arthur Dick's potlatch in Alert Bay's ceremonial house. Masked dancers perform before the assembled people at the climax of the festivities.

▽ The mysterious Easter Island figure. A series of clues from Sydney and St Petersburg finally linked it directly with Captain Cook.

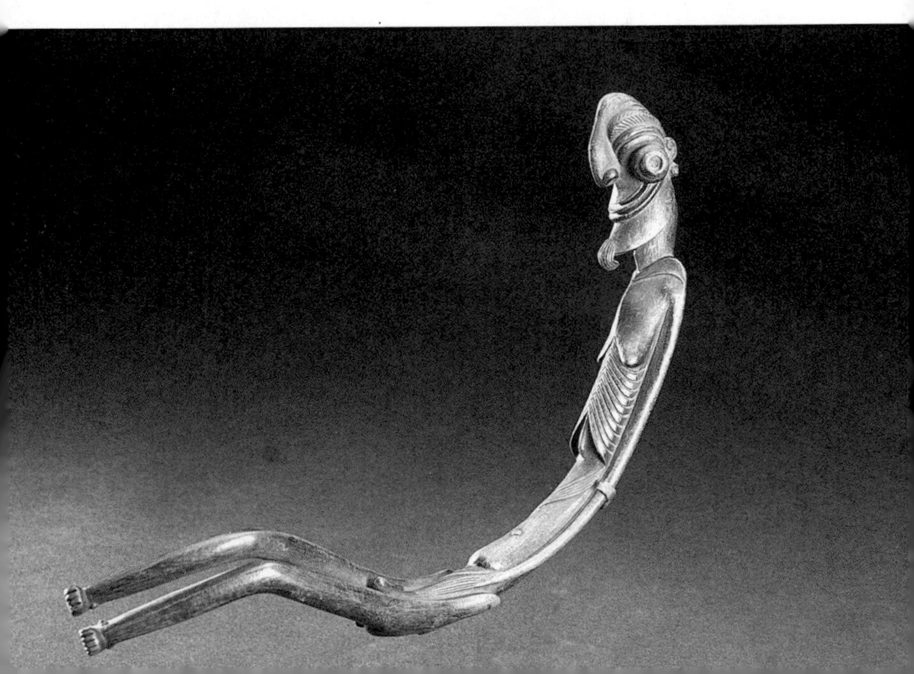

△ Dinosaur footprints at Purgatoire Valley, Colorado. Huge though they are, they did not show up well until filled with water.

The huge flowering of *Amorphophallus* in Sumatra. It has begun to close, but Mike Pitts has already filmed it. ▷

△ Paul Atkins, Alastair Fothergill, Trevor Gosling and I in the middle of a of king penguins' crèche on South Georgia.

▽ At night, a kiwi probes the sand for hoppers on a beach in New Zealand's Stewart Island. With poor eyesight the kiw relies on its sense of smell: the smell of rotting seaweed seemed to conceal mine.

conditions were like at these great altitudes – the extreme cold, the lack of oxygen, the silence. Why not film the sequence from a balloon? I raised the obvious objection that we could not catch spiderlings from a balloon, for we would be drifting with the wind and there would be no air current to drive them into a net. Adrian suggested that we should fit a fan on our net to create such a current and we could take some specimens that had been caught earlier. Eventually, I agreed to his plan, provided we made it clear that we had not ourselves caught the spiderlings that we showed.

There are not a lot of balloons that can ascend to twenty thousand feet. They have to be huge to go so high. In fact, as far as Adrian could discover, there were only two. But one of them was operated by Don Cameron, a highly experienced Bristol balloonist, and he agreed to fly it for us.

A high-altitude balloon demands very specific flying conditions. Take-off can only be safely achieved in total calm – and the best time for that is usually at dawn. There must not be too many clouds. And one has to know exactly what the winds are doing at high altitudes, for they are likely to be blowing in quite different directions and speeds from those lower down. We had to be sure that they would not take us out to sea or into the flight path of planes.

We stood at the ready for several weeks. Again and again the flight was postponed because conditions were not right. Then at last the message came for us to go. A strong air current was blowing in a north-westerly direction at high altitudes. If we took off from southern Scotland, we would drift towards the Grampians as we climbed and with luck would reach the right height for me to speak my piece and show off my pre-caught spiderlings before we left land altogether and drifted out across the Atlantic.

We started putting on all the gear we needed to be safe in an open basket at altitude. That meant blue space suits, high-soled moon boots, oxygen cylinders and mask, crash helmets with microphones inside so we could talk to one another, and the specially large parachutes that are needed if one is to be supported as one drops through extra-thin air. By the time we had got the lot on, we were all so bulky that we only just managed to squeeze into the basket. The huge balloon swayed and billowed above us. I was still trying to familiarise myself with the bits and pieces that had been hung around me when I glanced over the side of the basket and saw that, without my realising it, the ground had silently dropped away from beneath us. We were off.

The flight itself was rather uneventful. Don kept looking at various dials and told us how high we had climbed. It got colder. Adrian said that it was time we put on our oxygen masks. At 16,000 feet, rather less than Adrian had hoped, Don announced that we would have to descend. The winds were not blowing in the direction predicted. If we did not come down immediately, we would be blown out to sea. I hastily spoke my piece explaining that my spider-catching apparatus, devised by Adrian and based on a hair-dryer, had failed to collect anything but that I had with me some spiderlings that had been caught earlier. We started to lose altitude. We sank slowly and gently through several bands of cloud. When at last we were clear of them we saw far below, an empty expanse of moorland. We floated across a little valley with a small cottage at its head and on over a level moor. There were a few sheep wandering aimlessly through the heather. Ahead stretched a stone wall. Don gave a blast on the burner to try and lift us over it. The balloon rose slightly but not enough. The bottom of the basket just clipped the top of the wall so that it tilted and gently decanted us on to the ground.

While the others were gathering up the canopy, I set off down the hill to the cottage we had seen to try and telephone our recovery team. There were no mobile phones in those days. It was not easy walking through the heather in my huge moon boots and I was very hot in my space suit. I knocked on the door. A dour black-haired man opened it and stared at me suspiciously.

'We've just landed on your farm,' I said apologetically.

'Aye,' he said, narrowing his eyes, 'Ah ken.'

'I was wondering if we might use your telephone?'

He looked me up and down.

'Are yew from the tully-vision?' I nodded.

'The one with the animals?' I nodded again.

'It's ma wee bairn's burrth-day today,' he said. 'Would you wish her many happy returns?' I nodded again.

He disappeared and returned leading a small girl by the hand.

'This is David,' he said. 'He's come in a balloon to wish you a happy burrth-day'

His daughter sucked her thumb and stared at me wide-eyed.

'Happy birthday,' I said.

'The telephone,' he said, 'is over there,' and he and his daughter disappeared into a back room.

* * *

Our balloon flight had not really given us as dramatic a sequence as we had hoped. But Adrian had another idea for bringing a bit of drama into our programme about the upper atmosphere. NASA, America's space agency, had a specially adapted jet aircraft in which they accustomed potential astronauts to the experience of weightlessness. It flew in a series of parabolas. After climbing to around 35,000 feet, it tipped forward, following a carefully calculated curve, and started to dive. As it crested the top of the curve its passengers experienced zero gravity. It was like being lifted off your seat on a roller coaster but staying in mid-air for a very long time. The aircraft was known, with good reason, as the Vomit Comet. A pharmaceutical company had chartered it for use in field trials for sea-sickness pills. Adrian had persuaded them to let us join one of their flights.

The volunteers for the trials had been undergoing all kinds of strenuous and demanding physical tests in gymnasiums under the supervision of two charming blonde girls. The zero-gravity flight was their reward for having done so much hard work. We were already on board the aircraft when the latest batch of volunteers marched out to join it. At their head strode a six foot tall athletic-looking young man, grinning with enthusiasm. One of the flight attendants waiting by the aircraft door to receive them recognised him from earlier land-based experiments.

'Oh my Gard,' she said. 'If our tests mean anything at all, he is going to be as sick as a dawg!'

The first time we started on a parabola, we all looked fairly apprehensive. The climb was so steep it pulled us down into our seats. Then the engines cut back, the plane sailed over the crest of the parabola and we all floated up from our seats, our eyes wide with astonishment and all of us laughing with sheer delight. For a full thirty seconds – which in these circumstances is quite a long time – we floated gently around, nudging ourselves from the cushioned walls, bumping along the ceiling and revolving in mid-air. The attendants called a warning of the approaching end of zero gravity so that those of us who were still hanging upside down could propel ourselves back into our seats before gravity renewed its pull and threw us flat on the floor.

We worked out a plan for filming this extraordinary experience. I would learn my lines so that I could repeat them accurately with exactly the same timing. The pilots would give us a cue precisely twenty seconds before we went weightless. I was to begin sitting cross-legged on the floor, talking about the force to which every object on the earth, alive or dead, is subject – gravity. 'If that force were to be suspended,' I would say, 'the most extraordinary things would happen,' and on these words, without moving a muscle and still cross-legged, I should float up into the air. We would have plenty of chances to get it right for the Vomit Comet's programme required it to make forty parabolas on every flight.

By parabola number three, we got the timing just right. The sickness-tablet guinea pigs farther down the aircraft were having a tougher time. Some were labouring away at various exercise machines. Others were throwing up into paper bags. With my words delivered, I had a chance to relax and enjoy the experience. While we were still climbing, I filled a plastic cup with water. Then as we went weightless I lowered the cup, leaving the water hanging in mid-air, slowly disintegrating into floating globules.

By the time we got to parabola number fifteen, the enthusiastic young man who had led his contingent, having ridded himself of the contents of his stomach, was lying ashen-faced and prostrate on the metal floor, pleading with the attendants to ask the pilots to land. I too was beginning to feel that maybe I had extracted as much pleasure from the experience as I was likely to get. Adrian, on the other hand, had only just started to enjoy himself. Before we landed he had persuaded the crew to let him come along the following day and have the whole experience all over again.

* * *

The very first sequence in any big series is of crucial importance. It has to be visually arresting, to prevent people who might be wondering whether or not to watch it, from switching away to see what is on another channel. It also has to give some foretaste of the scope and editorial aim of the series as a whole. It then has to lead seamlessly to the main topic to which the rest of the programme will be devoted. We had already decided that the first programme of *The Living Planet* would deal with volcanoes, but we also had to make it clear in the prologue that the series was primarily about animals and plants and the way they adapted to differing environments. I suggested that we might achieve this by taking a three-minute journey up a Himalayan valley, seeing how the plants and animals changed character with increasing altitude and decreasing temperature. In its lowest reaches we would see orchids, sunbirds and perhaps tigers. Around 3000 feet, we would find rhododendrons and langur monkeys. By seven thousand feet, the vegetation will have shrunk to low sparse bushes and thickly furred animals should appear – snow leopards and Himalayan pandas. And at the valley head, around 14,000 feet, the plants would be tiny, and there would be very few animals of any kind. Ned Kelly knew just the place. The Kali Gandaki valley in Nepal.

The completed sequence, shot as we travelled up the valley, did exactly what I had hoped. At the end of it, I was seen walking into a small Sherpa hamlet, one of the highest human settlements in the world. We showed women working in the fields, some pulling heavy carts. Pursuing the series' theme of adaptation to the environment, I explained to camera that human beings were just as subject to adaptation as any other animal. The Sherpa people who lived here, I said, had become physically adapted to their conditions just as other creatures had. Their blood contains 30 per cent more red corpuscles than people born at sea level. Their chests and lungs are also exceptionally large so that they are able to take in more of the thin air with each breath than a lowlander can do.

I talked about this as I walked slowly alongside two Sherpa women who were dragging a plough and going rather faster than me. To underline my point, I panted a little as though struggling for breath. Perhaps I overdid it slightly. At any rate, when the series was finally launched in autumn 1984 one of the newspaper critics attacked the BBC on my behalf. It was unkind, they said, to send a presenter of my advanced age on such arduous trips. It was time I was given less demanding jobs more appropriate to my antiquity.

* * *

Andrew Neal, who had directed the last programme in the series, drove me back to London from Bristol after I had recorded the commentary. He asked me what I wanted to do next. I thought of the endless hours I had spent in airports during the last five and a half years since we had started work on *Life on Earth*, of the interminable nights sitting cramped up in aircraft seats, of being out of touch with my family for weeks on end, of the dreadful things we had had to eat in the back streets of remote villages.

'I don't know,' I said, 'but whatever it is, it will not take me more than a few hours away from home. I will always be able to phone my wife. There will be a reasonable chance that I can speak something of the local language. And there will be a decent meal every evening with a cheap but drinkable bottle of wine.'

'Like where?' he said. 'I can't believe you could make one of your series in any such place.'

'What about the Mediterranean? We could sketch its geological history, look at its animals and plants, and then make an overall survey of what humanity has done to them right up to the present day?'

Andrew did not hesitate for a second.

'You're on,' he said.

Now that I had said it, it seemed rather a good idea. One of my regrets about *The Living Planet* had been that it had devoted only one programme to the most recent of environments, those created by humanity. A Mediterranean series could be a logical and valuable sequel. If it were true that human beings first evolved in Africa and then travelled north to more temperate regions where they domesticated animals and plants and built the first towns, then the first landscapes in the world to be radically affected by them were the lands around the eastern end of the Mediterranean. They must therefore be the oldest humanised landscapes in the world. Before Andrew and I got back to London, we had agreed on a title – *The First Eden*.

I started research and quickly hit upon a highly dramatic way of starting the whole series. Only a few years previously, geophysical survey ships, boring into the bed of the Mediterranean, had discovered huge deposits of salt within the rocks five hundred feet down. They extended all over the Mediterranean basin. The deposits are so thick that the drills could not reach the bottom of them, but echo-sounding revealed that in some places they extend downwards for over a mile. Microscopic examination revealed that these salts contained the remains of algae. The conclusion was inescapable. At one time – some six million years ago – the Mediterranean had been dry, its waters having evaporated to produce these deposits. Then, around five and a half million years ago, the Atlantic broke through the rocky barrier at Gibraltar and a colossal waterfall was created, fifty times higher than Niagara and many miles across. Within about a year it filled the great empty valley to the east and created today's sea.

We did not try to create a visual image of this stupendous spectacle, as perhaps film-makers today, with computer imaging at their disposal, might be tempted to do. We did, however, go to the Dead Sea and filmed the salt deposits that surround it, to give some impression of what that vast empty valley must have been like, while I tried to describe in words how the Mediterranean came into existence. We filmed macaque monkeys on the southern shores of the Sea in Morocco and showed their cousins on the Rock of Gibraltar that are probably descended from pets first brought across the Straits by the Romans. And we chronicled the domestication and history of one of the greatest natural gifts that the Mediterranean has given humanity – the vine.

The series attempted to link zoology and botany to archaeology and ancient history, and to find the echoes of medieval myths in the customs and politics of the twentieth century. In doing so I became aware of continuities that I, at any rate, had never appreciated before.

The most impressive and convincing of these concerned the bull. Bulls, it seems, have been worshipped in Mediterranean lands without a break from the earliest times. Together with the horse, the bull is the animal most frequently painted by prehistoric man deep in the caves of France. It dominates the magnificent painted cave of Lascaux. The oldest known of these bull-paintings date from 35,000 years ago.

Bulls are also represented on the walls of a ritual centre in one of the earliest cities yet discovered, Çatal Huyuk in Anatolia which developed around 8400 years ago. They were the first animals to be worshipped as gods in the temples of Ancient Egypt a mere three thousand years later. A special bull calf, identified by particular markings, was selected by the priests as the mortal incarnation of Ptah, the creator of all. He was kept in a special temple at Memphis on the Nile. On his death, his body was mummified with all the care that was lavished on the corpse of a Pharaoh, sealed in a huge stone sarcophagus and interred in a special underground catacomb, the Serapeum.

The next great civilised centre to arise after Egypt appeared around 3000 BC in the island of Crete. There too bulls were worshipped. Athletes took part in extraordinary rituals during which they faced a bull and, when charged, seized the animal's horns and vaulted over its back. The Romans continued the cult of the bull with the worship of Mithras, a hero who slew the primordial bull and fertilised the earth with its blood. Today, the last relics of these cults survive in the bullfights of Spain and southern France. All that became the subject of our second programme. It made a more coherent and logical story than I had imagined when I first started researching it.

The third programme dealt with the growth of human population around the Mediterranean's shores and the Crusades, during which the armies of Christianity and Islam began the battles that still continue in that region even today. Indeed, a war was still raging around the valleys of the Lebanon while we were filming. Nonetheless, we were able to visit the most perfectly preserved of the Crusader castles – Krak des Chevaliers – and describe how both sides, Christian and Muslim, inflicted the most appalling atrocities on one another in the name of religion.

The last programme started in the sixteenth century, when Mediterranean peoples, as their number increased, sailed beyond the Straits of Gibraltar and discovered a new world on the other side of the Atlantic. They found species there unlike any in Europe and those they thought might be useful as food, such as the potato and the tomato, they sent back to their

Mediterranean home.

That process has continued ever since. In 1869 the Suez Canal was opened and the waters of the Red Sea started to mix with those of the Mediterranean. Slowly exotic fish, crustaceans and seaweeds strayed out of the Canal and spread westwards through the Mediterranean. One way to chart their progress was to research the appearance of Red Sea species in Mediterranean food. So I devised a final sequence to the series in which I had to sit in a succession of restaurants from Alexandria to Athens, Italy to France, and compared menus. I have never found a more delightful way to make a legitimate biological point – or to end a television series.

22

Extra Duties

Alasdair Milne, who had succeeded Huw Wheldon as Managing Director of BBC Television, summoned me to his office.

'We need a new producer of the Queen's Christmas message,' he said, in his normal abrupt way. 'The Palace wants you to do it. I told them that you were quite the wrong man. But they are insistent. Will you do it?'

'Do I have any option?'

'No,' he said, cheerfully. 'The job is yours.'

I went to see Bill Heseltine, the Queen's Private Secretary, to discuss treatments. My view was that it was a mistake to try and get a chatty, domestic, I'm-just-an-ordinary-person kind of feel. The whole point of having Royalty is that the Sovereign is *not* the same as other people. Wars were fought about that issue in the eighteenth century, and the nation still likes to believe to some degree in the divine right of kings. Equally, we ought to be a little more imaginative than simply to ask Her Majesty to sit behind an ormolu-decorated desk. What we needed was an occasion when the Queen could be properly queenly. After all, she had had some practice at doing that. Was there an event around Christmas time where she could deliver the message to a flesh and blood audience and at the same time address the world on television?

'There's a Christmas party in the Royal Mews she always gives for the children of the Palace staff where she hands out presents and wishes everyone a Happy Christmas,' said Bill.

The relevance of a stable to the Christmas story was not lost on me.

'Perfect,' I said.

There were a number of practical problems to be solved. The Master of the Horse pointed out that if the Queen walked down through the royal stables in the way I was imagining, all the horses would be facing away from her with their heads in the mangers. It would hardly do to have her approach down an avenue of equine buttocks. However, he said, that could be fixed. 'We'll turn the horses around so that they are facing her. It will look pretty silly to anyone who knows anything about stables, but I don't suppose that will matter too much.'

Come the night, there was a promise of snow which would make it all marvellously Christmassy. In the event it proved to be drizzle but one couldn't have everything. The Queen walked towards the camera between the lines of horses' heads, and was presented with a posy by one of the children. She smiled charmingly, advanced on to her marks, looked at the teleprompter and delivered her message perfectly. Overcome with relief, I started to tell her how splendidly she had spoken when the cameraman, Philip Bonham Carter, started plucking at my sleeve. I brushed him off and continued with my compliments. When he did so a third time, I thought I had better discover what his problem was.

'Tell her to go again,' he said.

'What's the matter. It was perfect.'

'Tell you later,' he hissed.

My thanks turned to grovels.

'I'm so sorry, ma'am. A small technical problem. May I ask for a repeat performance?'

She was surprisingly good about it and did it a second time, though perhaps not quite as well as the first. Then she went on to greet more of the children.

As soon as she had gone, I turned to Philip.

'What on earth was the matter? Not a technical problem for heaven's sake.' 'No,' said Philip, 'but just as she started to speak the big black horse peering over her left shoulder got that funny tickle of its upper lip that horses often get and started to waggle it. It looked as though he was doing the talking.'

I told Bill what the problem had been. He laughed and the whole evening was regarded as a success.

The Palace gave its approval to the final edited version. Preparing special versions – in some cases in foreign languages – for the little bits of the Empire that were still left at that time, and sending them off by air (it wasn't regarded as safe for such a crucial missive to be relayed by satellite), occupied the next couple of days. A lady translating Her Majesty's words into sign-language for the deaf had to be added in a vignette in the bottom left-hand corner. All seemed finished. But then Bill Heseltine rang up.

'The Queen,' he said, 'would like to see the Khartoum version.'

I broke out in a sweat. I hadn't prepared any such thing. Was Khartoum still part of the Empire? And why should it have to have its own version? But it was all right. Khartoum proved to be the name of the ventriloquising horse leaning over the Queen's shoulder. Bill had explained why we had filmed the message twice and the Queen wanted to share the joke.

The next year, 1988, it was felt that we could return to the routine desk setting. It was the three hundredth anniversary of the establishment of William of Orange on the British throne. Bill said that the Queen was rather interested in William-and-Mary. Only a few months earlier she had bought a diamond studded patch box that had belonged to that Queen Mary. Could we weave that into the message somehow? Maybe it could be on the desk and she could refer to it. That was agreed. The recording could be done in Buckingham Palace. Would I like to pick a room? A small one was found to which I gave my approval. And what was the Queen to wear? Well, what did she fancy herself?

'She thought she might wear the outfit she is wearing this afternoon for an investiture,' said Bill. 'See what you think.' So he and I lurked in a side corridor waiting for her to pass by, on her way to the Throne Room to hand out medals. A distant susurrus and the faint yaps of corgis told us she was on her way. A few moments later, she came past, her hand bag over her arm, leaning slightly forward as though into a heavy wind. She was wearing an outfit of searing acid green.

'No,' I said, 'that's altogether too dominant. And in any case, the wallpaper of the room we have selected is green.'

'Well, then,' said Bill, 'you had better go and pick something else.'

A couple of days later, I found myself inspecting a few hundred dresses. I picked out a few that seemed a little less strident than the outfit I had rejected. One was pale blue. Another a kind of mushroom colour. In due course, I was told that mushroom had been agreed.

Came the day and Bill and I awaited the Queen at the doors of the somewhat antiquated lift that connected the Queen's private apartments with the rest of the Palace. I was not quite so quiveringly nervous as I had been the previous year. The lift descended, clanged to a halt and the Queen stepped out wearing the mushroom outfit and looking like thunder.

'I hope this costume suits you, Sir David?'

'Yes, ma'am. It's perfect.'

'Really there is no pleasing you people from the media. I am told by one lot that I have to wear something colourful so I can be seen in a crowd; and then you come along and tell me that I have to wear something pale and nondescript. What was wrong with the outfit I was wearing for the Investiture.'

'Well ma'am, it wasn't just that it was bright, it was also that it was green, the same colour as the walls of the room where we are doing the recording.'

'The alternative was to re-paper the room,' said Bill, brightly.

'Re-paper the room!' said the Queen. 'Have you any idea what that would cost?'

'I think that was Sir William's little joke,' I said.

'Was it!' she said. 'In that case it was a very bad one' and swept off to take her place behind the desk.

* * *

The following year, we returned to the idea of recording the message at a big event and chose a huge military pageant in the Albert Hall. There was a certain worry at giving what was supposed, by convention, to be a highly secret message that must not be divulged before being delivered by Her Majesty. I thought that objection absurd. The message could scarcely be more anodyne. All we had to do was to ask the six thousand odd people in the Albert Hall not to spread it about unduly. That was agreed and so the programme had state trumpeters, massed military bands and all the pomp and ceremonial that should accompany the notion of royalty. It was, to my mind, one of the more successful of those that I handled.

The year after that, we returned to the desk. This time a copy of one of the drafts of the speech fell into the hands of one of the BBC news correspondents who leaked a few of its phrases to a daily newspaper. There was the most frightful row. Headlines. Scandal. Official reprimands. The correspondent in question was sacked from the Corporation's staff.

In 1991, I was 65. I discovered that 65 was the retiring age for permanent members of the Royal Household. That, I reckoned, should include me.

23

Vanished Lives

I wanted to make a series about fossils. I have been fascinated by them for as long as I can remember. Hitting a block of stone with a hammer, seeing it fall apart to reveal a coiled shell unlike anything alive today, beautiful in its shape and perfection, is an excitement that I first experienced as a small boy and one that still stirs in me just as powerfully as it ever did. The sun glinting on that mineralised shell is the first light to strike it for fifty million years, and yours are the first human eyes to see it.

The countryside around Leicester where I was brought up was easily reached by bicycle and as a boy, I spent much of my free time there. To the north-west lies the Charnwood Forest. I found out that the rocks there were among the oldest in the world, so unimaginably ancient that life at the time they were formed was only just beginning and had yet to leave its mark in stone. To the east, however, it was different. There you could find pits where a honey-coloured limestone was quarried for smelting iron. In that there were the loveliest fossils imaginable.

There were belemnites, six inches or more long and the shape of bullets. They were known locally as thunderbolts and were, in fact, the internal skeletons of an extinct squid-like creature. There were shells which, because they had a little spout just beyond the hinge, were said to resemble Roman pottery lamps and were therefore called lamp-shells. Paleontologists call them brachiopods. If one had lost part of its shell you could sometimes find inside it a scaffold-like structure of tiny ribbons which in life had supported the soft parts of the animal. Others, the size of hazelnuts, were like cockle-shells but with the junction of the two halves of their shells contorted into zig-zags. Best of all, there were ammonites, flat coiled shells. Some had narrow coils with multiple ridges, others only a few very broad relatively plain ones. When an ammonite was broken, you could see that it was divided internally into compartments. These had once contained air and had allowed the owner of the shell to control the depth at which it floated. If you discovered one where the outer shell had been delicately removed by rain and frost, grain by grain, the junctions between the walls of these internal compartments and the outer shell might be visible. These had the most extraordinary complexity, meandering in a series of tiny loops, each loop itself embroidered with minuscule petal-like shapes.

In 1933 when the British Association for the Advancement of Science held its annual conference in Leicester, my father, as Principal of the city's University College, invited the Association's President, Sir Frederick Gowland Hopkins, to stay with us. Sir Frederick had been the first to identify (and I believe invented the name for) vitamins, for which work he had been awarded the Nobel Prize. I remember him as a stooped, white-haired old gentleman. But I was dazzled by his daughter. She seemed to me to be the most beautiful lady I had ever seen. So enraptured was I that I asked my mother whether she thought Miss Hopkins would like to see my museum. My mother said she would ask Miss Hopkins. Miss Hopkins said that she would.

My museum was, I dare say, like lots of other seven-year-old-boys' museums. Its backbone was my fossil collection gathered from Leicestershire rocks. It also contained butterflies, birds' eggs (legal at the time), abandoned birds' nests (a long-tailed tit's nest neatly covered with fragments of lichen was a specially prized exhibit in this section), bun pennies, champion conkers, the shed skin of a grass snake, and a fragment of Roman brickwork picked up from near the City's Jewry Wall. Needless to say, for such an important visitor, new labels had to be written out by hand for each exhibit and I laid out the whole assemblage on a long shelf that went round the conservatory where I kept my fish tanks. Miss Hopkins descended from the guest room upstairs and allowed me to take her round, listening solemnly as I explained in considerable detail the identity and provenance of every item.

Many days later the postman arrived at the door of our house with an immense brown-paper parcel. It was addressed to me. I could not imagine what it might be. It was not my birthday. Neither was it Christmas. Those were the only two occasions, I thought, when one received parcels. I opened it, mystified. It was from Miss Hopkins. She said that she had enjoyed looking at my museum and wondered if I would like to add the enclosed objects to it. There was a pearly nautilus, a desiccated pipe-fish, some Roman tesserae and a medieval silver coin, a few grey sherds of Anglo-Saxon pottery, cowrie shells from the Pacific and pieces of coral. Each was packed separately. Each was a treasure. It was one of the most memorable days of my childhood.

Miss Hopkins in due course became Jacquetta Hawkes, the writer and archaeologist and subsequently Mrs J.B. Priestley. A quarter of a century later, she appeared in *Animal, Vegetable, Mineral?* and I reminded her of the occasion. She said she remembered it well, but she might have been being polite.

Her gift spurred me on as a fossil collector. I would leave home early in the morning on my bicycle, with special home-made collecting bags strapped to the carrier over the back wheel, and sometimes would not come back until after dark with the bags loaded with specimens, each carefully wrapped in newspaper for protection.

I showed my discoveries to my father. He had been a student teacher in his youth and while he earned his living in this way he had managed to win grants and scholarships to get himself to Cambridge – no easy thing to do in the early twentieth century. There he became a Fellow of his college and lectured in Anglo-Saxon, so he was not feigning ignorance when he told me that he did not know the names of my fossils. But I suspect he would have said that anyway for he was wise enough as a teacher and a father to know that pat answers are easily forgotten. He suggested to me that there were probably books that contained the information that I sought and that the City Museum might have named specimens on display with which I could compare my finds. Then he was duly impressed when I found the answers for myself and was able to explain to him how one could tell what the living animal had been like all those millions of years ago. So I came to know the names of the Leicestershire fossils. When I found out that one was an ammonite called *Tiltoniceras* because the first specimens to be discovered had been found near the Leicestershire village of Tilton where I myself had found ammonites, I decided that I must be living in one of the world centres of paleontological treasure.

As the time approached for me to take my first national examinations, I discovered that geology was among the optional subjects. My grammar school, next door to the University College, did not teach it but they were understanding and arranged for me to go to Nottingham University to eavesdrop, unofficially, on introductory lectures that were being given by the Professor of Geology. On the first occasion that I set off to do so I walked out across the empty playground and heard the classes chanting Latin verbs – and felt that I was almost playing truant.

I caught a slow train to Nottingham that stopped at every station along the line. I was just relishing my new freedom when the train stopped at a tiny station and a porter strutting past my window yelled 'Attenborough!'

I leapt to my feet feeling that I had been caught breaking school rules. But it was the name of the station. It was the first time I knew that there was such a place. I did know, however, that my father had been born in a little village called Stapleford where my grandfather, who died before I was born, had kept a corner shop. Back at home I looked at a map. Stapleford was only three miles from Attenborough. So it was there, presumably, that my father's family had originated.

* * *

The more I learned about fossils the more my interest in them grew and I offered geology as one of my examination subjects when I applied for a scholarship to Cambridge. When a telegram arrived saying that I had been awarded one, I ran down to find my father on the allotment where, as it was wartime, he grew vegetables for the family. I can still see his beam of pleasure.

At university, I read both paleontology and zoology. Once in television, however, I concentrated, understandably, on living animals that moved around and did things and I had little time for paleontological indulgences. The holidays, however, were different. Among the best of them were those that we spent in the late 1960s with Desmond Morris in his Maltese villa. Then both our families joined together to hunt for the spectacular triangular teeth of the huge fossil shark, *Carcharodon*, an activity which then, happily, was not prohibited by law as it now is.

I had managed to work a few fossils into the *Life on Earth* series. In the first programme that dealt with the very earliest life forms, we filmed at an extraordinary site in the Ediacra Hills in South Australia. Sandstones there are extremely ancient, around two thousand five hundred million years old, but they nonetheless contain the ghostly imprints of creatures that look like jellyfish and sea fans. They had only been discovered in the 1940s, but I had managed to keep sufficiently abreast of the paleontological world to know that among those at that site was one called *Charnia*. I asked the Australian scientist who was working on them and was showing us around if he knew how it had acquired that name. He didn't.

'It's because,' I said proudly, 'the very first of them to be discovered was found by a sharp-eyed fossil-collecting Leicester schoolboy called Roger Mason in the rocks of the Charnwood Forest where I grew up.'

How I wished I could have said that I had been the boy who had ignored official wisdom and had looked in the Charnwood rocks so assiduously and persistently that eventually he had discovered the extremely rare and barely discernible signs of life that they contain.

Later on that same trip, we wanted to film at a site in the central Australian desert called Gogo. There paleontologists from the Natural

History Museum in London had, some years earlier, found a rich deposit full of the bones of the earliest of all fish. They collected a great number of specimens and took them all back to London. There they devised special techniques to dissolve away the particularly hard Gogo sandstone and reveal the bones of skulls, perfectly articulated. Some Australian paleontologists had been a little put out at having lost this particular race. Permission for us to film the site was given with some reluctance.

'I don't know why you Poms still want to go there,' the scientist responsible for granting permits said. 'After your lot came over and pinched the obvious bits, we went in and did a proper job. There's nothing there now.'

All the same, we wanted to go so that I could describe the fossils on the actual site, and he came with us. We flew in by helicopter. He directed the pilot telling him exactly where to land. As I stepped out, I picked up a flat rock within a yard of the helicopter's landing bar. On it was a dark rectangular shape, over six inches long and beautifully patterned with a tight network of meandering lines. It was the huge bony scute from the head of an arthrodire, one of the biggest of the ancient Gogo fish.

I showed it to our scientist guide.

'You bastard,' he said. But he smiled as he said it and he had the decency to let me keep it.

* * *

The *Life on Earth* episode dealing with reptiles gave me the opportunity for further indulgence – a major sequence about the dinosaurs. We went to a small valley in Texas where there were some spectacular footprints of one of the great plant-eating sauropod dinosaurs. A tall young Texan ranger in a wide-brimmed hat and smart uniform showed us them. We said we would like to come back at dawn the next day so that we could film in the early morning light when the prints would be seen to their best advantage.

After we had done so, he invited us back to his base for a gigantic Texan breakfast. As we worked our way through great hunks of meat and odd-shaped bits of offal, he asked me what I thought of his State.

'Wonderful!' I said. 'I knew it would be even before I got here.'

'Howzat?' he said.

'Because your roads are all dead straight and run at regular intervals directly north to south and east to west.'

He was puzzled. 'How else would you build roads?'

'Well, you don't find many roads in England that run dead straight. If you did find one, it would be a Roman road and two thousand years old.'

'Two thousand years old!' he said, amazed. 'And you're still using it!'

It occurred to me that the parallel between the Texans and the Romans was perhaps closer than he suspected. Both had invaded a country and built straight roads as the shortest distance between two points with no regard whatsoever for the land rights of the native people who lived there. Both had left their political attitudes, as well as their tracks, fossilised on maps.

* * *

I had done two long series about living animals with only sidelong glances at fossils. Perhaps now I might try to make a series devoted exclusively to fossils. I suggested one, to be called *Lost Worlds, Vanished Lives*, that would show the way fossils were formed, the techniques that were used to free them from their matrix, and the clues that paleontologists used to deduce what long-extinct animals had looked like in life.

The series, happily, took me to some of the most famous and exciting fossil sites in the world – to Solnhofen in Bavaria where, a century and a half before, a quarryman had found *Archaeopteryx*, the earliest of birds with feathers on its wings but teeth in its bony jaws; to the Burgess Shales in the Rocky Mountains which at the time of their deposition had such perfect conditions for fossilisation that not only the hard shells but even the most delicate, soft tentacles of some creatures around 550 million years old had been preserved in astonishing detail; and to the Dominican Republic where miners dug shafts into the hills to excavate nuggets of amber within which were perfectly preserved ants, flies, spiders and even frogs, that were thirty million years old.

And of course, one whole programme was devoted to dinosaurs. I decided that we would not try to bring species to life with animations. At that time, in 1987, such visualisations were very expensive, clumsy and unconvincing for we had no computer-generated imaging. But in any case, I also believed that dinosaurs, like all fossils, grip our imaginations because their study is an exercise in detection. Clues have to be found, deductions have to be made. Only then can reconstructions be responsibly attempted. To present a reconstruction without the clues and the reasoning that justified it, seemed to me to be like disclosing in the first paragraph of a detective novel that the butler did it. Why read further? So I decided that our series would concentrate on the bones themselves.

We journeyed to a remote valley in Colorado in America where there were even longer trails of dinosaur footprints than we had filmed for *Life on Earth*. We also went to look for skeletons in the Sahara with a team including Angela Milner, the dinosaur expert from the Natural History Museum, London and Dick Moody from Kingston University. To my astonishment, Dick, driving through a remote part of the desert, had identified at least twenty sites where dinosaur bones were eroding out from the side of the low hills. Their debris formed a dark stain, like blood from a wound, trickling down the lighter coloured sandstone. Eventually the team decided on one as the most promising and we set up camp beside it. Every day we uncovered a little more of it as we dug our way along the animal's spine and across the side of the hill. In the evenings, when hard digging was over, and the light low and slanting so that it picked out surface detail, we walked across the level desert looking for the animal's little peg-like teeth. Disappointingly, there was no skull at the end of the spine when we finally got there but Angela had discovered enough to be sure that this was a species that had never been found in Africa before. Even so, the skeleton itself was so huge that the team could only manage to take away some of the diagnostic bones. The rest we reburied.

In the United States we visited Bob Bakker. He is a great authority on dinosaurs and a leading proponent of the view that many were warm-blooded. I interviewed him about the basis of his theories as he stood inside the mounted skeleton of a *Tyrannosaurus* wearing the cowboy hat that he keeps on his head at all times whether indoors or out. Most exciting of all, we went to the Museum of the Rockies in Boseman, Montana to visit Jack Horner, another great dinosaur expert. He took us to a place in the badlands where he had made one of his most thrilling discoveries – a place where hundreds of dinosaurs had nested in a great colony. Bones were few, but egg fragments lay everywhere. I crawled over the cliffs alongside him. As usual when fossil hunting, it takes some time to get your eye in. To begin with Jack picked up the tiny flat pieces, smooth on one side dimpled on the other, every few minutes while I saw nothing, but before long I too was finding them.

The series did not attract such huge audiences as *Life on Earth*, perhaps because of my rejection of reconstructions, but it did well enough. And when children wrote to me, in tones I recognised very well, asking if I had any fossils to spare, I was able to send them a bit of genuine dinosaur bone from the fossil debris I had shovelled into bags in the dunes of the Sahara. Though they could not know it, they owed those bits to Jacquetta Hawkes.

* * *

Some time after the series had ended, as I was walking through one of the main galleries in the Natural History Museum, I met Angela Milner again.

'Congratulations,' she said, 'on your plesiosaur.'

I didn't know what she was talking about. Apparently, Bob Bakker had just published a paper reviewing the whole of this group of Jurassic marine reptiles. They were huge crocodile-like creatures that had first been recognised from specimens collected by a redoubtable Victorian lady, Mary Anning, who earned a living by scouring the cliffs of Lyme Regis in Dorset and selling the fossils she discovered to gentleman naturalists. Since then a great number of other plesiosaur specimens had been found in Jurassic deposits all over the world. Many different names had been given to them. Some were attached to mere fragments of a skeleton. Others were given to specimens that, when their anatomy was better understood, proved to be the same as earlier ones. There was a lot of duplication and a certain confusion about which should be called what. A respected authority was needed to examine them all and pronounce on which names were valid and which should be discarded. Bob Bakker had taken on the task and had just published his decisions.

One species interested him particularly. It was very different from all the rest for it not only had the elongated neck typical of plesiosaurs but a long massive snout with huge teeth. The first and still the only known specimen of it had been found by Mary Anning who sold it to the vicar of Axminster, the Reverend William Conybeare. He was one of those Victorian clergymen who believed that since all animals, past and present, had been created by God, their study was itself a form of worship – and thereafter felt justified in devoting much of their time to it. His fossil was given the name of *Plesiosaurus conybeari*.

In due course, the Rev. Conybeare gave it to Bristol Museum. In 1940, Hitler's planes dropped a bomb on the Museum and *Plesiosaurus conybeari* was blown to smithereens. Fortunately, however, an excellent cast had been made of it by the Natural History Museum in London and that still survived. Bob Bakker had decided that it was so different from all the other plesiosaurs that it should be allocated its own genus. Without mentioning it to me, he decided that it should be renamed *Attenborosaurus conybeari*.

I asked Angela where this remarkable object could be seen. It was still on display, hanging on the wall of a gallery a few yards away from where we were talking. She took me to see it. I looked at the label and was, I confess, a little disappointed to see that it still gave the specimen the name of *Plesiosaurus*.

'I know,' Angela said. 'But this gallery was redecorated only a few months ago and we've used up all our budget. We'll put that right when its turn comes round again.'

So my paleontological fame, after all, rests on a label that is not likely to be put up for many years to come and given to a fossil which does not, in fact, exist. I wrote to Bob Bakker to thank him for the honour and explained that I often stayed with friends quite close to Lyme Regis. 'Before long,' I wrote, 'I shall take a few weeks off to try and find myself.' But I have never succeeded in doing either.

24

The Trials of Life

Zoology had changed since I was a student at Cambridge. Then the science had seemed to me to be largely laboratory-bound. We were taught about the anatomy of animals and peered into the entrails of crayfish, dogfish and rats. We sat in lecture theatres while the complexities of animal classification were explained and illustrated with skeletons and stuffed skins. We heard about painstaking experiments designed to find out whether pigeons could count and how quickly rats could learn to run the correct way through mazes. But there was no suggestion that we might ultimately, as qualified zoologists, watch elephants in Africa or crouch in a hide in the depths of a tropical forest watching some rare bird at its nest. That was what naturalists did. Not scientists. But that was what I really wanted to do.

Of course, this was the blinkered view of a 1940s undergraduate. In our ignorance, we did not realise that Julian Huxley had already laid the foundations of behavioural studies with his analysis of the courtship display of great crested grebes, that Konrad Lorenz had already identified the behaviour he called imprinting which he was to explain to me when I interviewed him on television a decade later, or that Niko Tinbergen had decoded the signalling behaviour of black-headed gulls, work that later still was to win him a share of a Nobel prize. It would be some time yet before such attitudes seeped down to the undergraduate level. By the time I realised that behavioural studies had become a recognised branch of zoology dignified with its own name – ethology – I was already immersed in television.

Even in the 1950s, the need to classify animals by studying their anatomy was still a powerful element in zoology. The London Zoo had been opened in the nineteenth century by the Zoological Society of London and was still run, not as a show-piece or a tourist attraction, but as a scientific institution concerned with taxonomy and the identification of species. That was why the *Zoo Quest* expeditions had been so concerned with obtaining species that the Zoo had never possessed before. When eventually the Zoo's exhibits died, their bodies were perhaps of still greater scientific consequence to the Society than they were when they had been alive. They were passed to a staff scientist, the Prosector, who dissected them and wrote detailed reports on their structure. The holder of that post in the 1950s had a particular interest in primates and his scholarly articles were illustrated by drawings prepared by his wife. I remember looking at her, as she walked by in the Zoo, awed by the thought that for the last ten years or so she had devoted her days to scrutinising the pickled genitals of different species of male lemurs which, according to her husband, were the key to determining that family's taxonomic relationships.

But by the 1980s, behavioural studies had become widespread. Indeed they were the most popular aspect of zoological science. There was scarcely a single species of large mammal that did not have a dedicated researcher following it around in the wild, recording its every vocalisation, gesture and interaction and then publishing the results in scientific journals. The accumulation of information was now so substantial that we could contemplate a series illustrating intimate and complex behaviour across the whole range of the animal kingdom. I drafted an outline which allocated a programme to each of the major stages of life that every animal must go through, whether it is a dragonfly or a frog, a cuckoo or an elephant – arriving into the world, finding food, defending itself, selecting a mate. I identified twelve of them – not thirteen as with *Life on Earth*, but still enough if shown weekly to occupy an annual quarter. I called the whole series *The Trials of Life*.

* * *

The first programme was to deal with the first of life's trials – being born. I could think of no more impressive way of illustrating this than with the aid of the crabs of Christmas Island. There are several islands with that name in the oceans of the world. The one the crabs inhabit lies a couple of hundred miles south of Java. Every November, when the moon enters its third phase, ten million female scarlet crabs emerge from their holes in the island's forest and begin to march down to the sea. Nothing stops them. A vast red congregation, so tightly packed that you cannot see the ground between them, floods down towards the beaches. They cross tarmac roads, clamber up walls and over boulders. Eventually they arrive at the edge of the sea and there they shed countless billions of eggs.

There was not time for us to arrange to get to Christmas Island by the first November in our production period. It was fixed for the second. The arrangements seemed to fall into place with surprising ease, bearing in mind how remote the island looks on a map. It is administered by

Australia. We could call there on the way back from our main Australian shoot.

It was while we were in Australia that their airline pilots went on strike. Our all-important opening sequence was at hazard. This was our only chance. By the following November, the series would have to be on the air. Alastair Fothergill, the director in charge of the programme, managed to get places on an Australian Air Force cargo plane that was taking stores to the island. It was a deafening, chilly journey, sitting in the bare hold, strapped to small metallic seats, but to our great relief according to the calendar we reached the island forty-eight hours before the crabs were due to appear. There were no signs of them. How could several billion crabs be invisible? The islanders assured us that they would, on the appropriate evening, materialise. We should not be impatient.

The appointed night came. We went down to the deserted beach at sundown, cameras at the ready, and waited. Darkness fell. Then, Alastair searching the cliff face at the head of the beach with his torch saw a small red dot at the top. Within five minutes there were a hundred or so. Slowly the scarlet curtain descended the rock face. It advanced over the beach. In spite of the noise of the waves, we could hear a vast chorus of clicks made by the astronomic number of legs clambering over one another. Alastair told me to seat myself on a rock at the edge of the lapping waves. Dutifully I did as I was told and allowed the red tide of crustaceans to surge around me. The crabs' compulsion to rid themselves of their eggs was so strong that the marching millions took little notice of me, but even so, I had to introduce the first of the *Trials of Life* with all the confidence and enthusiasm at my command and several inquisitive crabs climbing determinedly up my trousers.

* * *

You might think that a scientist would be reluctant to reveal to a television producer behaviour that has taken him years of sitting in some rain-drenched insect-ridden jungle to observe and understand. That proved not to be so. Every scientist we contacted, once they were convinced that their discoveries would not be distorted but presented as truthfully as we could manage, welcomed us enthusiastically. Some, admittedly, took a little convincing.

Christophe Boesch, a Swiss ethologist, was one of these. He and his wife Hedwige lived in the middle of a rain forest in Africa's Ivory Coast observing chimpanzees. Their animals behaved in a rather different way from the well-known chimp communities that Jane Goodall had been studying so famously in the open savannahs of Tanzania. What he had found gave him every reason to be wary of television. His chimpanzees were hunters – and they hunted monkeys. Our nearest relations, it seems, are killers. The subject would be only too easy to sensationalise. Alastair Fothergill, however, is not only very responsible but extremely determined. Eventually, he convinced Christophe that we would treat his findings responsibly and that they would be an important part of a serious review of animal behaviour. On that basis, Christophe agreed that Alastair might bring me, with recordist Trevor Gosling and cameraman Mike Richards, to join him for a week or so.

Christophe is tall, rangy, brusque and with a very firm set to his jaw. He had started his study ten years earlier. For the first one or two years, he caught only brief glimpses of his animals and often went into the forest every day for weeks on end without seeing them at all. The vegetation was too thick and the chimps were too wary. Gradually, however, they became accustomed to the presence of the tall silent primate that trailed after them wherever they went, that rested when they rested, that ran when they ran and fed when they fed. After five years or so, they seemed to be quite untroubled by his presence. He spent every day with them and could recognise each of them, even from a brief glimpse of an eyebrow or a nose. Every day he got up long before dawn so that he could join them before they had left their sleeping place and followed them wherever they went throughout the day, noting down whatever they did. He remained with them until they had made their beds of branches in the tree-tops and had settled down for the night. That was essential for he had to know exactly where to find them on the following morning. If they left before he arrived, it might take him hours, even days to find them again. Continuity of observation would be lost, vital statistics damaged.

His conditions for our joining him were strict and non-negotiable. We would all have to wear exactly the same clothes as he did – Swiss Army fatigues – for the chimps were used to their appearance and their smell. We must not talk except in whispers. He would take no notice of us because he had to devote all his attention to the chimps. If we did not keep up with him, he might not even be aware that we weren't there and he certainly would not turn back to look for us. We nodded, mutely. He assumed that the team were fit, but I was thirty years older than the rest of them. He didn't want a repetition of the trouble he had had a few months earlier with a visiting professor who had collapsed after the second day and had to be carried out. Was I fit? he asked, challengingly. I assured him, without much confidence, that I was.

I doubted it the following day. The chimps had spent the previous night in a very distant part of their territory. It would take us a couple of hours to reach them. So we left at four o'clock in the morning. Running through the forest in the dark behind Christophe was not easy. The air was heavy with moisture and within minutes sweat was streaming from every part of my body, drenching my clothes. I tripped. I stumbled into bogs. Thin, string-like creepers tangled with my legs or threatened to garrotte me as I ran into them full tilt. The others were having an even more difficult time than I was, for I was carrying nothing and they had awkward heavy loads, Mike with his camera, Trevor his recorder and Alastair with food and water.

As dawn was breaking, we finally contacted the chimps. They were sitting on their beds of bent branches, in the top of a tall tree, lazily stretching and scratching. Evidently they had had a good night and were fully rested. We flopped on to the ground beneath their tree to get our breath. Almost immediately one of the elderly males swung away through the branches, jumped into another tree and the whole group, sixty strong, set off for their morning jaunt. Wearily we hauled ourselves to our feet and set off after them.

This was to be our daily routine for the next week. The chimps were endlessly entertaining. Often they travelled on the ground, pacing alongside us and taking no visible notice. Periodically one of the males would let out a scream – a location call – to let others in the group know exactly where he was. More rarely one would circle around the base of a great tree, making a few preliminary grunts, and then suddenly sprint towards one of its flange-like buttresses, leap at it and beat out a rapid tattoo with both hands and feet, screaming alarmingly. A minute or so later, we would hear a distant answering tattoo from another group in the forest.

The chimps seemed to find food wherever they went. Up in the branches they plucked leaves and fruit. On the ground they gathered nuts and took them to special places where there were flat stones that they habitually used as anvils on which to crack them. Only the adults managed to do this with any efficiency, but they were watched attentively by the youngsters who were still learning. One morning, the group found an ants' nest in a rotten log. One of the adult males made a sudden dash at it, ripped the bark off the top with one hand and scooped out a great handful of white larvae with the other. He scampered off, cramming them into his mouth, while hopping up and down and flapping his other hand trying to get rid of the biting ants that covered him. Others did the same. Then the whole group sat down a few yards away, and started to pick off ants from their hair, one by one.

Even from where we sat, fifteen yards away, we could see that the ant colony was vibrating with fury. Clearly the chimps thought it wasn't worth multiple bites to get any more larvae. But one of the youngsters could not resist the idea. He climbed into a tree above the log and cautiously descended a liana that hung directly above the nest. And there he hung, a few feet away, trying to make up his mind whether to go any further. Suddenly he swooped down the liana, shot out a long skinny arm, scooped up another handful and rushed back up into the tree, beating his hands against the branches to get rid of the ants and clearly delighted with his extra mouthful.

But the behaviour we had come to film was the monkey hunt. A chimp stands little chance of catching a monkey by itself, for an adult weighs two or three times a colobus monkey, their main prey. So a monkey can escape along branches that will break under the weight of a chimp. Chimp hunters can normally only succeed if they work together as a team.

The odds against us seeing them do so were considerable. The group only hunted about once a week and more often than not, did so high in the trees well beyond camera range. But on our sixth day, we were lucky. In mid-morning, as we accompanied the chimps on their ramble through the forest, we heard the soft high-pitched grunts of a colobus troop feeding in the canopy not far ahead. Immediately the behaviour of the chimps changed. Half a dozen males who Christophe knew usually hunted together, started to walk silently along the ground in a determined and agitated way, staring intently up into the canopy, tracking the precise position of the monkeys ahead. The female chimps and the youngsters dropped behind. We kept up with the hunters.

Christophe was now dictating quietly into a recorder slung round his neck and from his words we were able to follow what was happening. The colobus were feeding in the foliage high above, moving in a leisurely way from tree to tree. After following them quietly for about ten minutes, the chimps' behaviour changed again. Perhaps they had decided that the monkeys were now in a place where they would be vulnerable. Two of the younger males swiftly and silently clambered up lianas into the canopy. They would act as blockers. A senior male sprinted ahead and disappeared. Two more of the young males then shot up a liana, hand-over-hand, at great speed. The colobus spotted them and jumped away through the branches. The blockers on either side of the troop suddenly screamed and the terrified colobus fled as fast as they could through the branches – straight towards the old male who was hiding in ambush ahead. He jumped out, grabbed one of the larger monkeys as it landed close by him and bit it in the back. Now the females had caught up with us. There was pandemonium – the colobus screaming in terror, the female chimps hooting with excitement. Mike Richards, miraculously, had his camera on its tripod with a telephoto lens focused on the male who was disembowelling his victim alive.

The male ran down the tree, bringing the dangling carcass with him and joined a few of the females in an excited scrum. It was only a small group – there is not much to eat on a small monkey. The carcass was ripped in two. A senior male took each half and divided it still further handing out pieces to the females. Christophe told us later that this was the way food was usually shared. The hunter who made the kill does not have priority or special rights. He may not even get a share.

It was, of course, as dramatic a sequence as we could wish for to end our programme about the various ways in which animals hunt. But it was not merely ghoulishness that led us to film it. As I said while crouching beside the feasting group, their blood-stained faces might well horrify us, but we might also see in them the faces of our long-distant ancestors. And if we are appalled by their mob-violence and blood-lust, we might also see in their actions the origins of the team-work that has, in the end, brought human beings some of their greatest achievements.

* * *

Rudiger Wehner, from the University of Zürich, was interested in a very different behavioural question. He wanted to know how an ant called *Cataglyphis* could find its way around on the shifting sands of the central Sahara.

Cataglyphis only emerges from its underground home in the very middle of the day when the dunes are so hot that it is painful to touch their surface with your naked hand. Then even permanent residents have to take shelter. There are not many of them. Sand geckos hide in the shadow of a bush or a stone. Insects too seek shade. If they fail to find it and remain exposed to the punishing sun, they are likely to collapse and die from heat exhaustion. Such casualties are what *Cataglyphis* lives on.

As soon as the temperature rises to a particular point in the middle of the day, the ants stream out from their tiny hole in the sand. According to Professor Wehner, *Cataglyphis* can withstand higher temperatures than any other insect. Workers run back and forth at high speed across the surface of the dune, in an urgent search for food. Once they find something, they must get back to their nest as soon as possible before they too succumb to the heat. Retracing exactly their zig-zagging trail would waste a great deal of time, even if they had been able to lay down a scent track on these oven-roasted sands. Instead, no matter how circuitous their search, they return along a straight line, directly to their nest entrance. How do they know where it is?

Professor Wehner suspected that they were able to navigate by the sun. To test his theory he had designed a special machine. It looked something like one of those lawn-mowers that hover above the surface of the ground and trim the grass with whirling blades. Its body was a square horizontal aluminium plate with a disc cut out of its centre, a castor under each corner, and a handle like a pram. Mounted on the plate were two vertical rods, one on each side of the circular hole. One carried an adjustable wooden disc and the other a mirror. The professor tracked individual ants with this machine, keeping pace with one so that wherever it went, he could see it through the hole in the aluminium plate. He could then adjust the wooden disc so that it blotted out the ant's view of the sun, and twist the mirror so that it caught the sun's beam and shone it down into the hole. In this way, he was able to bamboozle the ant into thinking that the sun was diametrically opposite its true position. And lo! the ant ran, not towards its hole but directly away from it.

The implications of his discovery are astonishing. It means that each time the ant changes direction as it rushes back and forth, it has to register both the bearing of its track in relation to the sun and the distance it has run along it. It must do that for every zig and every zag of its path until it finally finds something edible. Then it has to add up all its data and find out in which direction its hole lies. Once the professor had told us this, we could see the ants taking their bearings. At the end of each straight run, each ant lifted its head towards the sky and executed a little pirouette, presumably recording in its microscopic brain the statistics it required, before setting off in another direction.

Recording this on film was not easy. We provided the ants with tiny fragments of boiled egg yolk. To film them collecting it and rushing back to their nest carrying it in their jaws, we had to lie full length on the burning sand. At least we did not have to do so for long. Our subjects only came above ground for about an hour each day.

But perhaps the most engaging image of the trip was that of Professor Wehner himself, out in the midday sun, muttering statistics into a tape recorder hanging around his neck as he trundled his weird lawn-mower across the dunes. We filmed him as well as his ants, for we had decided to make an extra programme to go at the end of the series. The scientists who helped us with the series would make just as fascinating subjects as their discoveries.

* * *

Not all our jaunts were quite as successful. Sometimes we deliberately gambled a significant slice of our budget for a particular programme on a sequence that we were far from certain of filming but which, if we could succeed, would be a real thriller. Marion Zunz, who was directing the programme on the various ways that animals find their mates, had discovered a story in the scientific literature about Siberian hamsters.

Each individual, male or female, lives alone in its burrow as the breeding season approaches. When a female becomes sexually receptive, she issues a chemical invitation which is so pungent and effective that males converge on her in droves. It's a race. The first to reach her copulates with her. Normally this behaviour would be virtually impossible to film, for how could one know which individual female was going to come into oestrus in order to get to her even faster than the fastest male and have our cameras ready focused to film him as he arrived. But now a Russian scientist, working in collaboration with an American colleague, had managed to synthesise this irresistible perfume of the female Siberian hamster in his laboratory. This substance, a pheromone, was so concentrated and could be produced in such quantity, that if anyone uncorked a test-tube full of it on the steppes of central Asia, the ground would pullulate with sex-crazed male hamsters racing in from miles around.

Siberian hamsters did not seem to me to be the most exciting of animals – even when sex-crazed. One could, after all, buy them in pet-shops in London. But Marion was irresistibly enthusiastic. It was the one gamble she wanted to make in this programme. It would demonstrate more dramatically than any other example she could find, the power of a chemical attractant.

We flew to Moscow. The next day we took an immensely long flight to a city I had never heard of (Kyzyl), in a Soviet State (Tuva) that until then I did not know existed. A monument in the town, however, told us that we were in the exact geographical middle of central Asia. There we were met by Kathy, the American scientist who was working on the behavioural side of the project. We piled into her truck and she drove us off into the steppes. It was dark by the time we got to her camp.

The following day we met her team. As well as Alexei, her Russian collaborator, she had two dozen American volunteer helpers. They were a mixed lot, mostly elderly, many wearing baseball caps and vividly coloured clothes printed with slogans of various kinds. They were all unquestionably enthralled by the project on which they were engaged.

There was, however, a fly in the ointment. Things had not worked out quite as well in Alexei's laboratory as he had hoped. He had not managed, after all, to bring a test-tube full of the pheromone he had synthesised. Indeed, at the present time, he had not got any at all. We were astounded. Why had this news not reached us before we set out? Kathy, however, was consolatory. She had one captive-bred female hamster, whom she had been keeping in strict purdah in a cage in her tent and who was just coming into sexual receptivity. This female, after all, would produce the real thing. Not in quite so much quantity, of course, or perhaps quite as pungently, but males all around would soon be aware of her and display all the activity we would need. Kathy felt sure that we would be able to edit the film in such a way that the steppes would *appear* to pullulate.

Night fell and the research team assembled. Every burrow within a radius of half a mile of the camp had been identified and registered. Each member of the team had been allocated one. Each sat beside it with a torch, a compass and a tape recorder around his or her neck. As soon as a male hamster poked his nose out of his burrow, the researcher had to illuminate him with the torch. He or she then had to follow him wherever he went, dictating the compass bearings of his route into the recorder together with any other interesting details of its behaviour. Did he sniff, or scratch or hesitate? By an hour after dark, the steppes around the camp were studded with crouched creeping figures, each in a pool of torch light, muttering to themselves.

It was at this point that I heard the sound of horses' hooves. Out of the blackness galloped a group of nomadic Mongolian horsemen. They wore cloaks and wild-looking Astrakhan caps. Reining back their horses, they stared at the sight in front of them. The researchers paid no attention. The flow of data could not be interrupted. After gazing at us all for a minute or so, the horsemen pulled their horses' heads round and galloped off again into the blackness. I wondered what they would tell their wives and children when they got back to their yurts.

Now the moment had come for the sexually receptive female to be exposed. Kathy crouched with her in her hand and we waited for males to appear. Sure enough, one did – within seconds. The female must have known, for she gave a wriggle and jumped free. 'Catch her,' shouted Kathy and Alexei lunged forward to do so. He tripped and his torch hit the hamster on the back of her head. Kathy picked the little animal up. She lay upside down in her hand, her little legs twitching their last.

'Oh My!' Kathy drawled sorrowfully and added, scientifically observant even at such a time, 'and her little fleas are already evacuating,' which indeed they were, hopping off Kathy's hand on to the ground presumably hoping to find another hamster host as soon as they could.

We did get film of the males, scuttling through the stunted vegetation, and were able to give some impression of a pullulating steppe. Nonetheless, it was a very long way to have gone for a relatively humdrum sequence. Indeed, it was probably as frustrating for us as it would have been for the male hamsters, had they known what erotic super-stimulation they had missed.

* * *

It was not only the recent increase in scientific knowledge that made *The Trials of Life* a practical proposition. Television techniques had also improved dramatically since the days of *Life on Earth*. Now we had cameras no bigger than lipsticks that we could put inside a rodent's burrow or strap on a bird's back. We had portable light-weight jib-arms that could lift a camera thirty feet up in the air and allow the cameraman standing beneath to see the pictures it was taking on a monitor. We could put a camera in a radio-controlled model helicopter to take aerial shots or fly alongside birds. And, most importantly of all, perhaps, we had low light cameras.

Many mammals sleep during the day and are only active at night – a fact that in previous films we had tended to gloss over, for until about this time, the nocturnal activities of animals were virtually unfilmable without some artificial light. Even the most sensitive of film stocks required so much illumination that animals, when lit, seldom behaved naturally and more often than not would simply run away. Even if they stayed within camera range, the lighting tended to distort their behaviour.

But now manufacturers of film stocks and electronic cameras were competing with one another to see how little light their products needed. Both techniques had become so sensitive that with them we were able to record scenes we could not ourselves see with unaided eyes. This was a particular boon for *Trials*.

I wrote a sequence about fire-flies in the programme on communication. We went to Malaysia and drifted through the swamps to film tens of thousands of little beetles sitting on the branches of the mangroves flashing their tiny lights rhythmically and in unison so that the entire tree appeared to be turning its illuminations on and off as if it were decorated for

Christmas. The males by synchronising their timing were attracting females from far greater distances than a single male could manage by himself. The pictures on our electronic cameras were magical, capturing even the reflection of the display in the waters beneath.

Later in the summer of the same year, 1989, in a meadow in Baltimore, I was able to take part in such dialogues myself. North American fireflies are not closely related to the Malaysian kind but they also communicate with flashes. Several different species of them may inhabit the same meadow. The males outnumber the females by as many as fifty to one. They cruise in the air flashing their call signs. One species makes single sporadic flashes. Another makes two, a second apart. A third produces them every half-second in a long sequence. The female waits until she sees one of her own kind. Then she responds with a single flash and the male comes down to land beside her and copulate. I was able to impersonate a female using a small torch and succeeded in enticing several males to land on my finger. Neither of these sequences would have been technically possible only a year or so earlier.

* * *

For the programme on hunting, we borrowed a device from medicine. Surgeons were now commonly using endoscopes, fibre-optic rods that they push down their patients' throats in order to inspect the lining of the stomach. Just the thing we needed for army ants. Thirty years earlier in Sierra Leone, Charles Lagus had managed to film such ants by lying alongside their columns as they marched off to hunt. That was still worth doing, but the endoscope now gave us the possibility of seeing how ants behaved during another and more mysterious stage in their lives. Every two weeks or so, the whole swarm forms a bivouac. They gather together in a huge ball, around the base of a tree or under a log. The outer surface of this great bundle is formed by soldiers who link legs with one another and wait with their jaws wide open, ready to savage anything that might interfere with the colony. When the bivouac breaks up and the colony resumes its march, many of the workers reappear carrying eggs, so it is clear that the queen lays during her time in the bivouac. But how is the bivouac structured? No one knew. We oiled the surface of an endoscope's tube so that the ants would have difficulty in running up it and biting the cameraman on the eye. Then we gently inserted it into the very heart of the bivouac. There we saw, to the delight of the scientist who was advising us, that the workers within, by linking legs, had constructed living corridors and chambers along which other workers clambered as they tended their queen and carried away the eggs she produced.

* * *

Alastair Fothergill, who was working on the programme concerned with navigation, was anxious to film electric eels in the Amazon. Natural history producers like to get their human presenter as close as possible to the animal he is describing, particularly if he can be talking at the same time. They call it a 'two-shot'. The presenter, they think, will reveal, by his reactions and the way he speaks, something of the character of the animal beside him. Is it nervous, or deaf, or maybe even dangerous? The shot also gives some idea of the size of an animal if it is seen alongside a human being. At the very least, the two-shot should demonstrate to viewers that the presenter was actually there when the animal was being filmed and giving his commentary from first-hand experience. That way, a viewer can imagine to some degree what it would be like to be in the presenter's place.

A talking two-shot, of course, is hard to achieve underwater. The standard diving face-mask makes it difficult for its wearer to be recognised and impossible for him to speak, since he has to have a tube in his mouth in order to breathe. Alastair, however, had heard of a device called a bubble helmet that divers on the North Sea oil fields were beginning to use. It consisted of a transparent plastic sphere that goes over the diver's head and screws on to the top of his diving suit. This bubble is kept full of air and empty of water by the pressure of compressed air coming from the cylinders on the diver's back. That means that he need not have a breathing tube in his mouth and is readily recognisable through the plastic bubble. The helmet also has a microphone within it so that the presenter's words can be recorded as he swims nonchalantly through shoals of fish or in and out of coral grottoes.

Sensibly, Alastair felt that before confronting a giant electric eel and describing in words how exciting it was, I should give the bubble helmet a try-out in our hotel's swimming pool. Putting it on was more awkward than I had imagined. The distance between the back of my head and the end of my nose was clearly considerably greater than the opening of the bubble. Eventually, however, with much nodding, I managed to get my head inside it. I was more alarmed, however, by the time needed to get it *off*. The screws around its base had to be very tight, in order that the seal did not leak. Each took some time to undo. Then the bubble had to be carefully removed without taking the end of my nose with it. Supposing there was an emergency? Supposing there was a blockage in the supply of compressed air or it filled with water? How long would it take, with my head in the bubble, before I suffocated or drowned? I was not entirely reassured by the answer, but sufficiently so to get into the pool with it on my head to give it a first try. I submerged and started to swim down to the bottom. Immediately, I felt water swilling around my chin. I came up fast. Taking the helmet off took even longer than I had feared. Alastair spoke reassuringly. He was confident that by the time he had done this a few times he would be able to cut down the removal time very substantially. But why was it leaking? He inspected all parts of the device, including the fittings, and was unable to find any fault. Let's try again. He put it back on my head. I lowered myself into the pool again, but once again it started to fill with water. I came out even more quickly.

Alastair said that I must be doing something wrong. The helmet had worked perfectly during its trials. Maybe I was breathing in an eccentric way. He would show me how it should be done. I took some pleasure in screwing it down over his head and he waded confidently into the pool. To my relief, he too came up to the surface just as quickly as I had.

'There is a fault,' he said, as though announcing a new discovery.

The try-out was called off for the day and Alastair went up to his room to read the handbook, taking the helmet with him. I retired to the bar. He didn't come down to dinner. Late that evening, I went to his room. He had taken the whole thing to pieces. Springs, tubes, rubber sealing rings and lots of small shiny screws lay all over the floor. He was talking on the telephone to the designer of the device, a French diver who was, at that moment, sitting on an oil-rig off the coast of Scotland. The diver spoke no English and Alastair was having some difficulty in identifying in French the particular component he was talking about, a problem made the greater by the fact that nothing was in its working position.

That particular helmet never did function in Brazil and I, to my great relief, never had to describe to anyone how thrilling it was to be swimming with a giant electric eel.

* * *

Peter Jones, in overall charge of *Trials*, had heard of a revolutionary new lens. It had been invented by Jim Frazier an Australian cameramen and an old friend who had worked on *Life on Earth*. Jim specialised in filming very small creatures. Spiders had been the stars of some of his most dramatic films. One problem a cameraman faces when filming on this scale is that the depth of field – the horizontal distance in front of the lens within which everything is in focus – is very shallow. But Jim's new device allowed him to have a tiny object virtually filling the screen in the foreground, while at the same time everything behind it as far as distant horizon is pin-sharp. Peter could hardly wait to play with this new toy and he saw a chance of doing so in the programme I had written about animal partnerships.

One sequence I had suggested involved the caterpillar of the Oak Blue butterfly and green tree ants. The caterpillar has a small nipple on its back which when stimulated in the right way, produces little drops of liquid called by entomologists, perhaps rather romantically, honey-dew. The ants clearly relish it. One after another they run along the caterpillar's back, tickle the nipple and sip what it produces. They look after the caterpillar with all the solicitude of a farmer caring for a cow. In the morning they entice it out on to a leafy branch so that it can feed. In the evening they chivvy it back to a smaller version of their own nest – a little shelter they have made by gluing leaves together with a tissue of silk.

Described like that, it may seem that the ants are exploiting the caterpillar, but there are good reasons to suppose that the caterpillar actively welcomes the ants. For one thing, it summons them by erecting a yellow feathery plume at its rear end. This gives off a pheromone that the ants can detect from a considerable distance. And the ants, demonstrably, are crucial to the caterpillar's survival. There are plenty of creatures that eat or parasitise caterpillars. But tree ants are ferocious creatures with powerful stings and bites and they will attack any intruder, large or small, that invades their patch. The Oak Blue caterpillar, once it has attracted its ant customers, makes sure that it is recognised and not attacked by them by making a faint purring noise that you can feel as a vibration if you put the caterpillar on your finger. And a caterpillar placed on the branch of a tree that lacks green tree ants, does not survive for long. So instead of comparing the relationship to that of a farmer and his cow, it might be more accurate to see the caterpillar as a land-owner who recruits a private army to defend his territory and rewards them with regular rations of their favourite food.

Peter thought he might use Jim's lens to film a sequence about this remarkable arrangement. Happily the appropriate species of butterfly and ant occur in Australia and Jim was able to come up and join us. Jim placed the caterpillar within a few inches of his ingenious lens, so that its body extended right across the frame with its head on one side and its tail with its rear-end plume on the other. Behind it, in perfect focus, I could be seen as a small distant figure. I walked directly towards the camera, talking as I came and ended up pointing at the caterpillar with my finger only a few millimetres away from it. The shot was dramatic in the extreme, but it also warned us that there are dangers in trying to be too clever visually.

When we got back to Bristol, the *Trials* team viewed the rushes. Afterwards in the canteen, as we sat discussing them, I heard a girl who had recently joined to give us help in the office, telling her friend about it. She wasn't sure that she wanted to spend much time in the Australian outback. 'Do you know,' she told her friend, 'there are caterpillars there that are two feet long.'

* * *

Misleading the audience, deliberately as well as inadvertently, was becoming increasingly easy. Feature film directors had, for decades, been placing actors in a landscape by putting them in front of large studio screens on which they projected a photograph. That effect became even more convincing when they used film. Then in the background, waves could be seen breaking, trees waving, and traffic moving. Television drama producers inherited the technique. It then spread insidiously into factual programmes. Newsreaders, sitting in studios, appeared to be in some pent-house in central London with a view of Big Ben and the traffic beneath crawling round Parliament Square.

But now electronics enabled producers to combine pictures in much more subtle and elaborate ways. A presenter could be shown sitting, apparently without the slightest fear, within a few yards of a man-eating jaguar that he had never seen in his life. What is more, he could, with a little rehearsal, pretend to react to some movement made by the electronic jaguar so that the illusion of togetherness became even more convincing. Computer experts then made the final leap. They became so ingenious that they could create totally convincing electronic images of creatures that do not exist. So people, first in the cinema and then on their television sets at home, watched dinosaurs moving, running, urinating and tearing one another to bits. Viewers could have every reason to be baffled. What is true and what is false? Since anything can be invented, why believe that anything is real?

In this situation it seemed to me that a presenter might now have a new and additional function. Many of the things he talked about in natural history films were, after all, quite difficult to credit. Are there really such things as snakes that fly, fish that walk, or frogs that incubate their tadpoles in their stomachs – or are images of such things merely created by a computer programmer because someone believes that they do? If a presenter is known and trusted, viewers might accept that what he says and shows is indeed actuality. But if viewers are to continue to believe him, he has to guard his reputation for sincerity.

When *The Trials of Life* was at last broadcast in October 1990, the television critic in the *Sunday Telegraph*, A.N. Wilson, reviewed it. He clearly wished to make it plain to his readers that he knew all about the behind-the-scenes of television and could not be fooled by its trickery. His disparaging eye lighted upon a sequence in the very first programme. It concerned the mallee fowl.

These ingenious Australian birds build huge mounds of sandy earth in which they incorporate a certain amount of rotting vegetation. They bury their eggs in the mound and the decomposing vegetation provides the heat needed for their incubation. Even so, the birds must keep a close eye on things. To make sure that the inside of the mound does not get so hot that the eggs cook, the male bird returns every few hours and takes its temperature by poking his beak into the soil. If there is a risk of over-heating, then he will kick away some of the soil from the top of the mound, or alternatively, if the eggs are in danger of cooling, kick on a little more.

We filmed a mallee fowl mound whose owners were accustomed to seeing human beings and were almost as tame as robins in a London back garden. I crawled cautiously up to it and began to throw aside handfuls of soil from the top. Within minutes, the male bird appeared. He was outraged. He came to within a couple of yards of me and as fast as I threw soil aside, so he kicked soil back. He did so with such vigour and so fearlessly that some of it hit me in the face. It was, we thought, an entertaining and informative sequence.

Mr Wilson was having none of it. The encounter, he said, 'was almost certainly faked up with a shot of the bird kicking and another of Sir David getting sand in his eyes (probably thrown by the continuity girl). There is nothing less natural than natural history on TV.' I thought it important to defend my sincerity. Peter Jones, as producer of the series, wrote to the *Sunday Telegraph* to put matters straight. Mr Wilson was instructed by his editor to withdraw and put his name to a printed apology, saying how deeply he regretted having misled his readers.

Two years later, when I narrated a series about Antarctic wildlife, *Life in the Freezer*, Mr Wilson returned to the attack. In one sequence, a leopard seal was seen lying in ambush, waiting for Adelie penguins to take their first swim. Five cameramen worked for two weeks at the site. One of them, Doug Allan, actually went into the water and faced the huge nine-foot-long leopard seal, an action which many experienced Antarctic hands thought would be suicidal. They produced an extraordinary sequence documenting the way in which a leopard seal caught its prey. Having failed on one issue Mr Wilson now attacked us from another angle. He accused us of gross and deliberate cruelty. 'The chances of a cameraman being in the right place at the right time to catch such a sequence in the wild,' he wrote, 'are inconceivably remote. So this little piece of slaughter has been arranged for us by Attenborough's producer, Alastair Fothergill. One wonders how many live penguins they had to feed to the seal before they got the desired effect.'

This time it was Alastair, as series editor of *Freezer*, who responded. He did so with great vigour, threatening legal action. Mr Wilson had to write a grovelling retraction and we were offered a sizeable sum of money as an out-of-court settlement for libel. We accepted and asked the Editor, with a nice appropriateness I thought, to send the money to the Falkland Islands Conservation Society. It had just launched its Penguin Appeal.

25

The Private Life of Plants

The time had come for me to change tack. I had made three major series covering three main aspects of natural history. *Life on Earth* had chronicled evolutionary history; *The Living Planet* had examined how biological communities had colonised different habitats; and *The Trials of Life* had surveyed the many different ways in which animals behave in order to solve the problems of living and reproducing. But the fundamental basis of animal life on this planet had been largely ignored. None of us in the Natural History Unit, or indeed anywhere else as far as I knew, had dealt in any depth or detail with the very foundation of all life, the kingdom of plants. There were, of course, gardening programmes in the BBC's schedules, but they did not deal with the basic facts of botany, or explain how plants feed, how they reproduce and distribute themselves, how they form alliances with particular animals. The reason was only too obvious. How could you construct the dramatic narratives needed for a successful television documentary series if your main characters are rooted to the ground and barely move? Thinking about this, it suddenly struck me that, of course, plants do move and very dramatically. We tend to forget that they do because they operate on a time-scale that is very different from ours. But now, in the 1990s, we had the technology to transcend that difference more effectively and vividly than ever before.

Film-makers have been accelerating time since the very early days of cinema. Normal film is shot at twenty-four or twenty-five frames a second. If you expose only one frame a second and then project the resulting film at the usual rate, then you speed up the action twenty-five times. So shots of snowdrops breaking through the ground in springtime or beech trees bursting their buds and spreading their leaves in the sun, were among the great revelations of early natural history documentaries. But to produce such shots those pioneer film-makers had to keep the camera rock steady. They had no way of moving one, steadily and extremely slowly to keep pace with, say, the tip of a blackberry stem as it probes across a woodland floor, to follow a bud as it emerges from the ground and is then carried vertically upwards by its growing stem, or to track the tendrils of bindweed as they wrap round a host plant and eventually strangle it. Now, we had computer-controlled mechanisms that could be programmed to shift a camera along a track by a millimetre every five minutes over periods of days. We could make a camera creep with imperceptible slowness alongside a plant. We could pan and tilt, track in to a close-up or pull back to take a general view, and programme a camera to make these movements over a period of days or even weeks. Perhaps now was the time to try and make a dramatic series about plants.

No sooner had I written a brief synopsis of what I had in mind than I discovered that old friends in the Unit, Mike Salisbury, Keith Scholey and Neil Nightingale, with all of whom I had worked on previous series, were all thinking along the same lines. We met and without hesitation decided to combine forces. Our programme outline persuaded the Controller of BBC2 and even rather more commercially minded overseas broadcasters to back the project. But it is one thing to produce an enthusiastic synopsis that explains how exciting and thrilling such newly visible dramas could be; it is quite another to sit down and work out, shot by shot, how to solve the practical problems that such sequences posed. Thankfully, we were able to recruit two cameramen of near genius, Richard Kirby and Tim Shepherd. They welcomed the technical problems involved. The more ambitious the scripts we produced the more they seemed to delight in working out how to obtain the shots that we had asked for.

* * *

Tim lives in a small village in Oxfordshire and shoots his most technically difficult sequences in an old cowshed. One of the first problems we set him was to film the growth, flowering and pollination of that most celebrated and glamorous plant, the giant Amazon water-lily. The Royal Botanic Gardens at Kew grew specimens from seed every year and the then Director of the Gardens, Sir Ghillean Prance, had himself worked out the details of the pollination process so we had access to all the necessary horticultural and botanical expertise. But that was only a part of the problem. How was Tim to film the whole process on the varying time-scales and in the intimate detail that was needed?

He started by germinating some lily seeds in a small aquarium. Then he built a huge tank ten feet square and four feet deep and covered the bottom with a thick layer of cow dung. Into this he transplanted the young lilies. He installed heaters to keep the atmosphere and the water at a temperature comparable to the Amazon itself. That, however, caused the water to evaporate very rapidly in the comparatively dry Oxfordshire air, so the whole of the inside of the cowshed had to be lined with polythene sheeting, to keep the moisture in and the air humid. He had to replace the water as fast as it evaporated for if he didn't the water level in the speeded-up film would be visibly rising and falling. So he installed a pump on one side of the tank to keep it brim-full at all times and a spill-way at the opposite side so that the water level remained perfectly static. The cow dung, though it certainly proved to be exactly what the plant needed for swift and lusty growth, caused further problems. It generated gas in such quantity that every now and then a huge smelly bubble rose and broke on the surface, completely ruining the shot. Tim dealt with that by putting sloping panes of glass beneath the surface of the water so that the bellying bubbles were deflected sideways. That didn't entirely solve the problem because even though the bubbles surfaced out of the camera's view, their bursting created ripples that surged across the shot. He prevented that with floating strips of polystyrene held together with wire which dampened all the ripples.

Lighting created another set of problems. He installed daylight tubes to simulate the intensity of the Brazilian sunshine and fitted them with time-switches so that they turned off at night and on again the next morning. That suited the plant – but not the time-lapse camera. It had to take one frame of film every fifteen minutes, throughout both day and night, for the plant grew at least as swiftly during the hours of darkness as it did during the day. But the lighting during the day had to be exactly the same in intensity and quality as during the night if the finished shot was not to flicker unacceptably. So Tim put up blinds, operated by more time switches, that during the hours when the daylight bulbs were lit, rose and screened them just before the flash gear discharged and the camera exposed a frame, and then fell again.

As if these were not enough complications, Tim created further problems for himself. He decided he ought to get underwater pictures of the developing shoots, so he put a camera actually inside the tank. And he wanted the camera above water, on occasion, to move. He built a gantry that spanned the tank and carried rails along which the camera, mounted on a carriage, was propelled by half a millimetre each time the camera exposed a frame. So in a cowshed in Oxfordshire, with much clicking and flashing, with blinds rising and falling and cameras whirring, he began to record the growth of the magnificent Amazon water lily in detail.

As the plants developed, so leaf buds armoured with prickles rose through the murky waters. When they reached the surface, they unfurled and began to expand. In the Amazon they can be six feet across, big enough, as the Victorians were fond of demonstrating, to support a human child. Tim's tank was not big enough to accommodate such monster leaves, so he had to keep trimming off the edges as they pressed against the walls of the tank to make room for the next leaf rising from the middle of the plant.

And then the first flower bud broke the surface. As evening came the flower opened and exposed its cream-coloured petals. Twelve hours later, with the dawn, it closed. It remained tightly shut throughout the following day. The next night it opened again but now its petals had turned pink. The following morning it closed once more. Its stem contracted and the closed flower was drawn back down to the depths.

All this Tim recorded, varying the frequency with which he exposed frames according to whether the action he was filming took place over several days, as did the expansion of the leaves, or like the flowers, over a few hours.

But these shots by themselves were not enough to tell the full story. Viewers needed to see a whole carpet of the immense leaves covering the surface of a river, not with babies sitting on them but birds – lily-trotters – pacing across them with their enormously elongated toes. And what about the actual pollination? Such things could only be properly filmed on the Amazon itself. And there Tim went, to do so. Lily-trotters were easy. They were there all the time. Pollination was much more difficult.

Ghillean Prance had demonstrated that the agents that carried pollen from one flower to another were beetles and he also had explained why it was that the flowers changed colour. The beetles are attracted only to white flowers. As soon as the lily opens one of its flower buds in the evening, the beetles whirr in and clamber down to the base of the white flowers to chew on a circle of sugary knobs that lie in its centre. While they are busy guzzling, the flower slowly closes its petals, trapping the beetles within. And it remains closed throughout the following day. During their imprisonment within the closed flower, the stamens mature and begin to shed their pollen. The beetles rummaging around within, sticky with their chewed-up meals, get covered by it. On the next night of their imprisonment, the flower opens once more. Now, however, it has turned pink so it is no longer attractive to the beetles. Nor are there any white flowers on their erstwhile jailer, for the lily will not produce a fresh crop until yet another twenty-four hours have passed. So the beetles fly off to look for another plant with white flowers and cross-fertilisation is achieved.

To record these parts of the story, Tim spent days wading through the swampy backwaters of the Amazon, wearing rubber waders to minimise the risk of shocks from electric eels. The birds he filmed at normal speed, but to show the behaviour of the beetles he had to illuminate them with battery lamps and then slow down their movements by operating his camera at high speed. When these slowed-down shots were combined with the speeded-up shots of the lily's growth, when the delicately treading jacanas were cross-cut with the shots from the camera gliding smoothly across the water surface, when the beetles were shown elbowing one another aside as they burrowed through the petals to feast in the depths of the flower, a vision of the growth, the flowering and the pollination of the Amazon lily was produced that, I think, till then had never been equalled.

* * *

Richard Kirby had different problems. Most of the subjects allocated to him had to be filmed in the wild. We asked him to record in time-lapse the growth of meadow plants in spring. He found a suitable patch, rich with a variety of grasses and other small plants and built a neat enclosure of plastic sheeting around it. For three weeks his cameras efficiently clicked away. The grass grew tall. Dandelions and daisies, campions and lady's smocks were on the verge of flowering. One evening the sky darkened with clouds and that night there was a storm. A gale blew down the enclosure. Heavy rain flattened all the stems. The shot was ruined. It would be another eleven months before it could be set up again.

That same spring, he also built an enclosure in a bluebell wood. Once again, the shoots rose up from the ground and became swollen with flower buds; at which point, a family of wood mice found their way into the enclosure and neatly trimmed off every succulent stem at its base. And it was no consolation for Richard the following day to be able to watch the film and see them doing so at high speed.

He also got the job of filming the biggest flower in the world – *Rafflesia*. There is no way in which that could be filmed in a studio for it lives inside the body of a particular kind of vine that grows in the rain forests of Borneo and Sumatra. It is a parasite. The only time it becomes visible to the outside world is when it flowers. A bump appears on a section of the vine that lies on the ground. This bump grows steadily over a period of weeks, becoming more and more ball-shaped. Eventually, the outer coverings of the ball split and huge petals, dark red mottled with cream, unwrap themselves. A fully expanded flower of the largest species may be three feet across. Such an astonishment could not be ignored by a series seeking to survey the kingdom of the plants.

We contacted expert botanists in Borneo. They sent out search parties to find likely localities and in due course, Richard, Neil Nightingale and I flew out to Malaysia. After two nights with little sleep and several hours bumping along muddy tracks in a truck, we at last reached a rest house on the flanks of Mount Kinabalu, close to a locality they had recommended. And there we met our guide. He had news. It was hard to know whether it was good or bad. The good part was that he had found a *Rafflesia* bud. The bad bit was that he was sure it would open that very night.

I could not have blamed Richard if he had said that his time-lapse gear needed a lot of sorting out before it could be used, nor even if he had explained that he hadn't had a decent night's sleep in the last forty-eight hours and needed more. He said neither thing. He simply set about assembling his gear, there and then. Some local porters were found to carry it and as the sun was setting, he plodded off into the forest. He found the bud, as big as a cabbage, ballooning from the side of the vine and set up his gear. The lights began to flash and the camera to click. But even now he couldn't go to sleep. He couldn't even leave. Some wild creature could easily wander by during the night and knock over his equipment. So he spent yet a third night awake, sitting beside his flashing lights and camera. In the morning, the bud had still not opened a single petal.

It was not until the next night that the *Rafflesia* finally opened, peeling its fleshy petals upwards and stretching them outwards, to expose in its centre a circular plate supported by a pedestal and studded with spikes. Even then Richard had not finished. *Rafflesia*'s pollination also had to be documented. To do that, he cut a small hole in the base of the flower and inserted a lipstick camera to record pictures of small flies arriving dusted with pollen grains.

I am not one of those, like Aesop or Robert the Bruce, who readily derives moral precepts from the behaviour of animals, and I thought I would be even less likely to find them in the life cycle of plants, but *Rafflesia* did seem to me to provide a parable. One has to ask why this particular plant should produce the most extravagant and flamboyant of all flowers. It occurred to me that *Rafflesia* does not work for its living. The vine itself has to build leaves and stems to produce its food and ultimately construct its flowers. But *Rafflesia* does not concern itself with such practical matters. It simply absorbs all the food it needs from its host. Indeed there is virtually no limit on how much it can take and no curb to its extravagance. So it can build the most grandiose of flowers. It is the aristocrat of the tropical forest plant community.

* * *

I knew that Richard and Tim with their ingenious techniques would add spectacular and surprising dramas to the films, but even so, we needed a few animals every now and then to add their own kind of action. And there were good reasons to do so for animals, of course, play a major part in the lives of most plants – either as allies or enemies. One particular story appealed to me.

The great Indian rhinoceros is, in effect, a farmer. It has a particular taste for the fruit of the *Trewia* tree. After feeding on it in the morning, the rhinoceros habitually goes down to wallow in a river. Its bathe finished, it then stands about on a mudbank. It is safe there, out in the open and in no danger of being ambushed by a tiger. And there it defecates. *Trewia* seeds emerge with its dung and fall on to the highly fertile mud. So, over years, rhinoceroses create their own *Trewia* plantations conveniently planted alongside their regular bathing places. Mike Salisbury and I went off to Nepal to film this story, with Dickie Bird, once again, as recordist and a new cameraman, Stephen Mills.

The safest way to get close to an Indian rhinoceros is sitting on the back of an elephant. Mike's plan was that the camera crew should be on one elephant. I would be on another with a radio mike pinned to my shirt and connected to a small transmitter in my pocket that sent a signal to Dickie's recorder. When we found our rhinoceros, I would start telling the story and the camera would then pan from me to the rhinoceros which, with luck, would be either munching *Trewia* fruit or amiably and appropriately defecating its seeds. It seemed fairly straightforward.

The first evening we made a reconnaissance on elephant back and found plenty of likely places where a rhino might appear. That night it began to rain. It continued to do so for three days without stopping. The river we had crossed to get to our camp overflowed its banks and we became marooned on an island. On our fourth and last day, it stopped and we set out on our elephants. They patiently plodded through the marshes and the forest, creaking and swaying, shouldering their way through thickets of grass that came up to their ears, and wading through the rivers without finding any sign whatever of rhino. But then our little procession emerged from thick forest onto a grassy meadow and there on the opposite side we saw a huge bull rhinoceros. It was feeding on reeds, not the *Trewia* plant as we would have preferred, but we were overjoyed to have found it at all. Our elephants moved ponderously forward, mine a little ahead so that Stephen would be able to pan from me down to the rhino a few yards in front of my elephant. Just as I was about to speak, Dickie called out to me in alarm. Where was my mike? I looked down. It was no longer attached to my shirt.

'It must have fallen off,' I hissed back, despairingly.

'I guessed that,' said Dickie, somewhat testily. 'It's still transmitting and I can hear nothing but insects. It's on the ground somewhere. Get down and call to it. I'll be able to tell you when you get close.'

Cautiously, my elephant knelt on its back legs. I slid down its huge leathery rump and peered round its back end. I could just see the rhino beyond its front end, still munching the reeds. I looked around on the ground. The grass though short was quite thick. I'd never be able to find a tiny microphone and its transmitter in such a tangle unless I were guided.

'Hello, hello,' I said, as loudly as I dared and took a couple of steps out into the open.

'Warmer,' said Dickie, from the safety of his standing elephant. 'Definitely warmer.'

It took about three minutes of playing this somewhat desperate version of the warmer-colder game before I finally discovered the little microphone and its transmitter lying several yards back along the path made by my elephant. I picked it up gratefully and was about to clamber up again on to my elephant when Mike called out to me.

'Stay where you are,' he said. 'It's a much better shot with you on the ground. If the rhino charges you can always shin up that tree and we will rescue you.'

And that is why, in the finished film, it appears as though I have foolishly elected to approach a great Indian rhinoceros on foot. I admit that, judging from Stephen's shot the rhino appears to be many times farther away than I remember it. On the other hand, the tree Mike suggested I might climb looks, as indeed it certainly was, little more than a bush.

* * *

We decided to devote one programme in the series to plants that manage to thrive in extreme circumstances. That would bring the excitement of dramatic landscapes into the series and take viewers, vicariously, to little known corners of the world. It seemed a good idea, not least because it would take us there too. Where should we go? My mind went back forty years to the time when Charles Lagus and I had travelled up the Kukui River in Guyana, towards the vast mountain of Roraima. I remembered how frustrated we had felt at having insufficient time to try and reach it. Even though there are certainly no extraordinary animals on its summit plateau, as Conan Doyle had imagined in his novel *The Lost World*, it does have some bizarre plants that have evolved up there in isolation and now are quite different from those that grow on the savannahs three thousand feet below. That surely was excuse enough. Neil Nightingale felt the same way.

Richard Kirby and Trevor Gosling came with us. In Venezuela we were joined by Charles Brewer-Carias, a Venezuelan adventurer who had been up Roraima several times. There is a wide ledge that slants up one side of the mountain, giving relatively easy access to the summit, but walking up it and carrying all our equipment would take a lot of time. Far better to get whisked up to the summit by helicopter, Charles said. That seemed a good idea.

At least, it did at the time. I should have remembered that mountains the size of Roraima tend to generate their own weather. We sat on a small air-strip at its foot gazing at the bank of cloud on the horizon where the mountain should be, waiting for our chartered helicopter. Eventually a message arrived by radio from the helicopter's home airfield. There could be no flying around Roraima that day.

The cloud looked just as thick the following morning but to our surprise we were then told that a helicopter was already on its way. The pilot, when he arrived, surveyed our gear with gloom. If we wanted to take all that, he would have to make at least two trips and the weather was deteriorating so fast he was not sure there was even enough time for one. We would have to hurry. We threw in half the gear, Richard and I climbed in after it and before we could take breath we found ourselves careering through the sky and into the bank of cloud. As we roared upwards through the mist we knew we must be ascending fast because our ears popped and it got swiftly colder. I caught one alarming glimpse of an immense precipice of naked rock through a gap in the clouds and then once again we were travelling through dense mist. We started going round in circles. I felt sure we would have to return when suddenly the pilot spotted a small rent in the cloud. He swung round swiftly and plunged through it. We landed on a level expanse of bare black rock. The pilot's air of urgency had intensified into something close to panic. Yelling at us over the deafening sound of the engine, he hurled out all the gear and before we could discuss anything further, he had gone. In the sudden silence, we looked around at the scatter of baggage. At least we had had the wit, in the rush, to include a tent and some food. There was no guarantee that the helicopter was going to be able to get back.

We had come to Roraima in the hope of finding a strange landscape. We could hardly have done better. It was grotesque. The rock around us had been carved into the most extraordinary shapes. There were turrets and pillars. Some projections resembled swollen animal heads, others medieval castellations. Gigantic inverted pyramids teetered on narrow points and looked as though the slightest push would knock them over. All were as black as lava. But that colour was superficial. I moved a boulder and discovered that the rock beneath it was a pinkish grey. It was sandstone not lava. The blackness came from an alga that grew over every exposed surface. And it grew everywhere because everywhere was running with water.

The mist closed in. It began to rain quite hard. It was cold. We took shelter under a rocky overhang. To my relief, I heard the faint noise of the returning helicopter. It got louder and seemed quite close above us, but we couldn't see anything and the sound faded. I wondered how much food we had with us. The rain got heavier. There seemed nothing to do but to wait. After an hour or so, we heard distant voices. We yelled back. The pilot had not been able to find us so he had dumped the others and the rest of the gear a mile away.

Uniting took some time. Ravines at least twenty or thirty feet deep with vertical sides criss-crossed the rock. Some we could jump across. Others were too wide. If you slipped and fell into one there would be little chance of climbing out. Nor was it easy to find a place to put up tents because water was swilling everywhere. Guy ropes had to be tied to boulders or rocky projections, but at last we managed to establish a camp and made ourselves a meal.

During the next five days we explored the summit plateau. We had been deposited more or less in the middle of this huge tableland. Walking across it was not difficult in itself but any smooth area was almost bound to be covered with vast puddles deep enough to go over the top of your boots and the ravines made it virtually impossible to take a direct route anywhere. Eventually we found our way to the plateau edge and looked over the stupendous rock wall that fell vertically for over two thousand feet. Streams beside us, draining from the surface, spouted over the edge and cascaded downwards but before they had descended far the wind caught them and blew them away in a haze of droplets. Occasionally the clouds themselves blew away and then we could see the savannahs baking in the sun far below. But for most of the time we seemed to be in another world, a flat island of rock surrounded by a sea of cloud.

We had come, of course, for plants and Roraima is rich in them. Although so much of the landscape is naked rock, they grow in great variety in the crevices and beneath the overhangs. In some places sufficient peat has accumulated to enable them to sprout in such numbers that they form a thin sodden carpet. But the soil is little more than gravel and many plants have to find their nutrients elsewhere. They eat insects. There are several species of sundew, their leaves covered with long hairs, each glistening with a globule of sticky nectar at its tip which both attracts and entraps small prey. There are pitcher plants with leaves modified into cups in which their victims drown. Bromeliads, which retain the water in a vase at the centre of their rosettes, here on Roraima have developed an ability to dissolve and absorb the bodies of any small animals that tumble into their ponds. And – uniquely on Roraima – one of these carnivorous plants, a bladderwort, has become a thief.

Bladderwort is a water plant that grows in many parts of the world. Some of its leaves are modified into small capsules. It absorbs water from inside them thus creating a partial vacuum. If an insect or some other water creature blunders into one, a small door is unlatched and sucked inwards so that water rushes into the capsule carrying the insect with it. In other parts of the world, bladderworts live in ponds. On Roraima, they live in vase plants, trapping their victims before the vase plant itself can absorb them. The bromeliad's vase is quite small and can only accommodate a few stems of bladderwort, but its lodger grows by putting out a tendril that gropes across the rock to find more vase plants to rob.

Neil had decided that one of the plants we needed to film was a species of *Heliamphora*, a marsh pitcher plant which retains water by furling its leaves into long narrow tubes. It was unique to Roraima. Geological grotesquerie was all very well, he said, but we were, after all supposed to be making a botanical series. We searched every crevice and gully around our camp without finding a *Heliamphora*. Charles said that the only place he knew where it grew was on a promontory at the most distant end of the plateau. He was by no means sure that it was possible to find our way there through the maze of ravines. The helicopter had already been chartered to pick us up and ferry us down early on the morning of our sixth and last scheduled day on the mountain. There would be enough time for it to take us to this distant site and film a sequence before it finally lifted us all off.

That last day dawned clear. So clear, in fact, that the helicopter company, unknown to us, decided that they could deal with some of their backlog of commitments. Instead of dawn, it was nearly midday when the helicopter suddenly rocketed up over the edge of the plateau and settled down beside us. We told the pilot of the variation in our plan, piled in to his machine and surged off to our chosen locality. The *Heliamphora* was there just as Charles had said it would be. But it was not hugely impressive. Neil felt that in order that viewers might appreciate to the full what a privilege it was to see such a rare species, I ought to sit beside it and talk enthusiastically about it. Trevor pointed out, very truthfully, that he couldn't record me with the noise of the helicopter in the background and asked the pilot to turn off the engine. The pilot would have none of that. He might not be able to start it again at this altitude and then we would all be stuck. Charles suggested that, in that case, the pilot should go down and wait for an hour on the savannah below. Before anyone could think of another plan, the pilot was in his cockpit and away.

We managed to finish our filming within half an hour but we had no way of getting the helicopter back any earlier. Gloomily we watched the clouds build up. By the time the helicopter was due to return the clouds had descended and it was already raining. We heard it circling above us but unless the mist cleared no pilot could be expected to descend and land. And this one didn't.

We had brought a tent – a two-man tent – and there were six of us. By the time we had managed to pitch it on the bare rock, and tied guy ropes to boulders, we were drenched. All six of us lay in the tent in our sodden clothes, four of us in two layers, head to toe, like sardines. Trevor lay crosswise at one end. I was one of those at the bottom. Dawn was a long time coming but soon afterwards the helicopter reappeared. As we sat bleary-eyed in our soaked clothes, I wondered out loud whether anyone looking at our shot of a small aquatic carnivorous plant would have any idea of what we had had to do in order to film it. I need not have worried. The perspicacity of viewers was not put to the test. The shot never even appeared in the finished film.

* * *

Although *Rafflesia* is undoubtedly the biggest of all flowers, there is another plant that reproduces itself by means of an even bigger structure. It belongs to the arum family, as does the cuckoo pint, that strange little plant that grows in English hedgerows, and the arum lily (which is not, therefore, a true lily). Arums produce clusters of small flowers that grow around the base of a spike, called a spadix, which is encircled by a shroud, the spathe. Technically speaking, this structure is an inflorescence, not a single flower, so it should not be compared with *Rafflesia*. But those who are not troubled by botanical niceties, tend to regard the reproductive structures of the arums as single blooms and in that case, the biggest of all flowers is produced not by *Rafflesia* but by *Amorphophallus titanum*, the titan arum. Its spadix is reputed to grow to ten feet tall. Clearly it too should be in our series.

Filming the titan arum, we soon discovered, would not be easy. It grows in the Sumatran rain forest and first appears as a single tall shoot that rises like a pole out of the ground. When it is some ten feet tall, it branches out into three leaves. After a year it dies down, but the following season it sprouts again. Year after year it does this, building up a huge tuber in the ground. Then it dies down for a longer period. The stem rots away and after a few weeks all that is left is the tuber, hidden beneath the surface of the soil. Opinions vary about how long it remains like this. Some say ten years, others suggest that it may be even longer, but at any rate it is long enough for everyone to forget the exact place where the leaves once stood. Then unpredictably, the tip of the spadix starts to emerge from the ground. It grows with extraordinary speed – as much as three inches a day. By the time it is five or six feet tall the spathe is apparent, wrapped tightly around its base. Within a day or so this unfurls and spreads out like a wide trumpet. It only remains like this for a day or two. During this time, it is pollinated, though no one could tell us how or by what.

We started a little research. No botanic garden had a living specimen. Then we discovered that there was a doctor, Jim Syman, living in San Francisco, who had an *Amorphophallus* obsession. He had assembled every known record of it. He had a collection of every illustration of it ever published. And he had travelled thousands of miles in quest of a flowering specimen. He had seen plenty of the leaves. He had even seen a dead rotting flower. But he had never seen an example fully expanded and at its most glorious. However, during his travels through the forests of Sumatra, he had set up a network of *Amorphophallus* watchers who could swing into action whenever he had a chance to visit the island. He was planning another trip in a few months' time and we were welcome to join him.

We met in Medan, the biggest city in northern Sumatra. Jim proved to be as excitedly enthusiastic in the flesh as he was in his e-mails. He was large and heavily built. With him was Wilbert, a Dutch botanist who was studying the whole arum family and was as keen as any of us to see its star. We drove down to a small town in the south of the island where Jim's main agent, a local man with the unlikely Indonesian name of Darwin, was based. Darwin told us that he had found four specimens that were about to flower. However, people in a nearby village, he said, were very jealous that he should be making money out of their plants and had deliberately hacked them to pieces. I wondered about the truth of this but

Jim nodded sagely and assured me that it was the sort of attitude he would expect in this part of the world. *Amorphophallus* generated strange passions, he said. However, Darwin promised that he would spend the whole of that night combing the forest for more.

The next day, he arrived at our hotel triumphant. He had found one. The journey to it was not easy. Porters trotted along narrow paths through rice paddies and into the forest. We puffed after them, slipping on the mud, stumbling over roots. After an hour we caught up with the leaders who were sitting smoking waiting for us to arrive. In front of them lay a slimy rotting pile. Darwin beamed. Jim paid the promised reward but explained that a dead flower was not what we wanted. We would like to see one while it was still alive. Darwin nodded. He hadn't realised that, he said. He would try again. We wearily plodded back to our parked cars. But there we found another group of Darwin's men. They too had found one. Once again, we set off. After another hour, we slithered down a steep slope towards a stream, rounded a hump and there saw as astonishing a vegetable vision as I have ever witnessed.

Amorphophallus squatted in the forest looking like an alien. It was totally out of scale with every other plant around it. Its trumpet-shaped spathe, dark crimson within and pallid yellow on its outer surface, was at least four feet across. Its brown spadix, rising from its centre and tapered like the steeple of a church, stood at least nine feet high. I tried to lean across its huge spathe to see if I could smell the stench that was attributed to it. I could smell nothing. We set about filming. As we worked, we became aware that the flower was now beginning to produce a smell. Bad fish was not an inappropriate description, yet oddly, it was not offensive. And it came in waves. I would not have been surprised if I had seen the huge spathe visibly contracting and expanding.

Wilbert was on fire to cut it apart. He said he needed to take samples of its various tissues. He also wanted to establish the precise state of development of the flowers at the base of the spadix and to do that he would have to cut a great strip through the spathe. This seemed to me to be close to vivisection. It was certainly desecration. Happily Jim joined us in our pleas for him to spare the knife and Wilbert sat on the bank beside the flower with ill-concealed frustration while we busied ourselves with filming.

The following day we set off to return to the flower to see if it had changed and if so to film it before Wilbert slaughtered it. While we were preparing for the walk in, a villager arrived to say that he too had found a flower. We decided to concentrate on the one whose progress we had started to chronicle, but Wilbert saw a chance to practise his science unfrustrated by our art and eagerly left with this new guide.

Our flower certainly had changed. The spathe had closed up around the spadix, except in one part where its edge had drooped sufficiently for us to be able to peer into its depths and see the red female flowers deep below. Small sweat bees, in a steady procession, were flying down to visit them. So these were the pollinators! We caught a few to take back to Wilbert. It seemed very incongruous that such an immense structure should be needed to attract such tiny pollinators.

What is the explanation of *Amorphophallus*'s huge size? Perhaps it is connected with its rarity. If cross-pollination is to take place, two different plants have to be in flower at the same time. Since the plant only flowers once in seven years or so, and then for only a few days, two plants open at the same time will be a rare coincidence. The nearest flowering neighbours may be miles apart. Indeed, judging from Darwin's reports, they certainly were. So a flower has to be able to summon bees freshly loaded with pollen from another flower over very long distances. That cannot be done by sight which is the kind of signal that most insect-pollinated flowers rely on. It can only be done by smell. The flower generates its perfume by warming slightly in its depths. The odour must rise up through the gigantic spadix, which is hollow. I was being over-romantic in comparing it to a church steeple. It was functioning like a factory chimney.

* * *

Putting all our sequences together into finished films raised unusual problems. Ought we to make explicit the ways we had changed time scales? Should we point out that the rhythmic pulsing in some sequences filmed over a week or so was caused by the fact that plants grow at different speeds during the day and the night and that we had arranged for them to appear to be continuously well lit? And what should we do about sound? Watching plants making sudden vigorous movements without creating any sound whatsoever seemed odd. Music didn't seem to provide a totally satisfactory answer. In the end we added discreet noises. When the shoots of pitcher plants started to inflate themselves into huge jugs we added gentle creaks; and when the lid of the jug finally took shape and flipped open we added an audible pop. Dodder tendrils winding themselves around nettle stems and squeezing them were another opportunity, but in the end we resisted the temptation to add a faint strangulated scream.

26

The Lure of Birds

The British, notoriously, love birds. There is a royal society to protect them, merely a national one to protect children. Books about birds dominate the natural history sections in bookshops. Detailed field guides are there to help you identify them no matter where it is that you spend your holidays. And birds are, after all, the only truly wild creatures that many people see during their working lives. Alastair Fothergill, by now in 1997, Head of the Natural History Unit and himself an expert ornithologist, was very keen that they should be the subject of the next series I tackled.

But there was a snag. I am not one of those who can identify a bird from the merest glimpse of a silhouette. Nor am I expert in recognising bird calls. So I did not feel that I was the person to make a series about bird identification. On the other hand, I am fascinated by the way birds behave. How do they fly? Why is it that some have become flightless? How do they find their way around? How do they organise their social lives? A series that examined those things would be a very exciting one to make. For one thing, I should learn so much. Alastair agreed that I could make it along such lines. Sharmila Choudhury, a brilliant young scientist then working with the Wildfowl Trust at Slimbridge, was recruited to produce surveys of recent research and Mike Salisbury, who had headed the teams for *The Trials of Life*, the fossil series and *The Private Life of Plants*, once again took overall charge. There was no problem about finding directors to work on the programmes. Almost everyone in the Unit seemed to have a burning desire to try out some new and ingenious technique to record his or her favourite avian subject.

Pete Bassett took on the programme dealing with bird song. He determined from the outset that he would not be satisfied by filming birds singing and then adding the sound from recordings made at some other time. As he said, very correctly, you can tell. The movements of the singer's beak and the pulsations of its throat don't match the trill and swoops of the sound. There would be no such short cuts in *The Life of Birds*.

The dawn chorus in an English woodland was to form a major sequence. Pete determined that not only would every bird that appeared be heard singing its own song but he wanted its breath, warm from its tiny lungs, to be seen condensing in the cold air of first light. To film that the cameraman had to find a bird that, throughout the few weeks when the dawn chorus was at its height, used a singing post directly east of where the cameraman could put his hide so that its breath would be back-lit by the rising sun. There must not be any twigs between the camera and the bird. Equally importantly, there should not be too much vegetation behind it which might confuse its outline. Recording the sound made other demands. There must not be a motorway nearby, loud with the sound of cars. Nor a woodsman who had got up early to start work with his chainsaw. Nor an aeroplane droning its way through the sky. It must not be raining and the sun should not be hidden behind a bank of cloud but rise bright and clear. Andrew Anderson took on the job. There are only a dozen species in the finished montage but Andrew spent every spring dawn out in woodland before he felt he had got the quality of shots that he wanted. The finished sequence is little short of magical.

* * *

There is, of course, one very effective way in which you can attract a bird and encourage it to sing. If you can get a recording of such a song and play it in the territory of another individual of the same species, the rightful owner will very often appear and indignantly answer back. You can use the same trick simply to persuade a bird to come into the open within range of the camera. We decided to do so when we set about filming a lyre bird in the forests of southern Australia.

The male is one of the most accomplished of all avian singers. He not only produces a highly complex song of his own but he incorporates into his performance many of the songs that he hears made by other birds in the surrounding forests. One male was said to regularly imitate a dozen different species. As the outside world has come closer to their forests, so lyrebirds have started to mimic other sounds. We recorded one that gave perfect imitations of car burglar alarms, camera shutters, and even a shutter click preceded by the buzz of a timer. He even – sadly – sang of his own destruction, producing a perfect imitation of a chainsaw being used by men cutting down his forest nearby.

We decided to try and film this bird and me in the same shot. If I could precede him on to his stage, I could give a fittingly laudatory introduction to this most magnificent singer in the whole bird world. I would then retire and the bird would give his recital. Most of the forest in which this one lived was so dense that there was little likelihood of getting a clear view of him. But a fallen tree gave us a chance. I stationed myself at one end, the cameraman at the other. Now if we could persuade the lyrebird to hop on to the middle of the trunk, the camera would be able to get an uninterrupted view along the trunk of both the bird and me. The obvious thing to do was to put a loudspeaker beneath the middle of the trunk and play a recording of the song.

It worked too perfectly. No sooner had we started the playback than the bird appeared very close by me. Ignoring me, he jumped on to the trunk and rushed along it to try and find his rival who was producing a song as good as – indeed identical with – his own. But as he ran towards the camera and the sound, he passed the speaker. Suddenly he realised that the sound was now coming from behind him, so he turned back and ran towards me. Now, once more the sound was coming from the middle of the log. He turned again to face the camera and looked so angry and confused that we switched off the recorder before he had a nervous breakdown. His rival having apparently vanished, he flew away.

After a short rest to allow him to recover his composure away in the forest, we played him another very short burst. The camera started and I began, very softly, to speak my introduction. As I mentioned his name, exactly on cue, he took the stage, stalking haughtily along the trunk towards the camera – and since there was no song to be heard, jumped off again and returned to the bush confident, no doubt, that he had seen off his rival.

* * *

From Australia, we went south to New Zealand. Many birds here are flightless for the islands became detached from other lands so long ago in geological history that no ground-living predatory mammals were here to hunt them – before human beings arrived a thousand years or so ago. Among them is New Zealand's national bird, the kiwi.

The kiwi is a bird trying to be a mammal. Its feathers are almost hair-like, so that it appears to be furry. Its eyesight is very poor and it finds its food using its sense of smell, which the majority of birds lack, and long tactile whiskers at the base of its beak. Its wings have disappeared entirely. It is nocturnal and extremely shy but we heard of a lonely beach on New Zealand's most southerly inhabited fragment of land, Stewart Island, where we stood a good chance of seeing one.

We went down to the beach at dusk. As it got dark, a small hunched homunculus no more than a foot high, stalked cautiously out of the bush. It got bolder, advanced down the beach and started to walk along the line of stranded seaweed, probing into the sand for hoppers. Remembering this creature's mammalian inclinations I thought it would be a good thing to disguise my scent – and there was an obvious way of doing so. I lay down along the line of seaweed and draped some rotting fronds over myself. The kiwi worked its way along the tide-line, plunging its beak into the sand to tweezer up a sand-hopper and, with an audible snort, blow out the sand that had got up its nostrils in the process. Apparently it had no fear of the rather larger lump of jetsam that had been cast up with the seaweed and within five minutes I had the privilege of allowing one of the world's most extraordinary birds to pick up its food a few inches in front of me.

* * *

We needed to see birds in flight. Somehow we had to get a camera to travel alongside birds in the air so that we could see in detail the exact movements of their wings and give some impression of what it must be like to have the freedom of the skies. To do that we took advantage of the discoveries made by Konrad Lorenz which he had described when I had interviewed him in Alexandra Palace forty years earlier. As well as demonstrating his command of the language of greylag geese, he had also discovered that his geese would steadfastly follow the first objects they saw and the sounds they heard within a few hours of their hatching. In the wild, this ensures that young goslings follow their mother when she leads her brood down to the water and that they stay close to her for some time thereafter. Lorenz called the phenomenon 'imprinting'. If an experimenter hatches eggs in an incubator and substitutes himself for the mother goose as the nestlings break free of their shells, the young would also follow him or her, even when they were fully fledged. In later years it was discovered that many ground-nesting birds, and indeed many other animals, were profoundly and indelibly affected by sights and sounds of early infancy.

Nigel Marven, who was producing the programme on water birds, is an expert at this technique and imprinted himself on some mallard ducklings. One of them, a drake, became particularly attached to him and followed him wherever he went. Nigel decided to try and incorporate me into the shot of the flying drake. We went down to a reservoir in mid-Wales that had a long straight road running along the top of its dam. I sat in a dinghy on the lake surrounded by some tame mallard females, feeding them with bread to keep them swimming nearby. Nigel and the cameraman were in a small open car with the drake in a cage. The car started off along the dam road. The drake was released from his cage while Nigel called to him. The drake took off and flew within a yard of the camera as they drove across the dam at 30 miles an hour. Three quarters of the way along, the driver of the camera car suddenly accelerated. The mallard could not keep up, and veered away towards the water and there caught sight of the group of females so he swooped down and landed beside them and – coincidentally as far as he was concerned – me.

Nigel repeated the performance at a slightly faster speed. Then he decided he wanted to get a close-up of the drake's head as he quacked during the flight. For this he invited me to join them in the car. As we sped along the road, with the drake flapping alongside, the cameraman said that the drake was a little low. If he were to be flying a couple of feet higher, there would be a nice background of hills in the distance. I stretched out my arm, put my hand under the breast of the flying duck and pushed him gently upwards until the cameraman said that the shot was perfect. The drake, not in the least disconcerted, happily continued flying at his new altitude until once again he landed beside the dinghy and was put back in his cage.

Nigel was so delighted with his drake's performance that I think he would have continued flying him all afternoon. But after half a dozen flights the drake realised that there was *always* a group of females at the far end of the dam and that he could reach them more quickly by flying directly to them. So he decided to ignore Nigel's calls for the time being. It was time to take him home.

* * *

I was determined that one bird family in particular would star in the series, the one with which, I admit, I have an obsession – the birds of paradise. Alfred Russell Wallace's book, *The Malay Archipelago*, had started it in my boyhood. As well as its vivid plate of the birds being hunted, it also contained a graphic description of their courtship display. Wallace, in fact, was the first European ever to see and describe this dramatic sight. Ever since I read it, the birds had epitomised for me everything that is exotic, extravagant and spectacular in the natural world. The Indonesians had stopped me from going to the Aru Islands in my *Zoo Quest* days to look for the species that Wallace described, the Greater Bird of Paradise. On my second attempt, back in 1957, I had managed to film a close cousin to that species, Count Raggi's Bird in New Guinea, but it was a brief black and white sequence and in no way adequate. I had failed altogether when we had tramped through the headwaters of the Sepik in 1971. I was now seventy years old. This would be my last chance. So I decided that birds of paradise should be given a prominent place in *The Life of Birds* and that I would go to New Guinea once again to try and film them.

But could the series, with over nine thousand other bird species to choose from, provide enough time to document the family in the way I felt they deserved? I thought not. They needed a programme entirely to themselves. So I suggested that while I was in New Guinea, I should extend my stay and film material for an extra-long, extra-special programme devoted entirely to the whole family. I got together with Paul Reddish, a producer in the Unit whose obsession with the family rivalled mine. He agreed to be the producer and together we worked out a plan for a programme that would make a comprehensive survey of them.

There are forty-two different species. Many were scientifically described for the first time during the nineteenth century by German and Italian ornithologists. They thought they might get support for their expeditions if they named any new species they discovered after a royal patron. The names the birds received as a result are hardly usefully descriptive but nonetheless they are not entirely inappropriate for these aristocratic dandies of the bird world.

The family is extraordinarily varied. The males of the most famous kind, the Greater Bird of Paradise, which Wallace had illustrated on the frontispiece of his book, has plumes sprouting from its flanks and displays in groups in trees. There are half a dozen similar species – the Lesser, which like the Greater has yellow plumes, Count Raggi's which has red, and the Emperor of Germany's which has white. Another group within the family – the Superb, the Magnificent, the Parotia – have plumes of a quite different kind – iridescent breast shields, gorgets and capes. They clear special arenas on the forest floor and then dance solo to bewitch females who tour all the dancing grounds in a wide area before making their choice. Yet another group which includes Princess Stephanie's Bird, the Ribbontail, and the Sicklebill have greatly enlarged tail feathers and display singly on branches, fluffing out fans on their shoulders and breasts. In addition to all these, there are extraordinary eccentrics such as the King Bird, which is rich orange-red with a white chest and has two quills projecting from his tail each ending in a green iridescent disc; and the King of Saxony's Bird, which has a pair of notched enamelled feathers twice the length of his body sprouting from the top of his head and so is greatly favoured by the Wahgi men for their headdresses.

These decorations are so varied that it is hard to believe that all the different species belong to the same family. But they are only developed by the mature males. The females are drably dressed with brown backs and chests barred with black and cream but they are so similar to one another that when you see them you are no longer surprised to know that they are all closely related.

I started research for the programme by selecting a dozen species which, between them, would represent the family's extraordinary range. There was no way in which a single cameraman could film them all in the time available to us. Nor indeed could I hope to sit alongside a camera and witness, first hand, every one of them performing their displays. Instead, Paul and I decided that we would commission two of the most skilled and determined cameramen we knew to make two trips each to New Guinea. Each would go to different parts of the island, armed with the list of the half a dozen or so selected species that we knew occurred there. On their return, Paul and I would view all their footage. I would then write a script based on this material that made a coherent story about the family and its evolution. We would then be able to decide what gaps remained. The next season, all four of us would go to the island to try and film those species of which we needed extra material and those scenes in which I would have to appear. That way, we would produce a reasonably comprehensive film surveying the whole family – and I would, at last, get a proper view of some of the wonderful birds that I had dreamed about for so long.

We were lucky. Richard Kirby who had worked on the *Plants* series and Mike Potts who had filmed many of the bird sequences for *Trials* agreed to take on the project. They knew very well what the difficulties were. Wretched light for the most part, extremely uncomfortable living conditions, torrential rains most days, mud, leeches and days spent clambering up and slithering down steep muddy slopes in forests full of spines.

I particularly wanted to include the Blue Bird of Paradise – once known as Prince Rudolph's – for during his dance, the male spreads his ultramarine gauzy plumes into a huge semicircular fan, swings down on his perch and while hanging upside down and rhythmically vibrating his plumed fan, makes a noise like some malfunctioning electronic equipment. A captive specimen in a zoo had been filmed going through this astounding performance, but as far as I knew, no one had filmed it in the wild. We discovered that a young Englishman working as a teacher in a small village in the New Guinea Highlands, knew of a site where a male performed every day. Mike Potts went to film it.

The night before he got there, an inter-village feud erupted. Two men were murdered. Armed gangs were roaming the forest and waiting in ambush beside the tracks. Local people were quite certain that if a European insisted on sitting by himself in a hide surrounded by valuable cameras, he would certainly be attacked and robbed. Mike had to retreat.

Nonetheless, at the end of the first season, more than half of the species we had listed had been filmed. Paul and I discussed what was missing with Mike and Richard. The following year, they returned to New Guinea and this time, after a week or so during which they established themselves, Paul and I together with Dickie Bird as sound recordist, flew out to join them.

Phil Hurrell who had fixed the ropes from which I had swung about the forest canopy during the making of *Plants*, had gone out with Mike to help with the tree-climbing. They were working from a small village in the delta of the Fly River on the southern coast. Greater Birds of Paradise, the species which Wallace had seen, had a display tree a mile or so away in the forest. Its huge trunk, straight and smooth, rose vertically uncluttered by side branches for a hundred and twenty feet and there, above the canopies of the smaller trees that surrounded it, spread its girder-like branches. I could just see through my binoculars, half a dozen brown pigeon-sized birds sitting on the smaller branches that sprang from one of these. They were resting quietly. But I could also see yellow tufts on their flanks and knew they were male Greater Birds of Paradise.

The plan was for me to watch these birds perform their display from another huge emergent tree nearby. Phil had prepared it for me. He had fixed a pulley to the underside of one of its immense branches with a rope running through it. Attached to the rope's bottom end, on the ground, stood a frame, rather like a bicycle without wheels. I was to sit on a saddle at one end and look at a camera which was fixed to the place where the handlebars would have been. There were pedals for me to rest my feet on. By varying the pressure on these I could keep the frame on an even keel when it was hoisted into the air. At the other end, high above and immediately beneath the pulley, hung a kit-bag full of sand, exactly my weight and unflatteringly large. When Phil pulled on another rope, the kit-bag could be hauled gently downwards and I would rise smoothly upwards, talking to the camera as I went and describing what I saw.

I rose smoothly past the crowns of palm trees, up alongside dangling lianas, up through the lower canopy, up into the sunlight until at last I hung swaying immediately beneath the pulley. I was now on exactly the same level as the Greater Birds in the next tree. Through binoculars I could see them in intimate detail – their rippling emerald green bibs and the golden plumes that sprouted from their flanks beneath their wings. They seemed quite untroubled by my arrival. After I had been watching them for a few minutes, one shrieked. Immediately the rest of the group sprang into action, doubtless under the impression that the male that had called had glimpsed a female. Each ducked his head, threw up his plumes into a fountain over his back and rushed up and down his branch. One fluttered across and tried to settle on a branch that was occupied by another bird but was repelled and went back to his own. And then a female did appear. She flew to a perch in the middle of the group. The resident male was bowing, head down, his wings tensely outstretched, but as soon as she settled, he twisted round and approached her head-up. As he got alongside, he started to peck her head quite violently. She stood her ground. He started to flick her with his outstretched wings. Still she would not budge. Then the male jumped on her and within a few seconds had copulated.

Mike had already filmed every phase of this behaviour in close-up. To do so he had sat on a tiny platform a hundred feet above ground for several days. It wasn't too difficult, he said dismissively, because the females always selected the same male performing on the same perch. Once he had realised that was the case, he knew just where to look and focus his lenses. Clearly, the sequence devoted to this species was complete. It was certainly the closest approximation to the frontispiece of Wallace's book that I could ever have hoped for.

* * *

Some of the mystery and fascination that these birds have for me stems from the fact that their elaborate costumes are so difficult to interpret. Museum-bound naturalists when they first received specimens sent to them by collectors from the very edge of the explored world, were not sure what to make of these crumpled skins and bedraggled feathers. The very first to be seen in Europe were brought back by Magellan's expedition at the beginning of the sixteenth century. They were those of the Greater Bird which the expedition collected not from New Guinea but from Tidore, one of the Spice Islands, where they were regularly imported to be used as currency. The New Guinea people who collected them, five hundred miles away to the east habitually cut off the wings and the feet to expose the glory of the flank plumes. They were still preparing such trade skins in the same way in Wallace's time. They still do so today. Magellan's men, when they asked about these legless wingless creatures, were confidently assured by the Tidore traders, who of course had never seen the living birds, that they never did have wings. They did not need them for they floated high in the air – in paradise – feeding on dew. Neither did they need any legs, for instead of perching they suspended themselves from the bare curling quills that projected from their tails. Indeed, they were only found by human beings when they died and fell to earth. So it was that the scientific name given to the Greater Bird of Paradise, the first species to be seen in Europe was, and still is, *Paradisea apoda* – the 'legless bird from paradise'.

The species that were discovered in the following centuries continued to baffle ornithologists. Even as late as the late nineteenth century, the artists commissioned by John Gould to illustrate his magnificent monograph on the family were still getting things wildly wrong. They did their best to make sense of the strange iridescent epaulettes, crinolines and cockades. But how could they have known that the Blue Bird did not display perched upright on a branch, but hung below it; that the King of Saxony's Bird used a liana like a trampoline in his dance.

Even as we were making our film there were still mysteries to be solved. One species, called the Twelve-wired Bird of Paradise, is particularly strange. The wires referred to in its name are bare quills that project beyond its tail and then bend forwards. Naturalists who received the first specimens at the beginning of the nineteenth century understandably concluded that the bends in them must have been caused by the way the specimens had been packed. They therefore illustrated the species with these wires curling around its body in a fantastic aureole. When ornithologists at last saw the living bird they discovered that in reality the wires are naturally bent backwards and radiate from the birds' rear end like spokes of a wheel. But why? If they served as a visual display it seems odd that they had been reduced to mere quills. Mike Potts filmed the Twelve-wired's display for the very first time and revealed the answer.

The performance starts with the male calling from a vertical post, often a dead snag in the swamps. If a female alights on the top, the male starts his dance, circling around the bare trunk some distance below her, climbing closer and closer until, as he reaches her, he suddenly switches around so that his rear end is towards her. Then he flicks his twelve wires across her face. So we now know that the fore-play of birds of paradise can involve not only glamorous visions and eccentric sounds, but tactile stimulation.

Perhaps the most bizarre member of the family is a species that Wallace himself discovered and which now carries his name – Wallace's Standard-wing. He found it on Bacan, one of the most westerly of New Guinea's offshore islands. It has a pair of long white feathers attached to the front edge of each wing. Wallace wrote that they were erectile. The artist who illustrated his book showed them flaring outwards roughly at right angles to the body, though in truth the illustration is difficult to interpret. Gould showed them simply dangling. Neither version looks very convincing. No-one then seemed to know how they appeared in life. It was more than fifty years before another European naturalist saw the living bird and it wasn't until 1983 that there was a third sighting. An English ornithologist, David Bishop, found a population, not on Bacan but on the neighbouring larger island of Halmahera. We had map references for the site, we found guides and eventually the four of us, Paul, Richard, Dickie, and myself, together with David Gibbs, one of the major ornithological authorities on New Guinea birds, reached the patch of forest where the Standard-wings were to be found and there we made camp.

Wallace's Bird is one of the rarest species in the whole family, but here the forest was alive with them. It echoed with their calls. There must have been a hundred or so males clustered together in just one or two trees. They were bizarre rather than beautiful. Most of their body feathers were plain brown but each had a vivid green bib attached to its breast, rather like an extravagant made-up bow tie. And each had those extraordinary white feathers sprouting from the front edges of their wings. At rest, they dangled downwards, but when the birds started to dance they raised them so that they stood up like white masts. When they got really excited, they switched them into all kinds of positions like a frenetic semaphore signaller. The males were clustered thickly in the trees, each on his own stretch of branch. The whole group would suddenly be seized by a paroxysm of shrieking and frantic dancing, with all of them twisting round to show off their bow-ties and white standards. But they also had a speciality display of their own, used by no other bird of paradise – vertical jumping. Every now and then they leaned slightly forward and then shot vertically upwards, like rockets to a height of some thirty feet. Then back they fell, shrieking as they came, to land on the perch from which they had taken off.

* * *

Richard Kirby, helped by David Gibbs, had established a camp in the mountains in the far west of the New Guinea mainland. It was built on the side of a steep muddy slope in the forest and it was undoubtedly one of the most uncomfortable camps I have ever stayed in. Everything in it was either moist or wringing wet for heavy rain lashed the forest for at least half of every day and even when it was not doing so, the air was so laden with moisture that everything dripped. Our tents were simply large fly-sheets stretched over horizontal poles. Beneath them, platforms had been made by laying trimmed saplings side by side. These would keep our gear above the mud and serve as our beds.

But two hundred feet above us, on the very crest of the ridge, a Wilson's Bird of Paradise had his display ground. We climbed up to look at it. An area some twelve feet across had been meticulously cleared of every twig and leaf. A large sapling had fallen across it, but Richard said that this had not put off the bird and he had now incorporated it into his dances. Richard had built two hides, one on either side of the arena. He would film from one. I would watch from the other. I would wear a radio microphone around my neck and he would have head-phones so that if necessary I could speak to him. Only he and I would go up the following morning. The rest of the team would remain in camp. We dared not risk having more than two people beside the arena, in case the bird took fright and decided not to display.

The following morning, in the pitch dark, two hours before sunrise, Richard and I left camp and clambered up the side of the ridge by torchlight. We were in the hides well before the dawn. The bird, unless he had been roosting very close by, could not know that the two clusters of leafy branches among the lianas to which he had become accustomed over the past week, were occupied.

The leaves dripped. The forest began to pale. Mosquitoes started to bite. There were distant calls from birds down below us in the valley that I did not recognise. Then from right above my head came a piercingly loud whistle. On the other side of the arena I saw a very slow smooth movement among the leaves. Richard was tilting his camera slowly upwards to focus on something above my head. The call came again – and down fluttered the owner of the arena.

I knew, of course, what Wilson's Bird would look like for I had seen museum specimens and identification drawings. The male is one of the oddest of the whole family, for he is bald. Short jet black feathers, as fine as velvet, covered the front of his head and the base of his beak. His naked scalp was a rich ultramarine. His chest was a rippling glossy green and his back a deep red with what looked like a back-pack of red feathers with a special gloss, glinting like spun glass. A golden cape covered the nape of his neck. Two bare quills projected from his short tail, each curling into a ring, one on each side. So vivid were his colours that he seemed to glow. It was almost as though he had a light within him that gave his spectacular pigments an incandescence.

He noticed a small leaf that the cold dawn wind had detached from one of the branches above. It irritated him. He hopped across to it, seized it with his beak and with a sideways flick of his head, threw it into the nearby bush. He called again.

Then, although he had meticulously removed one leaf from the ground, he set about cutting others from the saplings that leaned over his arena, laboriously sawing through their stalks with his beak. Having cut them, he threw them into the surrounding bush. Perhaps he wanted to increase the light falling on his stage so that his colours could be seen to their best advantage. Wonderful though he was, I knew that we would not see his full thrilling glory unless a female approached. Only then would his accoutrements become apparent. Now he was like a theatre performer whose crumpled costume appeared to be somewhat over-sized and ill-fitting and would continue to do so until such time as he began the extravagant expansive movements of his dance. He continued to hop and call.

And then he shrieked. Simultaneously, he jumped on to the fallen sapling that slanted across his arena. The cape on his neck sprang up and became a glorious ruff. His breast shield lifted and projected at right angles to his back. He opened his beak wide and revealed a green enamel-like lining to his mouth. Only now did I see that a female had fluttered down from the branches above. He pirouetted in front of her, and hopped repeatedly between the sapling and the ground, jerking his head violently from side to side. I watched breathless. Slight movements in the hide opposite told me that Richard was recording everything.

Wallace, when he saw a species of butterfly that he had been pursuing for a long time, described the thrill of one of these long-sought-after visions. 'My heart began to beat violently,' he wrote. 'Blood rushed to my head and I felt more in danger of fainting than when I had been in apprehension of immediate death. I had a headache for the rest of the day.' I didn't have the headache but otherwise I felt much the same. Wilson's Bird made a last call, flew away and Richard and I scrambled and slid our way down the flank of the ridge to camp.

Dickie came out to meet us.

'We got it,' I called excitedly. 'We got it.'

'I know,' he said. 'I could tell the split second when the bird arrived. The radio mike was picking up your heart beat and I suddenly heard it double its speed.'

* * *

In February 1997 we were in New Zealand working on the *Birds* series, filming kea parrots raiding the nests of mutton birds and extracting the rotund fat chicks and eating them. That night I was woken by Mike phoning from a nearby room. He had just had a call from Bristol. Jane had been taken to hospital and was dangerously ill. I should get back to London as soon as possible. Trevor drove me through the night to Christchurch where I caught the morning flight. I landed at Heathrow and went straight to the hospital. As I stepped out of the lift, Susan, our daughter who had been with Jane since she was taken ill, was there to meet me. Almost simultaneously, the neighbouring lift doors opened and her brother, Robert, stepped out. He had come from Canberra where he was now an anthropologist at the Australian National University.

Jane was in a coma. She had had a cerebral haemorrhage. The doctor said that she might just be able to register my presence if I spoke to her as I held her hand. She did, and gave my hand a squeeze. She recognised that her two children were there too in a similar way. But she was sinking. We spent that night in the hospital. The following evening she died. It was the eve of our forty-seventh wedding anniversary.

The focus of my life, the anchor, had gone. Jane had always been there. She had been at the airport, unfailingly, every time I came back from a trip. She had supported me in everything I had done. Indeed, most of the things I had done were, in actuality, joint enterprises. I could not have led the life I had, were it not for her unwavering support. Now, I was lost.

The next filming trip for the *Birds* series was, of course, cancelled. But I could not simply abandon the series. Three quarters of the filming had been done. The film of some of the programmes had been edited but none of the commentaries had yet been written. There was a lot of work that I had to do – and I was grateful that this was so.

27

Mammals and Two-shots

I wanted to keep busy. *The Life of Birds* was transmitted in the autumn of 1998. The end of the millennium was approaching. In spite of the fact that the first year of the new one would clearly be numbered 2001, the BBC together with almost everyone else had decided that the terminal moment of the old millennium would be December 31st 1999 and that it must be marked with suitable pomp and gravitas. The Natural History Unit was to make its own specialist contribution to events and I was asked to make a pair of programmes assessing the condition of the natural world at this arbitrary but nonetheless meaningful moment.

I had been involved in the conservation movement since the early 1950's and had sat, as a very junior member, in the meetings during which Peter Scott and the other pioneer conservationists had created the first great international conservation charity, the World Wildlife Fund. I thus became involved, and increasingly so, with several voluntary organisations preaching the conservation cause and raising money for it. I believed, however, that people would not take action and give their financial support to conserve the natural world unless they cared about it; and they would not care about it unless they had seen something of its complexity and beauty and understood that we are not only part of it ourselves but are dependent on it. I therefore continued to focus my programmes on straightforward natural history. But I also made sure that I ended each series with a call for the protection of an increasingly endangered world. Now, however, I had the opportunity to publicise not merely my own concerns but the findings of ecologists, climatologists, glaciologists, geologists, zoologists and other scientists who were investigating the present condition of the planet and were alarmed at the increasing pace of degradation and the looming catastrophes of climate change. I took it gratefully.

The resulting two-part series, *The State of the Planet*, ended with a sequence shot on Easter Island. Archaeologists had shown that when Polynesian seafarers had first settled on the island it had been richly forested and capable of supporting a large population. But as their numbers grew, so the people cut down more and more trees. The thin soil from the cleared land became impoverished and started to blow away. When the people felled the last tree, they destroyed their ability to build new boats and effectively imprisoned themselves. Ultimately, they were unable to leave their shores even to catch fish to supplement the limited amount of food that they were able to grow on their devastated island. There could hardly be a more vivid and poignant example, in miniature, of the disaster that faces the entire planet unless we start to mend our ways.

* * *

Other broadcasting jobs continued to appear. I had been narrating half-hour episodes of the *Wildlife on One* series for over twenty years and these still continued. But now a bigger proposal appeared. Alastair Fothergill, with whom I had worked on *The Trials of Life*, had plans for a series on a truly epic scale that would deal with marine life worldwide. He was going to call it *The Blue Planet* and it would have seven fifty-minute episodes. By now the underwater helmet with which he and I had experimented during *Trials* had been improved so much that it was at last possible for a presenter to talk to a camera underwater without drowning. Even so, what with the inevitable change in the tone of voice and the gargling effect produced by intermittent streams of bubbles, presenters were undeniably less effective underwater than they were on land and Alastair had decided to do without one for his new series. Nevertheless he did need out-of-vision commentaries and I was delighted when he invited me to provide them. Both he and I recognised that as a consequence I was likely to get a considerable amount of reflected glory and undeserved credit but he was happy to accept that.

In the event, that prediction proved to be correct. And there was a bonus. When the series was sold in the United States, my commentary was replaced with one specially modified to suit American tastes and spoken by an actor. Overall, the series was so successful that Alastair was commissioned to make a sequel, *Planet Earth*. Once again I supplied commentaries and once again the United States distributors removed them, replacing them this time with a new version spoken by the Hollywood actress Sigourney Weaver. This time, however, the BBC decided that it would also put the original British version on videodisc in the U.S. Its sales were phenomenal. Within a year it had become the best-selling non-movie DVD of all time. My mind went back to the time when *Life on Earth* was initially rejected in the United States because of my refusal to allow my voice to be replaced by an American one. It had taken a long time, but it seemed that a battle had at last been won.

* * *

The next subject for a natural history series seemed obvious. Mammals. They do not, it is true, have as wide and dedicated a following as birds but they do have fur and that gives them a wide appeal. People are fond of animals they can pat even if they only do so theoretically, and we are, after all, mammals ourselves. The Unit needed little persuasion. Mike Salisbury was happy, once again, to take charge and he recruited some of the most inventive directors in the Unit. Some – Neil Lucas, Vanessa Berlowitz – I had worked with before, when providing them with commentaries for their *Wildlife on One* programmes. Others – Mark Linfield, Jonny Keeling, Chris Cole – I had not met before but if Mike had invited them, I knew we would all get on. As in previous series, we would do our best to include species that few people had heard of – pikas, tenrecs, yapoks and babirusa, fossas and uakaris. They might provide us with sights that we could truthfully say had never been filmed before. But if the series were to survey mammals in an encyclopaedic way, we would also have to include many of television's most familiar animal stars – lions and tigers, elephants and giraffes, orang-utans and chimpanzees. How could we film such well-known creatures so that sequences in which they featured would seem novel and exciting? The team put their collective heads together and concluded that the solution would be to include a lot of what is known in the trade as two-shots, that is to say shots in which a presenter (in this case me) appears alongside the animals in question. That would make it clear to viewers that they were getting new specially-filmed sequences and not something that an industrious researcher had discovered by simply rummaging through the Natural History Unit's archives.

We decided that each programme would deal with animals that shared the same diet. What an animal eats has a profound effect not only on the shape of its body but on its behaviour and even its breeding habits. Grass-eaters must have teeth that macerate their comparatively indigestible food and large stomachs in which to digest it. Gathering grass requires them to spend a long time out in the open. They therefore have had to develop good defensive strategies and many grazers find safety by congregating in flocks or herds. That, in turn, has an effect on their social systems and courtship rituals. Carnivores, on the other hand, have the lithe athletic bodies necessary for running fast, as they must do in order to catch their prey, and sharp-edged teeth with which to cut up their victims.

And they either live in small packs like wolves or are largely solitary like tigers.

The principle is not quite so clear-cut in other groupings – insectivores, omnivores and rodents – but such divisions approximate to the way in which science classifies the mammals and modern genetics confirms that this reflects their natural relationships. So organising our survey in this way also had some scientific validity.

The series would start with the most primitive of living mammals, creatures that still retain some of the characteristics of reptiles, the group from which all mammals are descended. The most dramatic of these concerns the way they reproduce. They lay eggs instead of giving birth to live young like all other mammals. Only two mammals do this today, the echidna and the platypus.

Filming platypus was going to be a major problem. We had tried to do so before, back in 1977 during the production of *Life on Earth*. None, then as now, were to be found in zoos outside Australia. Even on their native continent they had only been bred in captivity once – and breeding, of course, was the crucial characteristic that we wanted to feature. We had considered that illustrating this behaviour was so important for *Life on Earth* that we offered to finance a research project at any Australian university that would study the species, providing they allowed us to use their findings to work out ways in which we might film the animals performing their remarkable breeding techniques. Disappointingly, we had been unable to find anyone who thought the project sufficiently feasible to take it on.

We did, however, discover in an Australian government archive, some fragmentary film shots showing a tiny infant platypus breaking free of its parchment-like shell. As a film sequence it was a long way from satisfactory for it showed nothing of the nest and even the shot of the baby emerging from its shell only started when the baby was already half free. Nonetheless, incomplete though it was, it was an extraordinary sight and we included it in the finished programme.

Many years later I met the Australian government's official cameraman who had taken the shot. I thanked him for it but said how disappointed I had been that it was not more complete. Was the baby half-emerged when he found it? 'Ar, no,' he said with disarming Australian cheerfulness. 'It wasn't until I was halfway through filming the bloody thing that I realised I'd left the bloody lens-cap on!'

Now, however, we had another chance. Mike made a preliminary research trip to Australia and discovered a scientist from the University of New South Wales, Tanya Rankin, who was already several years into a study of a platypus in the wild. She had perfected a technique of implanting tiny radio transmitters beneath the skin of a platypus so that she knew exactly where it was – not only when it was swimming in the river, hunting its prey, but even while it was underground. She was coming to the end of her two-year project and with it the end of her funding. Mike immediately saw the possibility this offered us. We might be able to get one of those never-filmed-before sequences in our very first programme. He offered to provide money from our budget for the further year of her study, which she dearly wanted if, during that time, we could get her help in our own filming.

Our primary aim, of course, was to capture the extraordinary sight of a furry warm-blooded mammal cuddling, not a baby, but an egg. Witnessing the precise moment when the egg emerged from the female was scarcely possible. Even watching one hatch would require a very lucky piece of timing. But we might at least see the baby supping milk as it oozed from what are, in essence, greatly enlarged sweat glands buried in the fur on the female platypus' underside. But she only feeds her baby deep in her breeding chamber at the end of the long tunnel that she digs in a riverbank. Nonetheless, Tanya's familiarity with a whole population of platypuses in a secluded valley in New South Wales gave us a reasonable chance of getting such shots.

We were lucky to get the help of Mark Lamble, one of the most experienced and expert natural history cameramen in Australia. He filmed a platypus that Tanya knew well as it hunted for crayfish. It continued to do so even when I waded into the river and moved slowly towards it. So we were able to get yet another two-shot, even though my contribution was no more than a glimpse of my feet.

Mike and Tanya had already worked out how to film what was going on inside the nest chamber. They planned to use an endoscope, a flexible fibre optic rod originally developed to allow surgeons to inspect the inside of a living human body by, for example, pushing it down someone's throat in order to have a look at the lining of their stomach. Mark and Mike thought they might see what went on inside a platypus' nest chamber in the same sort of way.

The signals from Tanya's transmitters were sufficiently strong to travel through a yard or so of the soil so she could follow the female's progress as she travelled underground along her burrow. That burrow, Tanya had discovered, could be as much as twenty yards long. And when the source of the signals ceased to move, Tanya knew that the female had reached her nesting chamber at its farthest end.

Mark carefully plotted the position of the nesting chamber on the surface of the ground above. Then, very gently and exceedingly slowly, a millimetre or so at a time in order not to disturb the mother platypus, he started drilling a hole a centimetre across towards the nest chamber.

Platypuses do not burrow like rabbits, ejecting a telltale pile of soil outside the mouth of the burrow. Instead they force their way through the soft soil by sheer muscular effort. As a consequence, the soil forming the walls of their tunnels is harder and much more compact than elsewhere and Mark found that as soon as the tip of his auger came within a centimetre or so of the tunnel he could feel the difference in soil density and thus know when he was about to enter the tunnel. So he was able to make that last and possibly alarming breakthrough at a time when he knew that the female platypus was away swimming in the river.

So it was that for the very first time, he and Tanya were able to see and to film what no one had ever seen before – a mother platypus in her burrow. Mike and I joined them and together we watched the platypus as she crawled up and down her tunnel. We peeped at her as she slept curled up in her nest chamber. When she left it, we saw little movements in the leaves which covered its floor and at last we glimpsed a glistening, squirming, naked blob. Mark started the camera and recorded the first-ever shots of a platypus baby undisturbed in its nest. In themselves they were very exciting. But they were also something of a disappointment for they showed that, in this burrow at least, we were too late for a shot of the egg. Nonetheless, when the mother returned we were able to watch as globules of milk appeared on the female's underside and the baby suckled.

Within a week or so, Mark had as many shots as we needed of this very early stage in the youngster's life. It takes over three months for a platypus baby to grow sufficiently large and strong to emerge into the outside world and Mike decided that for the sake of both Mark and the platypus, there should be a pause in the filming. Mark would return in a few weeks' time when the baby would have grown fur and be visibly more developed.

But that return never happened. A week or so after Mark had left, in the Australian spring, there were sudden and torrential rains. The river rose several metres and the platypus' burrow was flooded. When at last the waters retreated, Tanya found that her platypus had vanished and the burrow was empty. So the sequence was not as complete as we had hoped it would be. It was, nonetheless, a first.

* * *

The endoscope enabled us to make another discovery. I had allocated a major sequence in the rodent programme to beavers. They are natural television stars, felling trees, hauling them down into specially cut canals that lead back to their lakes and above all, industriously and ingeniously constructing dams in order to create lakes for themselves. At one side of these lakes they build a lodge, a great dome of mud and leaves, often covered with logs and boulders, which they enter by means of an underwater tunnel that opens into the lodge's floor. We know that beavers do not hibernate during the winter for they can be seen swimming beneath the ice as they collect the leafy branches that they dragged underwater during the summer. But what did they do inside the lodge?

Mike deployed three endoscopes, one to convey a picture to our infra-red camera and two others to illuminate the spacious interior with infra-red light, invisible to our eyes and, we assumed, to those of the beavers. We watched the beavers slip into the pool at one side of the lodge's floor that was the mouth of the tunnel leading out into the lake and saw them bob up in it to drag out a leafy branch that they had stored in the lake during the summer and on which they now all fed. But as well as the adult beavers, there were also two small furry creatures in the lodge. At first we thought that they were very young beavers. But they were extremely small and lacked the wide flat scaly beaver tail. They were a completely different species. They were musk rats. Whether they had contributed their fair share to the beaver's cold store beneath the ice we did not know. Nor could we tell whether or not the beavers knew they had lodgers living alongside them, in the pitch darkness. But it seemed that the beavers were not as insensitive to infra-red light as we had supposed, or at any rate that they could sense the slight warmth produced by the lights for eventually, as we filmed, they plastered mud over the end of the endoscopes and returned their lodge to total darkness.

* * *

We were doing well with assembling entirely new shots of little-known animals. But what about the two-shots of me talking alongside some of our better known stars? The radio-tagging technology that had helped Tanya Rankin helped us here too, though on a much larger scale. The blue whale, as everyone knows, is the biggest of all living animals, weighing 200 tons and growing to a length of 100 feet. Indeed, it is almost certainly the biggest animal that has ever existed, being twice the size of the largest dinosaur yet discovered. But its size can only be properly comprehended visually if something familiar is placed alongside it for comparison. And what better than a television presenter. Sadly, however, a two-shot in this case seemed a virtual impossibility.

Blue whales, unlike most whale species, travel in groups of only three or four or by themselves. They can stay submerged for half an hour and can swim at speeds of up to 15 miles an hour. Getting a shot of me alongside one seemed out of the question. I had once swum with humpbacked whales in Hawaii but that was back in the 1970's during the filming of *Life on Earth* and now, in my seventy-fifth year, I didn't think I could manage to keep up with one even if the whale was idling.

At that time, little was known of the movements of blue whales through the oceans. But Bruce Mates, a professor of zoology in the University of Oregon, working off the coast of California, has developed ways of tracking them. After prolonged searching he manages to get alongside one as it surfaces and then shoots a small radio tag into its flank. Once he has succeeded in doing that, he is able to track it wherever it wanders in the vastness of the ocean. Every time it surfaces to snatch a breath, the tag sends a signal to a satellite several miles high in the sky which is then transmitted by way of the internet back to Bruce's computer wherever he may be.

We joined him south of San Francisco on the coast of Half Moon Bay. He was optimistic. A whale he knew well was approaching from the south about twenty miles out from the coast. Bruce has a friend, Steve, who is professionally a pharmacist but whose abiding pleasure is to fly in his small Cessna plane. He was more than happy to cruise up and down high above Half Moon Bay spotting whales with the very genuine justification that he is helping scientific research.

We set out in a rubber Zodiac driven by a powerful outboard engine. At the tiller is Barb who has been working with Bruce on the whale research programme for several years and is marvellously expert in anticipating where and when a whale is likely to surface – once she has located it. She is talking to Steve in his Cessna five hundred feet above and ahead of us. He has seen a whale and is telling her which way to steer in order to be near it the next time it surfaces. I am tied to a wooden construction in the bows so that I can lean over to see what is in the water beneath us without too much of a risk of falling overboard, even though we are travelling at great speed and bouncing vigorously on the waves.

Then I see it – a huge pale shape, as big as the wing of a small aeroplane, moving slowly up and down in the blue depths beneath us. Barb, driving at speed, keeps us alongside it. It is the tail of the whale and with the top of each upward stroke it seems to be getting closer to the surface.

And then twenty yards ahead of us, so far away that it is difficult to believe that it belongs to the same animal whose tail is beneath us, a spout of mucus-laden spray blasts into the sky. An immense shining flank, slate blue mottled with cream spots, slides up at speed and breaks the surface. 'A blue whale,' I shout triumphantly to the camera. It was not just a blue whale but a blue whale two-shot. I never dreamed that I could be so lucky.

* * *

For the most part, natural history filming is not a dangerous occupation. It can be made so by bravado, of course, but our kind of programme aimed to show animals behaving naturally as though the camera – and the cameraman – was not there. And for most of the time our subjects are indeed unaware of our presence or at any rate indifferent to it.

Bears, potentially, are very dangerous indeed. Jeff Turner, however, has spent many years with them in his native Canada and in Alaska and he knows them as well as anyone. He agreed to shoot specific natural history sequences of them, but he also undertook to try to get two-shots of me reasonably close to them. Together with director Huw Cordey and cameraman Gavin Thurston, with whom I had filmed for decades, we went to Kodiak Island off the south coast of Alaska, where the grizzly bears are reputed to be the largest in the world. They can be nine feet long and weigh three-quarters of a ton. They are not only big, they are fast. They can run faster than you can.

Jeff already knew many of the bears here individually. He recognised them by the shape of their ears and nostrils. And recognition is very important. Bears have markedly individual characters. Some are placid. Some are bad-tempered. There is nowhere to hide on the open gravel flats around the estuaries of Kodiak Island. Walking behind Jeff as, step by hesitant step, he approached a particular bear, one had to hope that he hadn't misremembered a nostril.

The information I had to impart, when I came to address the camera, was that an ability to eat almost anything is essential in such a bleak place as the Alaskan tundra. The bears can survive here because they can collect and digest a varying succession of foods as the changing seasons make each available. We were there in August. The bears at low tide had been digging up clams, showing remarkable dexterity and delicacy in extracting the flesh from the protective shells. But now, salmon were beginning to swim into the rivers from the sea and were making their way up to the headwaters where they would spawn. As we watched, a female bear, whom Jeff said he knew, stopped digging in the sand and lumbered away from us towards one of the shallow streams that snaked across the wide gravel flats. We hurried back towards the shore, waded across the stream up which the fish were swimming and returned towards the sea so that we could get a front view of the bear across the stream as it approached. Gavin was enthusiastic. The bear walking straight towards us made an impressive sight. Huw stood me close to the bank and asked me to start my piece about seasonal diets. To do so I had to look at the camera. And that meant I had to turn my back to the approaching bear. I could only hope that Huw would tell me if she showed any sign of preferring fresh meat to fish.

I looked earnestly at the camera. 'Bears have a menu that changes with the seasons,' I said. 'And top of the menu right now is – salmon.' I turned round, anxious to see where the bear had got to. She was within a few yards of me on the other side of the stream. As I spoke the word 'salmon', she reared into the air, plunged head-first into the water and lifted her head with water dripping from her shaggy fur and a four-foot-long salmon flapping in her jaws. And she lumbered off.

We were jubilant. If anything, her timing was too perfect. These days, electronic trickery makes it easy enough to take a shot of a bear catching fish and then to combine it with one of a man talking to camera. Perhaps viewers would think that our wonderful shot was a fake. On the other hand, I suspected that people might realise that the combination of delight and relief that showed in my face was something that would take a better actor than me to simulate.

* * *

Huw Cordey had not finished with me as far as two-shots were concerned. I like most animals. Even the least prepossessing – naked mole rats, for example, that resemble white sausages with little more than two pairs of curved teeth at the front end – have their fascination from a scientific point of view. I do however have one wholly irrational dislike that amounts almost to a phobia. I do not like rats. The kind that infest houses, that is. Perhaps it is because they carry so many human diseases and flourish in sewers. Maybe it is because, unlike most animals, they deliberately seek out human habitations in order to collect what they can of our food. Whatever the reason, the appearance of a rat in a room makes me want to jump on the nearest table.

Nevertheless, when I came to write the shooting script for the programme that would deal with opportunists, animals that are prepared to eat almost anything they find such as pigs and racoons as well as bears,

I could hardly leave out rats. They are the most successful opportunists of all. And I thought it would be cowardly, when scripting the sequence that dealt with them, not to include some to-camera pieces that would require me to be in the presence of rats. There was one obviously dramatic place to film such a sequence. Not far from the Indian city of Bikaner, in Rajasthan, there is a temple where rats are worshipped.

Huw and Gavin went there ahead and met me and Trevor Gosling, who was to record the sound, at the railway station. Everything was fine, said Huw. The authorities had agreed that we might film in their temple. And there were certainly plenty of rats. 'Never mind,' I said with somewhat forced jocularity, 'I will just wear my heaviest boots and tuck my trousers into my socks'. 'I'm afraid you can't,' replied Huw unsympathetically. 'Remember, this is a Hindu temple. You can only enter if you are barefoot. But I will make sure that there is a tall stool on which you can sit while you do your to-camera pieces and so keep you well above them.'

I was not feeling well. It was years since I had had any kind of digestive upset but something in the Indian cuisine had not agreed with me and as soon as I had swallowed a welcoming drink at our guest-house I felt a need to retire swiftly to the elegant little pavilion to which each of us had been allocated. As I sat on the lavatory, feeling considerably relieved if exhausted, I sensed activity beneath me. I looked down and saw a large and somewhat bedraggled rat leap out from between my legs. It shot across the room and disappeared into my bed. Gavin heard my shrieks and helped me to chase it out of the building and into the surrounding fields where doubtless it would somehow find its way back to the sewer from which it had come. But it was not a good augury.

The following day Huw and Gavin went ahead to the rat temple to set things up. Trevor and I followed half an hour later. The temple was indeed spectacular. Those who worship there believe that the rats are sacred and must therefore be protected. Accordingly, netting covers the open courts that surround the main shrine to keep away birds of prey, and a guard stood at the door to repel dogs and cats. Reluctantly we took off our shoes and socks and entered. Rats swarmed all over the stone floor collecting the grain that was liberally scattered for them. They ran in long processions along the tops of walls. The absence of any predators of any kind meant that some rats that were so diseased, crippled or senile that they would never have survived in the outside world, were here still limping around among their younger more agile relatives. A worshipper sat cross-legged and motionless beside the entrance to the inner courtyard with a large brass tray loaded with grain in front of him. Rats ran across his shoulders, over his head and beneath his beard as he, together with the rats, took mouthfuls of grain.

Trevor and I gingerly picked our way through the rat swarms towards the central and most holy chamber where we expected to find Huw. And there he was, bending over the stool on which I was going to have to sit in order to keep me well above any inquisitive rats. I was outraged to see that he was smearing bananas up the stool's wooden legs. He was deliberately laying a trail that would tempt the rats, once I was seated, to run up my trousers as I was trying to speak my lines. He did have the grace to look abashed. 'Well,' he said. 'We all agreed we needed two-shots.'

* * *

We went to considerable trouble to devise new ways of filming the familiar. Justine Evans nobly and courageously sat with an infra-red camera for night after night at the far end of a long tunnel-like cave on the flanks of Mount Elgon in Kenya while elephants cautiously felt their way past her in the pitch blackness to dig for salts. Mark Linfield used special cameras that slowed down the speed of flying bats so extremely that he was able to show how they could pluck a spider from the middle of its web and then reverse in midair to avoid becoming entangled in the sticky silk filaments. We even managed to produce a visual answer to that hoary conundrum 'How do hedgehogs mate'. The conventional answer of course, is 'With extreme care.' A more scientific description was hard to come by. Not so long ago, it was believed by many authorities that the pairing couples managed to avoid spearing one another by the female rolling over on her back so that they could mate belly to belly. But there seemed to be no film to settle the question.

I did, in fact, have a hedgehog living in my garden at the time, but he was by himself so Mark located a wildlife refuge that had a large number of hedgehogs of both sexes, including a number of young females. One evening, he brought twenty of them to my garden, in which we had installed low but nonetheless revealing lighting. He released several of the females and gave them a little time to settle down. Then one at a time, he released the males. It was past midnight when at last a young male encountered a young female that appealed to him. Round and round the garden they trotted, the female puffing and snorting, the male following immediately behind, until at last both seemed to consider that the proper moment had arrived. The female lowered her haunches to the ground and flattened the spines on the rear half of her back. The male put his forelegs on her back and pushed his pelvis forward. The truth was revealed. He was so splendidly equipped that, distant though the female's vent was, he was able to reach it without any risk whatever of doing himself an injury.

So after three years of near continuous work that included, cumulatively, some twelve months of filming overseas, *The Life of Mammals* was completed and started its weekly transmissions in November 2002.

28

Undergrowth

With the completion of the Mammals series, I could fairly claim to have dealt with the more popular groups of animals. Those I had so far failed to survey were mostly invertebrates, animals without backbones. They were certainly more unfamiliar – some might say, more unattractive – but that meant that they were also, in television terms, largely unexposed. Broadly speaking, they can be classified by the number of their legs. Insects have six. They are by far the largest group. Indeed, they constitute half of all known animal species. Science has named about a million of them and there are probably two million still unlabelled. Then there are spiders which have eight legs, centipedes that can have about forty, millipedes that can have hundreds and slugs that have none at all. There are also groups that virtually no one except professional zoologists have ever heard of such as the lobopods and the amblypygids. One way and another, there would be plenty for us to have a go at. So why not make a series about all land-living animals without backbones?

The idea, I thought, was a good one, but I had been long enough in the game of pitching programmes to know that controllers and commissioners like a title. Trying to sell a programme idea without one is to give yourself an unnecessary handicap. They also like a title that harks back to past successes, if there have been any. 'Son of....' is Hollywood's standard way of doing this. In our case, the first half of the title should therefore contain the word 'Life'. *The Life of Terrestrial Invertebrates*, however, was not, I thought, a title likely to enthuse them. On the other hand, I did not want to descend to the language of the nursery, as many museums do, and call it *The Life of Creepy Crawlies*. I spent a long time brooding on this until one evening, in the bath, the words came to me. *Life in the Undergrowth*. And it was with that name that we managed to sell the proposal to the Controller of BBC1.

I felt confident that now, in the first years of the new millennium, we would be able to film small animals better than ever before. In the past, the amount of light required by even the most sensitive film made it necessary to flood a small animal such as an ant with such an intense light that it could scarcely be expected to behave in a normal way. At worst, it could well be shrivelled and fried. The need for maximum light also meant that the iris diaphragm in the camera's lens would have to be opened to its fullest extent to admit as much as possible. As a consequence the depth of focus of the image would be minimal. So if the ant's head was sharp, its body would be no more than a blur. But now the latest hypersensitive electronic cameras would allow us to work with very little additional lighting. So both problems would be minimal.

But there was another obstacle to be overcome. If we really wanted to take the viewer into the world of small creatures, we would have to get down to the level of our subjects to look at them in the face and move alongside them at their own pace. A normal camera, simply because of its bulk, cannot do this. But one exceptional cameraman has spent much of his professional life working out solutions to these problems. His name is Martin Dohrn. He is a big burly man and when working, he travels with a great number of gigantic bags. His arrival in a hotel lobby can make even the brawniest porter gulp a little. But worse, from the hotel's point of view, is to come. The huge bags, when eventually lugged into his room, disgorge a huge number of strange implements and specially assembled mounts. Martin likes to be versatile.

He starts by assessing a filming problem – how to get inside a flower to see the probing tongue of a drinking insect, how to glimpse the inside of an occupied wasps' nest. He then is likely to explain to the rest of us with a mixture of gloom and barely concealed enthusiasm that the subject can only be properly filmed by a completely new camera set-up. This will involve major modifications to cameras, camera mounts, optical cables, lenses and much else. He then starts work. Eventually a new contraption is revealed which is then given its own name – Antcam, Spidercam, Flycam. In due course, a special case is made for it and another jumbo bag joins the gear. The truth, of course, is that Martin enjoys building new bits of kit, especially if in the process it is necessary to turn what was once a smart hotel room into an engineering workshop.

* * *

His skills were soon tested. The climax of the programme dealing with colonial insects would be those formidable warriors of the tropical rain forest, army ants. These are the New World equivalents of the driver ants of Africa that I had encountered on my first excursion to the tropics to Sierra Leone in 1954. Then they had invaded a hut containing our collection of several dozen snakes and attacked them so aggressively, sinking their secateur jaws between the snake's scales, that several died. The army ants of the New World, though only distantly related to the African species, have very similar habits.

If you go out at night, these insect armies can be a serious nuisance, even to a human being. It is only too easy to step into their path as they fan out across the forest floor in search of prey. Scorpions, centipedes, insects of all kinds, even small vertebrates like lizards are quickly seized, killed and carried back in pieces to the queen and the rest of the colony that are bivouacked under a rock or between the buttresses of a tree. If you inadvertently stray among them as they hunt, they will certainly swarm up your trousers, but they usually delay the moment of attack so that they may be everywhere in your clothing before you are aware of their presence. The first thing to do then is to run and get out of the way of the marauding army. But that is not necessarily easy for in the darkness you cannot be certain of the right direction in which to go. Once you are clear of the main army, you have to strip and pick off the ants, one by one. During the day, the menace is not so great. At least you can see them.

Every few days, a colony will abandon its bivouac and set out to find new hunting grounds. The workers run in a column several inches wide, carrying their cocoons with them, guarded on either flank by the giant soldiers with their great caliper-like jaws open and at the ready. Filming them, day or night, is just the sort of challenge that Martin relishes and Antcam is part of his standard apparatus. The trick is to ensure that no part of the equipment physically touches the path of the racing army. Martin achieves that by fitting his camera with a tiny lens on the end of a fibre-optic cable and mounting the whole assembly on a jib arm. This enables his camera not only to hover above the racing columns but to track alongside it, keeping pace with individual ants as they race along over pebbles, under leaves and over roots.

* * *

The first insects that appeared on land, around 350 million years ago, were wingless, like their living descendants, the silverfish and bristletails. In due course, however, wings appeared. It seems likely that they first developed as pairs of pouches on the back that were stretched taut by veins pumped full of liquid. I planned to represent this dramatic event in evolutionary history by filming the unfurling of the wings of what is considered to be the most primitive of winged insects alive today, the mayfly.

There are some two and a half thousand species of mayfly in the world, but the biggest, with wings almost three inches across, lives in a few small rivers in the middle of Hungary. The wingless larvae, like all the mayfly clan, spend the first year of their lives burrowing in mud or gravel on a river bed. They spend most of their lives in this way but eventually, in a final and dramatic burst of activity they rise to the surface of the water, moult into the winged form and take to the air. They do not feed. Indeed they cannot do so for they have no mouth. Their mission during their brief flight is solely to find a mate and, as far as the females are concerned, a suitable place to deposit their eggs. And then they die.

Most species emerge, to the excitement of fly fishermen, in a series of risings throughout the summer months. The giant Hungarians, however, coordinate their emergences and only do so on relatively few occasions at the height of summer. In consequence, when they do rise they do so in huge numbers. And what is more, the dates on which they appear are said to be predictable.

When we arrived, however, Jozsef Stzentpeteri, our local expert, explained that this year, it was different. This year there had been an unseasonable amount of rain. The river water was therefore colder than normal. That would probably delay things.

We went down to the river and waited. Martin Dohrn inspected the site and announced that, he would not, after all, need to construct a special Mayflycam. Mike Salisbury and I sat on the river bank, sometimes in pallid sunshine, sometimes in drizzle and waited.

After three days, the weather began to warm. At last Martin spotted a dimple on the mirror surface of the river with at its centre a wriggling insect struggling to free itself from its larval skin. The mayflies were starting their rise. The first individuals to appear were all males. Within a few minutes the air was so thick with a blizzard of wings that the opposite bank of the river was almost invisible. Thousands of insects were fluttering into the air and flying across to bushes on the bank, there to hang while, for reasons that no one properly understands, they performed a second moult. It took only a minute or so before the adults emerged with slightly longer twin filaments at the end of their abdomens and wings that, instead of being transparent as in their previous incarnation, were now a delicate blue. The process of emergence was so urgent and occupied them so completely that I was able to sit in a dinghy and allow them to settle on my fingers and describe the details of their transformation as they performed it.

Then the females started to appear. The males, when their bodies had dried, flew low over the surface of the water looking for them. As soon as one emerged gangs of males pounced on her and struggled among themselves to mate with her. Once that had happened, the female fluttered into the air and flew off to find an appropriate place to lay her newly fertilised eggs. But that is not straightforward. Mayfly eggs take several weeks to hatch. During that time, the flow of the river will almost inevitably carry the eggs downstream so that if they are laid in exactly the same stretch of the river where the females themselves hatched, the next generation would find themselves farther down river. If that happened year after year, then the whole population would doubtless end up in the sea. To prevent this, the females, once they have finished mating, start to fly upstream in long processions. Somehow or other they select a place which will ensure that when eventually the eggs hatch, they will have been carried back to the precise stretch of the river where their parents emerged.

* * *

We also planned to film the emergence of the seventeen-year cicada. This North American insect, like the mayflies, spends the bulk of its life away from human eyes, though in this case not under water but under ground. It lays its eggs beneath the bark of a tree. As soon as they hatch, the young drop to the ground and burrow into the earth. There they find the roots of the trees, fasten their mouth parts on to them and start to drink the sap. And there they stay for seventeen years, stolidly drinking, regularly shedding their skins as they grow. They are, by far, the longest-living of any insect.

At the end of that time, they start to dig their way back towards the surface. When they reach the top of their burrows, they wait for a day or so. And then, they all break out into the open air and climb up the trunk of the tree that has nourished them throughout their long existence. And there the males among them start to sing.

The noise is deafening. Local people shut themselves away in their houses, just to get some relief from it. The females respond to the songs – not by singing themselves but, mercifully, by flicking their wings. That action makes a relatively faint click but it is at a much lower frequency than the song of the males and sufficiently loud for the males to detect it in spite of their own resounding chorus. Having located such a clicking response, a male stops singing and moves towards the female making it. She encourages him with a double click. He moves closer and finally the two mate.

Our plan was that I should try and replicate the female's encouraging signals by clicking my fingers and see if I could persuade a male to clamber on to my hand. After some research, we decided that the most suitable place to attempt this was in a large cemetery that surrounded a nunnery outside Cincinnati. It was pleasantly rural with wide expanses of lawn interspersed with a few low trees. When we arrived, the air was throbbing satisfactorily with the cicada chorus. The trunks and the branches of the trees were coated with ranks of male cicadas, shoulder to shoulder, all industriously singing away. Rod Clarke, the cameraman, selected one on which the insects were particularly thick. I stationed myself beyond it and prepared to stroll forward explaining how I was proposing to bamboozle one of the sex-starved males. All was ready. I started to walk and talk, shouting quite loudly to make myself heard over the din. It seemed to go rather well. My words ended neatly with my arrival at the tree and I prepared to summon a male with a click of my fingers. Rod interrupted me. 'Stop, stop, stop.'

A small white minibus had driven into the camera's view, a hundred yards or so behind me. We turned and watched it as it stopped slap in the middle of Rod's picture. Out of it came a procession of a dozen or so elderly ladies, each holding in front of her a long thin shaft. They were now wandering among the trees, sticking their wands vertically into the ground, rather like golf caddies planting flags on a putting green. We went over to ask if they were going to stay long. Not long, they said, but what were we doing? We told them. 'And you?' we asked. 'Well,' they explained amiably, 'we have spent our lives in this nunnery and now we are selecting our graves.' So, bizarrely, while we had been filming insects resurrecting themselves out of the earth in the foreground, there were other beings in the background preparing to enter it. The nuns were greatly entertained by the thought and tickled by the idea that they might be appearing on television. But we decided that, all the same, we had better re-shoot the piece.

* * *

Our techniques of working as far as possible without bright lights and getting our cameras down to the level of our subjects were working well. And we had one further technical refinement up our sleeve. Lenses can enlarge an image so greatly that a three-inch twig can appear to extend from one side of a television screen to the other. But lenses do not similarly extend time. So an ant which walks along the twig in a couple of seconds appears to cross the television screen at frantic speed. Slow-motion technology, however, makes it possible for you to slow down time in the same way as you can enlarge space. So you can use it to show ants and other small creatures moving around their miniature world in a properly measured manner. You can watch the way two ants, on meeting one another, establish whether they are strangers or old friends by exploring one another with their antennae or exchanging little blobs of spittle. This technique, together with our emphasis on getting down to the level of our performers and following them around at their own pace, meant that we began to see our subjects as individuals. And to our surprise, we discovered that many of them, far from being mindless automata, had not only complex and subtle behaviours but individual personalities.

Kevin Flay took on some of the spider sequences. I caught up with him in the southern United States. He was filming bolas spiders. They hunt during the night in a particularly ingenious way. Instead of constructing a complex web, they spin a single filament of silk with a sticky blob at the end. They then hang beneath a leaf holding the filament with one of their long front legs. When they sense the approach of a moth, they use their leg to whirl the filament with its blob of silk like an Argentinian gaucho whirling his bolas. At the same time they release a chemical scent, a pheromone, that attracts the particular species of moth that is flying at the time. The moth approaches and more often than not is hit by the whirling blob, which sticks to it. The spider can then haul in its catch and eat it.

Kevin showed me his filming set-up. He had ten of the spiders, each on a leafy twig placed in an empty milk bottle. They were all females, for, as is usual among spiders, it is the females that are the larger and more active of the sexes. He introduced me to them individually. This one, he explained, disliked the light and simply would not hunt while a lamp was on it. The next one would be put off hunting by the slightest noise. And this one was, as far as he could discover, either fully satisfied or bone idle for she simply sat there no matter how quiet or dark it was and would not hunt. But the last one was a darling. She would whirl her bolas, light or no light, in noise or in silence, and irrespective of whether she had just had a meal or not.

Our tiny performers, we were discovering, might only have a microscopic brain, but they had individual personalities like many other animals a million times their size.

* * *

Chris Watson had the job of recording all the natural sounds for the series. He is a man obsessed with sound and has the most discerning ears of anyone I know. He maintains, for example, that he can tell the difference between the sounds made by Pacific and Atlantic waves, though since he is also a Yorkshireman and has a very well developed sense of humour, one can never be absolutely sure that he is being serious.

When we filmed termites, using endoscopes to peer inside their huge mud fortresses, Chris assumed we would need recordings of natural sound to go with the video sequence even though, as he lugubriously pointed out, when we came to edit the finished programme we would almost certainly drown his delicate sounds with that of a full orchestra. Nonetheless, he lowered a miniature microphone down one of the chimneys on the top of a termite hill and discovered that the hot air rising from the centre of the nest created a ghostly sound like that of the wind blowing up a chimney – which of course was exactly what it was. He also could hear the sound of the soldier termites tapping their heads on the walls of their galleries and snapping their jaws together. Having spent a long time totally absorbed in recording these other-worldly noises he eventually pulled up the microphone to discover that some of the sounds had been caused by the termites scissoring off the protective cap of foam rubber with which the microphone is normally covered. 'Fifty bloody pounds, that,' he said, assuming his music-hall Yorkshireman persona. 'There'll be no new shoes for the little ones this Christmas.'

Termites, in southern Africa, are regularly raided by Matabele ants. This species is named after a famously war-like African people, because they regularly emerge from their underground nests and set out in a column several hundred strong to raid nearby termite colonies.

Matabeles belong to an ant family that, unusually, possess a special noise-producing structure. It is placed across the joint that connects the ant's abdomen to the rest of its body. There is a comb-like line of spikes on one side and a rough scraper on the other so that when the ant waggles its abdomen up and down there is a faint stridulation. Chris, crouched beside a battalion of the ants as they marched out from their nest, could hear nothing. But when they reached the termite hill and poured into one of the entrances he could hear a pandemonium of rustlings, snappings and clickings as battle was joined within. The Matabeles eventually emerged, most carrying the bodies of the slaughtered termites in their jaws and to Chris' delight stridulating loudly, as if in triumph after their victory.

His ingenuity was further tested when we asked him to make what I suspect may well be the first field recording of earthworms. In Gippsland, in southern Australia, lives a giant among the earthworm family. Stating how long they are is not easy, for collectors eager to claim a world record for one of their captures, may well pull the poor earthworm to reach a length that it does not normally achieve (and break it in the process), but there are claims that they may reach the astonishing length of seven feet. The species was hardly likely to be a star performer for us but nonetheless the giant earthworm was an un-filmed character and we thought it deserved a place. The problem for Chris was that its tunnels seldom open on the surface. The animals live some inches beneath, where the soil is perpetually moist. As one inches its way along its tunnel, munching rotting vegetable matter, it occasionally contracts some of its segments and hauls part of its length up the tunnel. The squelching noise this creates is so loud that as you walk through the South Australian meadows you may hear, immediately behind you, what sounds like someone flushing a lavatory. Since there is little warning of when this is going to happen and no sign on the surface to suggest where an earthworm might be deep beneath the turf, recording them is not easy. But Chris indefatigably walked for hour after hour through the meadows with his recorder and eventually collected a most convincing recording of these phantom lavatories.

* * *

Filming our invertebrates took two years. By the beginning of 2005 we had finished our last foreign trip and were starting the absorbing and pleasurable task of fitting all the disparate pieces together to make coherent narratives. It was while we were in the middle of this process that Mike, sitting in an editing suite in Bristol, received a call from a lady in the Television Centre in London who introduced herself by saying that she had special responsibility for the overall presentation of programmes. She would like to discuss a title for the series. She had assumed that *Life in the Undergrowth* was used as a working title. It could not possibly be the final one. She had set up a number of focus groups, and had tested it on them by word association. She had thus established that the word 'undergrowth' was linked with darkness, gloom, squalor and general unpleasantness. It reminded some people, she said decisively, of underwear. Its use was clearly out of the question. She realised that the word 'life' would have to be included and she accepted the team's view that it would be unscientific to use the word 'insects' since other groups also appeared in the series. But something had to be done. She had therefore decided to set up a 'brain-storming session' at which ideas would be 'kicked around'. Mike would be very welcome to join it.

Mike, calm as ever, explained that as far as the production team was concerned the title was *Life in the Undergrowth*. If she would forgive him, he would stay in Bristol as he had quite a big enough job on his hands – making the programmes. He would wait with interest to hear the result of her brainstorm. It arrived two weeks later. She and her group after an afternoon of debate and head-scratching had decided that the best title would be *Life of the Multilegs*. Mike dutifully filed her note and continued as before. *Life in the Undergrowth* duly reached the television screen in the autumn of 2005.

* * *

As the series was in the course of being transmitted, it dawned on me that I had now made series about all the major groups of land-living animals except for the amphibians and reptiles. Together, the two would certainly provide enough variety to sustain a complete series. If we were to make it, our survey of terrestrial life would be complete. And suddenly that became the one thing I wanted to do more than anything else. Grouping them together in this way, however, raised a problem. What title could we give such a series that would not only be snappy but scientifically correct?

And then I had a telephone call from Miles Barton. He had worked on the *Birds* series but had always told me that his real passion was for reptiles and amphibians. As a lad, he had kept many species of both groups. He was, in short, one of those people who call themselves, rather unattractively, a herp – that is to say a herpetologist, a word derived from Greek terms meaning someone who knows about creeping things. How about making a series about them, he said, using as a theme the fact that they lacked the warm blood of mammals? I thought it a splendid idea. 'Why not' I said, 'call it *Life in Cold Blood*?' And for a short period thereafter I genuinely believed that I had invented that title myself before I realised that Miles, of course, had used the standard technique, when selling an idea, of allowing a buyer to believe that the idea he is accepting is his own.

Within days, the proposal was accepted. I had no time to waste. And in May 2006 Miles and I were sitting surrounded by marine iguanas on the Galapagos Islands. Where better to spend my eightieth birthday?

29

Completing the Set

I had visited the Galapagos before – first during the making of *Life on Earth* and then for the series that dealt with birds, but the excitement of arriving on the islands was undiminished. To see wild creatures that have no fear of humanity – mockingbirds that pull hairs from your head because they need them for their nests, land iguanas that, when they finally rouse themselves at your approach, nonchalantly amble away over your boots, sea-lions that swim beside you and challenge you to cavort through the water with them – such experiences give you the feeling of having wandered into the Garden of Eden. There were it is true many more people there than I had seen on my earlier visits. However, the fact that visitors are necessarily divided into small groups and put aboard boats in order to move between the islands makes it easy for their tours to be scheduled in a way that gives each group the impression that they are almost alone.

And this time, my advanced years brought me some travelling concessions. This time in order to reach wild giant tortoises, I did not have to spend two days labouring up the flanks of the Alcedo volcano carrying on my back all my gear together with two gallons of water. This time, the team and I were flown up to the summit by helicopter and deposited beside one of the Conservation Department's huts on the rim of the crater. From there we were able to stroll among the tortoises as they lumbered about, cropping the sparse grass with sideways snatches of their heads and copulating with the pained and sustained roars that I remembered so vividly from our previous visits.

The marine iguanas still assembled in great herds to sunbathe on the lava rocks beside the sea. Now, however, we were able to film them in a more informative and imaginative way than we had done before for we had a new technological trick. We had a thermal camera. This device, developed largely for military purposes, it seems, uses not light to produce an image, but heat and it indicates the intensity of that heat with different colours. A human face, which radiates a great deal of heat from the naked flesh of the cheeks, appears as a golden yellow rising to a bright white around the eyes. A lump of cold rock, on the other hand is shown in blue. So a patrolling soldier, equipped with such a camera, is able to pick out the shining face of an opponent a considerable distance away even if that face is visually perfectly camouflaged.

Heat – or rather the lack of it – was, of course, the unifying theme of our series, as we had made clear by our title, *Life in Cold Blood*. Amphibians and reptiles, unlike birds and mammals such as ourselves, cannot generate heat within their bodies. They must draw most of what they need directly from the sun. The marine iguanas when viewed by a thermal camera provided us with a splendid demonstration of how they did it. In the morning they emerge from the cracks in the lava rocks where they had sheltered during night. The thermal camera shows them as deep blue. But then, as they sunbathe, they warm up and their colour changes to red, orange and finally bright yellow. Fully energised, they swim out to sea and dive down to graze on seaweed growing on the sea floor. But the water robs them of their heat and after a few minutes, they are so cold they have to return. As they clamber out of the water, they can be seen to be a chilly blue and they have to bask once more to restore their energy.

The series not only took me back to places like Galapagos to film creatures I had filmed before, it also gave me a chance to film animals that I had failed to find on earlier trips. In Madagascar in 1960, Geoff Mulligan and I had filmed many spectacular animals such as the indris and the little mouse lemurs that had never before been seen on television. But we did not find any members of a family of small reptiles that I had hoped very much to see – dwarf chameleons. These, judging from their scientific description, seemed to be most extraordinary creatures. Unlike most members of the chameleon family, they do not clamber about in bushes but live entirely on the ground. Down there, they no longer need the muscular tail with which other chameleons grasp branches. Their tails instead are short and stumpy. But most remarkable of all, they have become very small indeed. Some species are scarcely bigger than a large grasshopper. They were high on my list of desirables during our Madagascar *Zoo Quest* but what with one thing and another, we never made a determined search for them – and I had regretted missing the opportunity ever since. Now I had a chance to put that right and I firmly wrote them into the *Cold Blood* script.

Miles was as enthusiastic about the prospect as I and managed to contact Bertrand Razafimahatratra, a skilled Malagasy naturalist who said that he would certainly be able to find some for us. We had other things to film on the island, in particular the full-sized chameleons of which there are many spectacular and highly coloured examples. But the little brown dwarfs were the ones I really wanted to see. One night Bertrand led us into the forest. Together we crawled around on our hands and knees, sifting through the leaf litter with our fingers. Within minutes Bertrand had found one. It was even smaller than I had imagined, only an inch or so long from its little snub nose to the end of its stumpy tail. It sat on the end of my finger in the light of our torches while Bertrand told me about its habits. It fed, he said, on fruit flies, which are scarcely bigger than particles of dust. He spoke with such joy and reverence that when the sequence was shown, he captivated viewers and made the midget reptile one of the more unlikely stars of the series.

* * *

Life in Cold Blood, of course, had to have its statutory examples of species that had never been filmed before. The stump-tailed chameleon was one, but enchanting though I found it myself, we also needed something that was rather more obviously dramatic. James Brickell had no doubt what that should be. He was the snake expert in the production team. No one, he told me in dramatically lowered tones, had ever filmed a venomous snake actually catch its prey in the wild. Of course, some viewers might think they had seen such a thing, but they had been deceived. The sequences they had watched were fabrications.

No one in the BBC had ever filmed a captive snake killing a live chick or mouse. The Corporation's house rules prohibited arranging such a thing. Most zoos have similar regulations. And the chances of seeing a kill out in the wild were negligible. Most snakes will lie motionless in one position for days on end, waiting for a suitable meal to pass by. No one could wait with a camera focussed and ready to film until such a moment arrived. In any case, most kills were made at night. Indeed, James knew of zoologists who had devoted their entire professional lives to the study of venomous snakes and had never seen one feed naturally. So producers who felt it essential to have such a sequence in their programmes had first taken shots of a snake striking at the camera's lens, provoking it to do so by waving something in front of it. Then they filmed a mouse away from the snake but running around in a similar setting. And finally they fed the snake a dead mouse. The shots had then been intercut so skilfully that viewers had the impression that they had seen a kill.

However, James continued, he had worked out a way in which, for the first time, without transgressing any of the BBC's very proper rules, we would be able to get a completely genuine, un-manipulated sequence of a rattlesnake hunting in the wild. He had contacted a team of scientists working in the woodlands of the eastern United States who had put electronic tags on a dozen or so rattlesnakes in their research area so they knew exactly where each one was and had a good idea of what it was up to. They would be able to select one of their study animals that they knew had not fed for many days and was showing every sign of waiting in ambush. He and the camera team would then set up electronic cameras and infra-red lights connected to motion detectors, much like those used in anti-burglar systems, so that any significant movement around the snake would start the camera. After the action came to an end, it would record for a further two minutes and then switch itself off. That way we would be virtually certain to get a recording of this spine-chilling event.

His plan worked perfectly. He and cameraman Mark McEwen had spent only four days with the scientists in the woodlands of upper New York State before one morning, when they checked the equipment, they discovered that the recorders had indeed run during the night. They started the playback and watched pictures of a small wood mouse trotting confidently down a dead branch lying on the ground clearly oblivious of the rattlesnake lying within inches of it. Surprisingly the snake did not move. But after the mouse had vanished, the snake slightly adjusted its position and then once more relapsed into immobility. The camera, seeing no further action for two minutes, switched itself off. James thought that he had had a near-miss. But the camera switched on again to reveal what seemed at first to be a replay. The mouse – or it could have been another one following the same route – appeared once again at the top of the branch and set off down it. But this time, in one electrifying fraction of a second, the snake struck. The mouse, with a convulsive jerk of its hind legs, shot into the air, landed on the ground and immediately bounced up again in a high parabola, soaring away from the branch. This time it landed in a patch of withered grass where it was out of sight. Once again, James was momentarily disappointed, for he needed to see the snake actually feeding if his sequence was to be complete. But once again, the camera having switched off, switched itself on. This time the snake was slowly uncoiling itself and crawling into the grass patch. It disappeared from sight among the withered stems. But within seconds there was more rustling and the snake reappeared. It turned and looked straight at the camera holding the body of the dead mouse in its jaws.

And then, facing the camera and with chilling deliberation, it slowly swallowed its prey.

* * *

With a rattlesnake kill in the can we already had a spectacular success. But it was not enough for James. There are other venomous snakes that behave in a different but equally dramatic way. The spitting cobra, for example, uses its fangs not only to kill prey but to repel enemies. If alarmed, it rears up, opens its mouth, swivels its fangs forward and squirts venom from their tips. It is particularly interesting, James explained, because the snake always aims for your eyes. If just one drop of venom gets into them, you will be blinded, certainly temporarily and probably for life. His plan was that I should demonstrate such an attack. Of course, he said solicitously, we will give you a plastic visor. We will coat it with a special powder that will turn bright pink if it comes into contact with the venom, so after the snake has spat, we will then be able to inspect the visor and see how wonderfully accurate its aim is.

In due course, James, a camera team and I arrived at a zoo in northern South Africa that specialises in reptiles. Donald Strydom, who runs it, had several cobras which, he was confident, would obligingly spit at me, given half a chance. We took one of the biggest out into the bush and placed it on a suitable rock. The camera was set up on one side of it and I walked towards it from the other side of the snake, wearing my visor and explaining what it was I was about to do. I stopped when I was about six feet away from the snake and leaned towards it in a provocative way. The snake performed exactly as it should have done, using its fangs like water pistols, squirting twin jets of venom at me and turning its head as it did so, with the result that droplets of venom struck my visor in a neat horizontal band directly in front of my eyes – and duly turned pink. James was delighted. I was relieved.

The series was transmitted early in 2008. The newspapers at the time were having a field day, baiting the BBC for fakery and dishonesty, a spasm of self-righteousness occasioned by the fact that the Controller of BBC1, when promoting his forthcoming programmes, had shown shots taken from a documentary about the Queen that had been assembled in the wrong order. Instead of her being mildly irritated at having to appear in her heavy regalia for a session with a stills photographer, they were so edited that she appeared to have been irritated by the session itself and to be leaving in a huff. Anxious as ever to turn a story into a campaign, one newspaper discovered that *Life in Cold Blood* had filmed a spitting cobra that was not discovered in the wild but had been taken from a zoo. More fakery, they screamed. I explained as patiently as I could, that the sequence had indeed, as they put it pejoratively, been 'set up'. That was perfectly proper. No one had pretended otherwise. I did not normally wander about the African bush wearing a plastic visor coated with a pink dye unless I had arranged to encounter a spitting cobra.

That explanation resurrected, yet again, the question of how much of what is shown by natural history films is specially arranged for the camera. Happily, I had no need to be defensive. The Natural History Unit had decided a long time before that it should make no secret of its filming techniques. Indeed, we had made a special programme about them called *The Making of the Living Planet* which was shown at the end of that series when it was first transmitted back in 1984. Since then, I had given a televised lecture with the title *Unnatural History* in which I explained the various techniques we had to use if we were to record the pictures and sounds needed to create properly informative programmes about the natural world. That might well involve filming animals under controlled circumstances. If we wanted to show the mating dance of scorpions, as we had done in *Life in the Undergrowth*, it was not unreasonable to collect scorpions in the wild and then introduce a male to a female in a place where a camera could see in detail what was going on - and that could well be on a table so that a camera could easily observe them from ground level. If we wanted animals to appear out in the open so that the camera could see them clearly, it was not improper to entice them with some judiciously placed food. We might also film several courtship displays of a particular bird and edit them together to give a complete picture of a performance. And if we needed to film down a burrow or in an underground den, it might be better not to use fibre optics but to build such a den and persuade an animal to make its home there. Such techniques could be legitimately used in order to present the most complete – and therefore the most truthful – picture of an animal's life. And it was neither necessary nor desirable to give a running commentary throughout the finished sequence detailing the various techniques we had used to get particular shots.

Even so, the issue is not simple. Editing is becoming increasingly skilful and computer imaging is now so versatile that it is possible to make an animal appear to do something that it does not in fact do. A zealous producer researching his subject may read a report of his chosen animal behaving in a certain way. Yet no matter how patient he is and how hard he tries, he does not see it doing so. He therefore edits his film shots in a way that makes it appear to be behaving in the way he has read about. He may even try to stimulate the animal in some way to perform the action he hopes to see. But it could be that the original report is mistaken. And then, of course, the film-maker is in trouble. Sometimes his error, although indefensible, is inadvertent. Sometimes, culpably, he may deliberately set out to deceive.

Today, viewers can have little chance of deducing from the screen what is the truth and what fiction. If computer animation can show dinosaurs rampaging across the land in a wholly convincing way, it can certainly without any difficulty whatsoever modify some detail in the behaviour of a real live animal. Ultimately, a viewer will only be able to judge whether a sequence is truthful or fictitious by considering the integrity and record of the people who made the film and the organisation that broadcasts it.

* * *

In over fifty years of programme-making, I have never filmed an animal that has subsequently become totally extinct. We had made sequences about several species that are now extremely rare. In *Life on Earth*, I had sat with mountain gorillas, of which the world population had then numbered only a few hundred. During the series on birds, we had filmed the kakapo, a giant flightless parrot that at one time had been reduced to seven individuals of which only two were females. In each case, thankfully, devoted conservationists had managed to bring the species back from the very brink of extinction. But for *Cold Blood*, we filmed a species that by the time the series was transmitted had almost certainly disappeared entirely from the wild.

It is a tiny frog, smaller than my thumb, known scientifically as *Atelopus ziteki*, that once lived in a small patch of rain forest in Panama. In colour it is a beautiful almost luminous gold, spotted with a few variable blotches of black. Traditionally, it was revered as sacred by local people. In centuries past, they made images of it in gold and wore them as talismans. In modern times, living specimens have been collected intensively and sold for high prices to the pet trade. That and the steady destruction of its wild habitat has meant that it has become increasingly and dangerously reduced in numbers.

We were interested in it, not just because of its rarity but because it had a most engaging way of signalling. It lives on the banks of small swiftly running streams. Most frogs, of course, in the breeding season communicate by croaking. But that is not always effective for the golden frog because the sound of rushing water in a nearby stream may make its voice inaudible. So it reinforces its rather feeble calls with gestures. It makes a slow jaunty wave with one of its front legs, rather like a dandy expressing his good humour by touching the brim of his hat.

But frogs all over the world, in recent years, have been afflicted by a fungus that infects their moist skin. Although frogs do have lungs, they also breathe through their skins. The fungus, as it spreads over their bodies, eventually prevents them from doing so and they die from suffocation. The advance of this fungal disease across Australia, Africa and South America has been recorded in detail. When we started filming *Cold Blood*, it was moving along the Isthmus of Panama steadily and inexorably. No antidote, no spray, has been discovered that can stop it. At the rate it was travelling, it would reach the golden frog's territory in 2007. The decision had therefore been taken by conservation scientists to marshal all available local help and collect every golden frog they could find. They could then be taken to zoos in Panama and elsewhere, where they could be kept and bred in captivity until such time that either a cure can be found for the disease or the fungus has disappeared in the wild for want of hosts – if, in fact, either of these two things ever happens. In captivity, however, the frogs do not wave to one another as they do in the wild. So if we wanted to record that behaviour, we had only one season in which to do it.

We were guided to the last area where the species could reliably be found by Erik Lindquist, a biologist who has studied the species for many years. It was he who had worked out the details of the species' gestural language. To do so he had used a small but accurate model of one attached to the end of a long pole with which he could make it move its front legs in just the way a live one does. With that we were able to persuade one of the frogs to make its last wave before Erik and his team collected it and took it away to safety.

* * *

Life in Cold Blood completed our picture of how the natural world had appeared to a film-maker at the end of the twentieth century. But the assembled series could also be seen in another way. Arranged, not in the order I had made them but in that in which the various groups had appeared during evolution, it could provide an outline of the history of life. And, almost inadvertently, I had made one series that could be seen as the climax to such a sequence.

Back in 1987, with *The First Eden*, I had traced the way in which humanity had changed from being one species of primate among many into a particularly ingenious and uniquely successful one. After the first human beings had learned how to make stone tools, they migrated out of Africa and settled in the lands around the Mediterranean. There they began to modify the natural world in an unparalleled way. They domesticated both plants and animals and by selective breeding started to mould natural species to suit their own particular purposes, creating animals that barely resembled their wild ancestors. They cut down the forests in order to build their growing settlements and cleared fields in which to graze their herds and grow their crops. And there, around the Mediterranean, they inflicted the first man-made ecological disaster on the earth. The lands around the southern shores of the sea, were so fertile that they were able to produce vast quantities of grain for the burgeoning Roman empire. But the demands the farmers made on them eventually became so great that the soil blew away and the lands became the deserts that they are to this day. The Mediterranean people were also the first to interfere with the complex ecological communities of their lands by introducing animals and plants from other continents – from Asia, from farther south in Africa and finally, as they became more ingenious technologically, from the Americas. Eventually they even cut through the link of land separating the Mediterranean from the Red Sea so that the fauna of the two seas began to mix. In short, they foreshadowed in microcosm a process that before long would spread across the globe.

Since the six series cohered in this way, it occurred to me that we could perhaps transfer the whole set, forty-one programmes in all, to videodiscs and in the process equip each with electronic markers. We might then produce a separate printed index so that it would be possible for a viewer to look up a species, or some aspect of natural history such as courtship or aggression, and then, with a few clicks of a remote control, summon that sequence to their screens.

That, to my great satisfaction was done and the six series were bundled together in a box and given the title *Life on Land*. The survey they provide is of course, superficial and sketchy but I wonder, considering the way that television stations are now proliferating and the audience consequently fragmenting, whether any broadcasting organisations in years to come will have the finance and the patience to commission anyone else to make such a survey of terrestrial natural history.

30

Voice-over

What next? I was 82 but even so, I didn't feel like putting my feet up. On the other hand, to propose that I should write, present and narrate yet another six-part series that would take three years to make did seem like tempting Providence. And would the BBC wish to join me in doing that? On the other hand, making a single film or recording a commentary for someone else's, would keep me happily busy, if such opportunities came my way.

And they did. Writing commentaries is, in fact, quite a pleasurable business. One of my earliest tasks in television, back in 1954 had been to produce them to accompany the amateur footage shown in the *Travellers' Tales* series. Ideally, the travellers themselves should have spoken them but in practice few were able to read scripts in a convincing way. So in the end I not only wrote them but spoke them as part of my normal producer responsibilities. 'Staff : No Fee' as the BBC's form-filling jargon of the time had it.

There are, of course, many different styles of commentary. The one I favoured was quite demanding. Spoken words, I thought, should be subsidiary to picture. They should not say what viewers could see for themselves. They should be kept to a minimum, avoid the flowery, and be so accurately timed that a particular word would be delivered at the precise point in the picture when it was most appropriate and had the maximum effect. Achieving that required a kind of split-second verbal carpentry which was at its best when listeners were unaware of it.

Spcaking the words also had its own expertise. In the days of *Travellers' Tales*, a commentary was recorded on hefty rolls of magnetic 35 mm film that ran on a separate machine linked to run synchronously with the picture. We were not allowed, in the interests of economy, to cut or edit these rolls. Each one ran for ten minutes. So if you fluffed a word nine minutes in, you had to start again. That usually produced enough adrenalin to enable to you to keep your concentration right to the very end. Today, it is possible to stop and start after every sentence, but I still think it preferable to record an entire commentary in one run, even if that lasts an hour.

That way, apart from anything else one can get an overall feeling for changes of pace. You can then go back, where necessary, to correct and replace glitches or fluffs, which is easily done with modern technology.

Selecting a speaker to deliver commentaries for natural history films is not easy. Ideally the voice on the film should be that of the filmmaker or scientist whose work is being featured. But not all such professionals are clear speakers, so actors are sometimes engaged instead. They after all, are trained to speak clearly and also to take instruction on the particular emphasis a director may want for his words. Back in the 50s, one independent television company, imagining that it would add to their film's attraction if the narrator was not only an actor but a star, engaged the great cinema heart-throb of the time, James Mason, to speak the words. It was not a success. Viewers knew perfectly well that, however experienced James Mason might be at seducing heroines on-screen, he knew little about non-human animals. So thereafter, while they maintained their policy of using actors, they chose ones whose voices were not readily recognisable.

Fortunately for me, my voice, having been heard on and off since *Zoo Quest* in the 1950s, was already identifiable as belonging to someone who had some knowledge of the natural world. And I was also sufficiently experienced as a film-maker to be able to speak, revise or even, if necessary, totally rewrite a narration script.

At any rate, when I left administration and set off down the freelance road, Chris Parsons, then the Head of the Natural History Unit, invited me to read all the commentaries for a series of half-hour films that he was proposing to call *Wildlife on One* – 'One' being a reference to BBC1, the channel that would show them. Each would be made by a young, up-and-coming natural history cameraman or director. The series would thus be a valuable place where newcomers in natural history film-making could get experience. The series, however, should have some continuity and that, he thought, might come if the same voice narrated them all – and he invited me to take on the job. So once again I became a narrator of all kinds of natural history films, not just those that I originated myself.

In 1998 the role had brought me an extraordinary reward. Alastair Fothergill – who had early in his BBC career contributed a film on bee-eaters to the *Wildlife on One* series and who later nearly drowned me in a Brazilian swimming pool during *Trials of Life* – was starting work on a new and extremely ambitious series he had devised about the underwater world. It was to be called *Blue Planet* and it was, without question, a somewhat risky proposal. Underwater filming is as complicated and as expensive as natural history film-making can be, so it would need a huge budget. But would viewers want an eight-part series that showed almost exclusively underwater life, week after week and hardly ever, as it were, came up for air?

There was another problem. Who should be that character who was increasingly seen as an essential for a major series – an in-vision presenter? At that time, people in diving gear talking under water sounded very odd. Their voices were distorted by the water pressure and interrupted by noisy gasps for breath. They themselves were, in any case, barely recognisable since their faces were largely obscured by the masks. Alastair decided to dispense with them altogether. Instead, he would rely on an off-screen narration and to my great delight he asked me to provide it. I would be given an outline script prepared by the director of an episode, revise it as I thought necessary and then speak it. A narrator could have scarcely asked for anything more exciting.

I was at this time filming my own series, *The Life of Mammals*, but there was nothing for me to do for *Blue Planet* as it was still being filmed. The main part of the work was being handled by four very experienced and expert underwater cameramen. Thirty or so others were given particular short-term individual assignments. Many were asked to secure shots of species that had never before been filmed, behaving in ways that no one up to that time had ever observed in the wild. Progress was inevitably slow – but Alastair had budgeted for that. Every now and then I joined him to watch batches of material as they started to arrive from all over the world. Sometimes one of the marine biologists who were giving us specialist advice on particular species would join us. Occasionally they were even more astonished by what they saw than we were, for few had ever had the time or indeed the underwater skills to do what Alastair's team of expert underwater cameramen were doing as they pursued their particular subjects day after day.

The series was an immense success. Alastair's pertinacity in staying with stories until they finally yielded the sensational result he had had in his mind when he first devised the series, produced some astounding sequences. George Fenton composed a richly orchestrated tuneful score for the series with a memorable few bars for the titles. In Britain, the series topped all the ratings. Overseas, television stations competed with one another for the rights to show it. Some sixty or so people – cameramen, sound recordists, and directors had made significant contributions to the series. And because my voice was heard throughout each episode, my name got disproportionate prominence in the reviews.

Alastair knew well enough what his next move should be after *Blue*

Planet – to propose another series with a similar if not an enhanced scale of ambition. This time he would survey the whole planet with particular emphasis on its terrestrial life. He had to admit that some such thing had been done before. In truth, I had had a go at doing it myself with *The Living Planet*, the successor to *Life on Earth*. But Alastair had two technical advantages. The first was that this would be the first major natural history series to be shot in high-definition, so that all its shots would have a crispness and detail hitherto unmatched. He had also discovered a newly invented and revolutionary camera mount. That may not sound much, but its effect was huge.

Only too often, a cameraman making a wildlife film is stuck inside a hide, waiting for his subjects to come to him. Sometimes, if he is very intrepid and the animals are fearless, he may be able to follow them on foot. Very occasionally, he may chase after them in a Land-Rover though doing so is more than likely to make them stop doing whatever it was that he wanted to film. This lack of mobility can be very frustrating. How can he hope to film a sequence showing, let us say, the tactics used by a pack of hunting dogs when pursuing an antelope. An aerial shot might seem to be the answer. But if a helicopter flies low enough to see what is going on the ground, the noise of its rotor will cause the animals below to stampede away in panic. Even detailed shots of a landscape are likely to be unacceptable because the down-draught from the rotors flattens the bushes. You might suppose that the solution would be for the helicopter to keep sufficiently high in the sky for the animals to be unaffected by it and to put a long-focus lens on the camera to get reasonably close-up shots of what is happening below. But long-focus lenses have to be rock steady. The whirling blades of a helicopter's rotor make the whole aircraft vibrate and only wide-angle shots are sufficiently shake-free to be useable.

Alastair, however, had heard that someone in Hollywood had produced a revolutionary new system called Cineflex. An electronic camera, gyro-controlled and set in gimbals, is mounted beneath a helicopter's fuselage. The cameraman sits within the helicopter behind the pilot watching the camera's output on a television screen using a joy-stick to pan it in any direction he might wish, and zooming in on any detail that takes his fancy.

One of the first sequences to be shot in this way followed a pack of wolves in Alaska, pursuing a herd of buffalo. The helicopter was flying so high that neither wolves nor any buffalo showed any sign of being aware of its presence. The Cineflex mount, however kept the camera perfectly steady so that Mike Kelem a cameraman who had specialised in operating this new equipment nonetheless was able to fill the screen with a close-up of the head of a wolf and keep it in frame as it ran at full speed.

A second camera crew on the ground was following the hunt in a Land Rover. The director, high in the sky and sitting alongside the helicopter pilot, was able to direct them so that as the chase wound through the forests, the ground crew were every now and then able to take a short cut and almost get ahead of the chase. And they were virtually alongside when the climax came and the wolves dragged a young buffalo to the ground and tore it apart. Then by combining the aerial and the ground coverage, it was possible to produce a complete and vivid picture of the action and the wolf pack's complex tactics in a way that had never been possible before.

Planet Earth exploited this new device to the full. Not only did it give the most vivid pictures imaginable of animals in action, even landscape shots were transformed. So a camera that seemed to be skimming only a few feet above the surface of a river torn by rapids could suddenly sail onwards through the air while the river itself plunged over a precipice and fell a thousand feet in an immense cascade. Cineflex gave the whole *Planet Earth* series a completely new feel. Once again Alastair had a huge success. Once again I bathed in reflected glory.

Undeserved credit of this kind was to come my way for yet a third time. Alastair had another itch. He is a big man, with a marked taste for physical adventure and a great talent for persuading others to join him in bringing his great plans to fruition. Like many such people he is fascinated by the polar regions where human beings have to pit themselves against the most hostile environments the earth has to offer. So now he suggested a series he called *Frozen Planet* which would survey both the Arctic and the Antarctic.

Once again there were editorial doubts on the part of planners and commissioners. Might six one-hour programmes about snow and ice get a little monotonous? Would it be possible to retain viewers' interests over such a long haul? Alastair once again stuck doggedly to his guns. The series would, he explained, be very varied for the Poles are very different. The North is covered by a frozen ocean and populated by creatures straying up across the ice from the lands that extend north across the Arctic Circle – Canada, Siberia and Scandinavia. So there you find bears, foxes and indigenous people. The South, on the other hand, lies in the middle of the vast frozen continent of Antarctica which is isolated from the rest of the world by the encircling storm-torn Great Southern Ocean. So the only animals that have been able to reach it, before humanity arrived were those that could fly or swim. The series would, as Alastair explained, at last enable everyone to understand why polar bears could never, in the wild, hunt penguins. The planners and network controllers nodded sagely and approved Alastair's budgets.

Alastair decided that I should appear at least briefly, once at each Pole. I was delighted by the prospect. In fact, of course, one flat snowfield stretching unbroken in all directions looks very like another. Not only that, but a human figure, trussed up in all the bulky cold-resistant kit that would be necessary, would look exactly the same, whichever Pole he was standing on. Nevertheless, veracity was paramount so arrangements were put in hand to get me to both places.

First was to be the South Pole. Several countries around the world have established bases on the Antarctic continent, in order – in theory, at least – to conduct scientific research. Britain and Norway were among the first to do so at the beginning of the twentieth century. Antarctica's nearest neighbours, in the southern hemisphere – Chile and Argentina, Australia and New Zealand – have also, very understandably, done so, as have others who lie much farther away such as Japan and even Belgium. But only one country has the technological and financial clout needed to establish and maintain a base at the Pole itself – the United States. There the year round and especially in the long black Antarctic winter, it is possible to conduct research in astronomy, meteorology, particle physics and other sciences more effectively than anywhere else on the planet. The Americans generously named it the Amundsen-Scott Base after the first two human beings – non-Americans – to reach this point on the globe.

To reach Amundsen-Scott you have first to go to the United States main base at McMurdo on the coast of Antarctica. The United States Air Force runs regular flights to it from Christchurch New Zealand which is where, in due course, Alastair and I presented ourselves.

Sitting in rows in a vast hangar-like building, we surrendered our civilian accoutrements and freedoms and were taken up into the maw of the military. Our own clothes were bundled into bags, labelled and removed to await our return. In their place, we were issued with proper cold-weather equipment, from goggles and woolly hats to thermal underwear. We were given mini-lectures about the Base where we would be going and told what to do in the various emergencies that might conceivably overtake us. Even now, we were told, we could not be sure that we would reach our destination. We might unwittingly be on what was termed a 'boomerang flight'. Antarctic weather was very changeable. If it was not good enough for us to land when we were within an hour or so of the base, we would have to turn round and return to Christchurch, for we would not have enough fuel on board to enable us to hang around in the air waiting for the conditions to improve. And with that we filed out and clambered into the immense C17 Globemaster transport aircraft that stood awaiting us on the tarmac.

Packed together on rows of metal bucket seats, surrounded by enormous piles of cargo, in the dimly lit, virtually windowless interior we awaited take-off. The plane's engines rose into a deafening roar and we were away.

It was as well we had been warned. After six hours, the captain told us that we had indeed become a boomerang. Conditions at McMurdo were bad and landing would be impossible. We would have to return to Christchurch. Far from depressed, I was slightly exhilarated to discover that even today, the South Pole, so long the ultimate in terms of remoteness and inaccessibility, should still be able to repel humanity.

The following day, after a repeat of all the lectures and educational videos for the benefit of one or two more passengers who had now joined us, we set off again. And this time, we landed.

McMurdo Base is by far the biggest human settlement on the Antarctic continent. In the summer, as now, it has well over a thousand human inhabitants. Over a hundred separate buildings, scattered in a seemingly haphazard way, stand on a rocky snow-covered hillside. Between them they accommodate dormitories, canteens, workshops and laboratories. There are two permanent airfields out on the sea-ice and a helipad closer to the main settlement. This being summer, a passage had been cleared through the sea-ice to allow ships to dock and bring in bulk supplies. The last to leave from here at the end of the season will take with it all Base's waste, including human effluent, in accordance with international treaties aimed to avoid polluting the continent.

Gavin Thurston and Chris Watson were already here, working out on the ice getting extraordinary sequences of the killer whales which were patrolling the ice edge and rearing out of the water to take a good look around, as they certainly do when hunting seals. Vanessa Berlowitz, who was directing the aerial sequences, and Mike Kelem, the Cineflex cameraman, were here too.

They were taking aerial shots using one of the Base's helicopters when it was not needed for other work. Alastair and I, travelling in another helicopter, joined them and I was deposited on the very summit of spectacular vertical-sided rocky towers to provide some kind of scale for Mike's shots. I sometimes wondered how viewers would imagine I had managed to get to such clearly inaccessible places. But there was a childish thrill to be gained from standing where no human being had ever stood before – and was unlikely to do so ever again.

Our task at the South Pole would be a simple one: to record the first and very important sequence in the series. I would appear in close-up, suitably muffled against the bitter cold, preferably with hoar-frost bespangling the furry rim of my anorak hood, and say 'I am standing at the South Pole. All around me is emptiness.' I would then spread my arms out wide. This would be followed by another shot showing me as a lonely figure standing in a vast field of snow-covered ice. Gavin and Chris would take the close-up. Vanessa and Mike in the Otter would then take an aerial shot which would pull back and reveal white emptiness until I was so small I disappeared.

Apparently simple sequences are often not simple in practice. The first problem was to get the Cineflex to the Pole. It obviously could not be mounted on the giant transport plane that would take me and Alastair there. But equally it could not be shot from the McMurdo helicopter that and Mike and Vanessa were using, because the Pole was far beyond its range. So it was eventually arranged that the Cineflex would be mounted, for the first time as far as Mike knew, on a fixed-wing aircraft – a quite small machine called a Twin Otter that was based at McMurdo and had a much greater range than a helicopter. They would then fly to the Pole to join us, taking aerials of the interior of the continent as they came.

After several days, Alastair and I, together with Gavin and Chris, left as arranged on a scheduled flight to take us the last eight hundred and fifty miles to the Pole. Once again, we took our places in the twilit belly of a giant military transport and thundered southwards, and once again we emerged into the brilliant white light of the Antarctic summer.

Amundsen-Scott Base, tiny compared with any of McMurdo's bigger buildings, sat on the ice sparkling in the sunshine and looking as out of place as a space craft on the surface of the moon. It is supported by stilts so that during the winter, snow driven by the ferocious gales does not pile up against the walls but is blown beneath the building and away. Some nonetheless does accumulate but the stilts are, in fact, powerful jacks that are able to raise the entire building – each season, if necessary – so that it remains clear of the snow.

In the winter, the block is home to some two-dozen people, many of whom may not leave its walls for months on end. A minority of them are scientists. The others are technicians of one kind or another needed to maintain this tiny capsule of life in the empty vastness of Antarctica. Now, however, in the continuous polar daylight, there were several hundred people here, many of them, like ourselves, short-term visitors with their own limited agenda.

Vanessa and Mike arrived in the Otter soon after we did and we all started to plan the shot in detail. First there was the problem of where exactly I should stand. There was one obvious answer. Twenty or thirty yards in front of the Base's main building there was a neat semi-circle of flags belonging to all the nations which have Bases in Antarctica. This was where visitors always stood. But fluttering flags would not really suggest the extraordinary remoteness that we wanted to convey. Equally – and perhaps even more importantly – they were not standing on the site of the South Pole.

The Antarctic continent is covered by ice. Here at the Pole it is almost three kilometres thick. And it is in motion. It is sliding down towards the coast at a rate of about ten metres a year. So the flagposts and the Base have been travelling away from the precise site of the Pole since the flags were first erected. Now they are several hundred metres from it. Perhaps the authorities once considered re-siting them but eventually decided that visitors would only want to walk a short way from the shelter of the Base to get the statutory shot of grinning handshakes in front of a flag. At any rate, the flags now stand much closer to the Base than they do to the Pole.

A more geographically accurate site of the Pole is, in fact, marked by a small post with a stainless steel globe on the top. This is moved annually. But even that can only be accurate on the day it is planted. So in the end we decided that the best compromise would be a spot on the ice midway between these two points where there was nothing whatever to be seen but snow and ice.

That agreed, the Otter took off, with Mike and Vanessa. I, with a mobile phone in my pocket with its speaker set at its loudest, duly marched away from the base in a direction that Mike said suited the light. Vanessa's muffled voice in my pocket told me when I had gone far enough. The Otter appeared over the horizon. I, cued by Vanessa, started to speak and spread out my arms. The Otter disappeared. And then we did it again. And again. I was beginning to feel quite cold. It was after all 35 degrees below freezing. And there was a biting wind lashing my necessarily exposed face. Viewers should at least be able to see my lips moving – even if there was nothing else identifiable in this trussed-up figure on a snow-field.

The Otter reappeared once more. Once more, I mouthed my words and spread out my arms. At times like this, all kinds of uncharitable thoughts flicker through the mind. Again and again, the Otter disappeared – only to return. But at last Vanessa's voice came through to say that she had what she wanted.

The words, 'I am standing at the South Pole' may not have been the most imaginatively phrased of the statements that I might have made under the such circumstances. But at least they were accurate.

* * *

Six months later, we filmed a similar sequence at the North Pole. Its location, of course, is very different from that of its southern equivalent. Instead of being three thousand feet up in the centre of a continent, it lies literally at sea level – on the surface of the Arctic Ocean. It is best visited in the few weeks between the time when the winter begins to loosen its grasp in early spring and a few weeks later when the sea ice starts to break, creating leads – long jagged corridors of open water.

A Russian organisation arranges trips to the North Pole. It is a very different operation from the American one at the other end of the earth. You fly first to the northernmost town of any size on the planet, Longyearbyen on the island of Spitzbergen, part of the Norwegian Arctic territory. From there a Russian plane will fly you seven hundred miles further north to a camp within seventy miles of the Pole.

The camp itself could hardly be more different in appearance and feel from the somewhat clinical super-efficient Amundsen-Scott Base. It is, of course, only temporary and has to be dismantled altogether before the spring melt is at all far advanced. So instead of a super-efficient modern building, full of high-tech scientific apparatus, there is an untidy cluster of immense canvas tents. Each is in effect, double – an inner one, with a kind of quilt attached to its walls and ceiling, is pitched inside a slightly larger one – so that there is an insulating layer of still air between the two. Each double-tent has its own heater, a roaring fan sitting in its own small tent outside the main one which blasts a continuous stream of hot air down a short canvas tunnel and into the main tent. Inside there are a few camp beds. Any additional comforts are those that have been brought by the visitors themselves.

The camp's inhabitants were also a rather more motley group than those at the South Pole. There were, doubtless, people with serious scientific programmes of their own. But there were also private adventurers who were planning to walk the remaining seventy kilometres to the Pole for no better or worse reason than that they wanted to do so. The whole was organised by a small group of Russian Arctic specialists as a purely commercial enterprise.

Our task was to produce a sequence to parallel the one we had taken in the south. I would once again say where I was, spread my arms and reveal nothing but snow and ice. But once again, if I were to say that the icy emptiness where I was standing was the site of the North Pole, that had to be true.

We had arranged for a Russian helicopter to take us there. It stood on the ice a few hundred yards away from the camp and close to a line of black plastic bags filled with ice that marked the edge of the runway, a long avenue of specially cleared and smoothed ice. Even to my technologically unsophisticated eye the machine seemed somewhat antique, with two huge engines and long drooping rotor blades. But it would not move until the meteorologists back in Longyearbyen declared that the weather was suitable. Meanwhile, we must wait.

We sat on our camp beds reading, telling one another stories of filming trips that we had already told many times before or simply staring at the canvas ceiling. Every few hours we tramped across the ice to the mess tent to have another helping of borscht, Russian beetroot soup. And every twelve hours or so, in the continuous daylight, we decided that we might as well pretend that night had come and try to go to sleep.

After some hundred and fifty hours – about seven days' worth – our Russian host brought us good news. The weather at the Pole had suddenly improved. We could go. Suddenly everything became urgent. We grabbed all our filming gear and clambered aboard the helicopter.

The Russian crew were cheerful and well-disposed but their English was scarcely better than our non-existent Russian, so our communications were not as explicit or as unambiguous as we might have hoped. But they cheerily agreed that the North Pole was indeed our landing place of choice. That after all was where all their other passengers wanted to go – and off we went.

Helicopter interiors are, at the best of times, noisy places. But in any case, our communications were largely by signs and after half an hour, their nods and beams seemed to suggest that we had arrived. Landing an aircraft on snow or ice is not easy. It is almost impossible to judge your height above the ground because there is nothing beneath that is of recognisable size. The Russian crew had their own way of doing that. As they hovered, they opened the helicopter's door and threw out an old, very worn car tyre. It landed almost immediately, from which I deduced that we must be only few feet above the ice.

I scrambled out. Alastair, Vanessa, Gavin and Chris followed. They quickly set up the gear. I looked at the camera, spoke my words and spread my arms gesturing at the white emptiness. We did it a second time 'just in case' as we always say on these occasion but without specifying what that case might be. Then, back into the helicopter and within the hour we were back in the camp supping another bowl of borscht. And there was good news. A plane would soon be leaving Longyearbyen with another load of visitors and would be able to take us back. We packed our gear, lay down on our camp-beds and shut our eyes to make one last attempt to pretend that it was night.

It is an odd feeling camping on floating ice. For most of the time you forget that the snow beneath your feet is not lying on solid rock. But then you may fancy you felt a certain unsteadiness, a slight shift, that makes you wonder whether the ocean only a few feet below you was making an infinitesimal heave. Then you may notice a hair-thin line zigzagging across the snowy ground. A crack.

One such crossed the path cleared through the snow that led to the airfield. The Commandant had laughed dismissively when I pointed it out. Cracks like that were continually opening and closing, he said. All the same, I could have wished that this particular one did not run between our tent and the airfield.

Promptly on time, the Russian transport plane thundered down on the ice. We loaded gear and a few hours later, we were sat drinking beer in the pub in Longyearbyen. It wasn't until the following morning when we were gathering for the return flight to Oslo and London that we heard the news. Soon after our plane had left the Russian camp, the crack across the path to the airstrip had suddenly started to widen ominously. The Commandant ordered an immediate evacuation. Tents were struck, bags half-packed, were dragged over the crack. As it visibly increased, ladders were brought to bridge the gap. A few minutes later and those on the camp side could no longer stride across the gap but had to jump. Within an hour or so the crack had become impassably wide but by that time, most of the baggage and all of the visitors were safe, huddled beside the airstrip awaiting a rescue plane. Our trip to the North Pole was, as it turned out, routine and uneventful. But it might not have been.

The *Frozen Planet* series was finished and transmitted in 2011 and like its two predecessors was a global triumph, completing a trio of magnificent series devised and brought to fruition by Alastair Fothergill. But now he had even greater ambitions. He wanted to bring his skills in creating visions of the spectacular on to a bigger screen – the cinema. It seemed that my good fortune of working on such big scale and prestigious television productions had now come to an end – for the moment, at any rate.

31

New Images and a Third Dimension

Now, in the twenty-first century, television was changing faster than ever. When I joined the BBC it was a monopoly. No one in the country understood its intricacies except for those of us who were lucky enough to find ourselves in that charmed and privileged company. That mystique still persisted even after commercially based Independent Television was introduced. It too, like the BBC, recruited a large in-house production staff. But as the years went by, electronic equipment became more versatile, cheaper and more reliable, and people began to realise that, in truth, there was no mystery about making television programmes. Anyone could do it. Small independent production companies appeared. They too wanted a share of the business. The Government encouraged this diversity and in 1990 it told the BBC that 25% of its programmes had to be made by independent producers.

Although I had hardly worked for any other television organisation in Britain, I had left the BBC's staff in 1973, when I resigned from my administrative job and started to work as a freelance. But before long invitations came to me to contribute to programmes that, although they were destined for the BBC, were indeed being made by independents. The first arrived in 2009 when *Frozen Planet* was still in production. It came from an ex-BBC producer, Anthony Geffen, who now had his own independent company, Atlantic Productions.

He had been commissioned by the BBC to make a film about the discovery in Germany of a very exciting fossil, now owned by the Natural History Museum at Oslo University, which was arguably the earliest known creature on the branch of the primates that eventually led to mankind. The film was accordingly being entitled *The Link*. Distinguished scientists from Germany, Norway and the United States had already filmed interviews for it. Would I narrate it?

The narration did not take sides but allowed the scientists to speak for themselves. It was a well-made if scientifically somewhat controversial film and I gladly recorded the narration. Its particular interest to me, however, was the way in which it brought to life animals from the far distant past using CGI – computer-generated imaging. This had been done by a small but extremely expert unit working within Anthony's Atlantic group called ZOO. I thought the images were the most convincing I had yet seen of any fossil species.

Soon after this, I received a letter from an Australian palaeontologist, Professor Patricia Vickers-Rich who was working on some of the most ancient of all known fossils. Some looked like tiny jellyfish, others appeared to be segmented worms. There were striated discs as small as coins, and others as big as door mats. The biggest of all were feathery organisms that resembled sea pens and grew to over a metre long. They were first discovered in the Ediacara Hills, part of the Flinders Ranges north of Adelaide. I had already filmed some of them for the first programme in the *Life on Earth* series back in 1979. Since that time, however a great deal more had been discovered about them and Pat wanted advice as to who could create video images of these mysterious creatures, showing them as they must have appeared when alive.

So remembering the computer images produced by Atlantic for *The Link*, I put her in touch with Anthony Geffen. Thereafter, one thing led swiftly to another, as it tends to do when Anthony is involved, and in no time at all I found myself writing an outline script for him that would make use of such images.

We would call the programme *First Life* and it would start in Australia with Pat Rich's Ediacara fauna. The programme would then go on to trace the history of life as it evolved into the myriad invertebrate groups that still survive today – sponges, jellyfish, corals, segmented worms, sea-urchins, molluscs – and finish with creatures that, although now totally extinct, were in their time hugely abundant, the trilobites. They disappeared some 260 million years ago. That would be a good point to end. Up to that time, life had been confined to the waters of the planet. Thereafter it spread to the land, which until then had remained sterile and empty, and eventually gave rise to dinosaurs, birds and mammals. But that was another story. Anthony was convinced and before long *First Life* had been commissioned.

I now had the chance, in effect, to write myself a ticket, with all expenses paid, to all the most spectacular and important fossil sites in the world. Filming the Ediacaran fossils in detail would please me greatly, because, as I have related earlier, one very important species belonging to this group had been found in the ancient rocks of the Charnwood Forest where I had spent a lot of time as a boy. Then I had missed my chance of palaeontological immortality by not finding it myself. But now, at least, I could film it.

Our first trip was to Newfoundland where the rocks are of about the same age as those in Ediacara but contain specimens that are, if anything, even more dramatic and spectacular than their Australian equivalents. It was my first journey with Anthony Geffen. He is a man of action. At Oxford he had represented the University at athletics, hockey and golf. In the BBC he had quickly acquired a somewhat swashbuckling reputation. He was filming in the Presidential Palace in Bucharest at the very time that Ceausescu was murdered and in Tiananmen Square in Beijing, with cameras rolling, when Deng Xiaoping's tanks were halted by a lone student protester.

I had never met anyone with his degree of non-stop energy. Having flown to St John's, the main town on the island of Newfoundland on Canada's eastern coast, we set off in a hired car, early in the morning, for our location at Mistaken Point, several hours away. Anthony started his day's work by doing five things simultaneously. He was, of course, talking on his mobile phone (one). He had an iPad on his knee on which he was checking some business documents (two). In between tapping that with his forefinger, he was picking up a camera to snatch a quick picture of the mountainous scenery outside the car (three), talking to me (four) and telling the driver (five) that he had taken a wrong turning.

The animals whose fossils we had come to look at had been entombed at the bottom of a deep sea by heavy and repeated falls of volcanic ash some 586 million years ago. These ash layers had been compressed and turned to rock and then bent, tipped and folded by later movements within the earth's crust – but not so vigorously that all traces of the animals within them had been destroyed. Quite to the contrary. The sudden falls of fine ash had preserved the creatures in extraordinary detail. Now the sea was eroding the rocks, working along the lines of weakness between the strata and exposing what was once the sea floor. So you can crawl on your hands and knees over an area the size of a tennis court, albeit one that is steeply tipped, and examine in detail each of the tiny organisms that had died there. You can see how a current sweeping across the sea floor had aligned many of the bodies and occasionally trace a wandering trail to find at the end of it, the remains of its creator. It was a thrilling beginning to our filming.

We worked on the programme for several months. We visited Charnwood where I was given a privileged view of recently discovered specimens still in place in a secret site. We went to the Burgess Shales in the Rocky Mountains to record their riches in more detail than I had been able to do when I first went there to film for *Lost Worlds* thirty years earlier. And finally, as a last treat, we went to Morocco to film trilobites.

The basic body pattern of a trilobite is a little like that of a woodlouse. Its back is covered with a jointed carapace beneath which are numerous pairs of long jointed legs. The entire family is now extinct and has no living descendants, but in their heyday trilobites had dominated the seas. Some were no bigger than the nail of one's little finger. Others were almost a metre long. Some swam, others crawled and some burrowed deep in the mud. Many species could curl up like tiny armadillos. A few had eyes that were the first organs on earth capable of producing a sharply focussed image. One species had a three-pronged projection on the front of its head the function of which is still completely unknown. Some of the earliest occur in the rocks of Wales and in Morocco you can find specimens in their final and most baroque complexity.

Our guide to these astonishing creatures was an old friend, recently retired from London's Natural History Museum and a world authority on the whole trilobite family, Richard Fortey. He not only took us to the best sites but talked expertly to camera explaining the anatomy of these strange creatures. He is also a highly expert all-round naturalist.

Anthony, on this as on every subsequent trip we have made together, had decided not to make just one film but two simultaneously – the second being a record, made by a solo cameraman-director, of how we made the first. They are known in the trade, somewhat clumsily and ungrammatically, as 'Making-Of's'. We had stopped in the Moroccan hills for a sandwich and a drink. I was sitting in the open door of our car. Richard had wandered off looking at bushes, turning over pebbles, picking up bits of rock and examining them with his hand-lens. David Lee, who was filming the Making-Of, approached me with his camera to his eye, its recording light flashing.

'What's Richard doing over there?' he asked.

'Well, ' I said, 'he is a naturalist. He's looking at everything. I'll bet you that in a minute he'll come over here, produce something from his pocket and say "I've just picked this up. What do you think it could be?"'

David continued recording but retreated slightly to get a wider shot. Richard reappeared, just as I said he would, looking slightly abstracted. He sat down beside me, reached into his pocket and handed me a small oddly-shaped stone.

'I've just picked this up,' he said, 'What do you think it could be?'

He knew perfectly well, of course. It was a small beautifully preserved fossil coral. Such are the lucky moments for a *Making-Of* cameraman – and the pleasures of travelling with a great naturalist.

* * *

First Life, for me, was revelatory not only because it brought me up to date with the new extraordinary discoveries that had been made since I had studied palaeontology at Cambridge, but because of the perfection of the reconstructions that ZOO produced. When I had made *Lost Worlds, Vanished Lives* back in 1989, affordable animation was still very unsatisfactory. No one watching the images of dinosaurs hobbling and jerking their way through picture postcards of Jurassic landscapes could have imagined that they were anything other that clumsy constructs. But *First Life* contained images that were so convincing that one could truly have thought, unless informed otherwise, that they must be living species that somehow or other had hitherto escaped one's notice.

ZOO was headed by James Prosser, a master of computer invention. He co-ordinated the work of two dozen or so programmers each of whom sat in front of a large computer screen, with specialised palaeontological publications at his or her elbow. The first creation they showed me was of a small three-inch long species called *Opabinia,* fossils of which have been found in the Burgess Shales. It had a segmented body somewhat like a shrimp's, flattened plates on its tail, a head with five stalked goggling eyes and, beneath, a long extendable proboscis armed with an array of sharp spines with which it presumably grabbed its prey. It was so bizarre, so unlike anything alive today, that when the palaeontologist working on the specimens first showed his reconstruction of the species at a scientific conference, his audience had burst out laughing.

But I did not laugh at the image that appeared on ZOO's computer screens. I was transfixed. The little creature swam into picture, its eyes glinting with iridescence like a living shrimp's, sculling itself slowly forward using flaps on each side of its body, probing the mud of the sea floor with its proboscis. Then, with a snap of its tail, it accelerated away. It was clearly a fascinating and totally believable little animal even though it was fundamentally different from anything alive today.

The process of creating such an image is a long and laborious one. First a featureless uniformly grey model is constructed on the screen that matches the general outline of the animal's body but with meticulously correct proportions. It is complete in all three dimensions so that it can be revolved and twisted to be viewed from any direction the operator chooses. Then one by one, details are added – scales or feathers or hair. Sometimes the original operator will do this work. Sometimes this early model is sent over the ether to a computer operator sitting on the other side of the world – perhaps in India – who is known to be particularly expert at creating very convincing textures such as scales or wrinkled skin. And as each stage is completed, the image is also sent instantly to one of the scientists who is expert on this particular species asking for criticisms. Would it have moved like this? Is the colour a reasonable guess? Then, at last, the completed creature is inserted into a separately produced picture of the environment it inhabited, through which it then saunters until – perhaps – it is grabbed by something even bigger.

To me, the results were breathtaking. The reconstructed animals not only looked totally convincing, they moved and behaved like creatures I know. They hesitated. They reared back. They pounced. And they weren't just superimposed into their environmental setting, they reacted with it. So tiny particles in the water drifted by, caught by the currents created by an animal's beating gills and little flurries were raised by the impress of its legs on the sea floor. Such details could only be envisioned by someone who was not just working from textbooks but who had a true feeling for the natural world – and one that had existed millions of years ago.

I wasn't to discover until several productions later who that was. It was James Prosser himself. Many months later when we were location together in Borneo, he mentioned that at university he had studied vertebrate palaeontology but soon after graduating, had found that his computer skills were more marketable than his paleontological scholarship. But I did realise as species after species was added to the cast list of *First Life* that it was now possible to bring fossils to life visually, in a way that was so convincing that it opened up a whole new area of programming.

* * *

It was while we were in the Rocky Mountains filming the Burgess Shales sequences for *First Life* that Anthony, over a beer at the end of a hard day's filming, started talking to me about 3D television. The cinema had made several attempts over the years to show feature films in three dimensions. And they succeeded pretty well. Images appeared on the screen that could have such depth that audiences, sitting in their plush seats, could be made to duck or raise their arms in defence as an aeroplane or a bird appeared to fly out of the screen and over their heads. But to see images in this way, spectators had to wear special spectacles and none of these productions had had more than a temporary success.

Now television was developing its own new improved 3D technology. Viewers were still required to use glasses, but the three-dimensional effect was greatly improved and completely convincing. As ever, Anthony was fascinated by this new technological advance. He thought he knew a broadcaster who was keen to try out this new area of programming. Supposing we were asked, he said, what subject would we choose for a 3D film?

I was captivated by the thought. I had started my television career using the first cameras to produce a regular, publicly available service of 405-line pictures. I had been in charge of the network that introduced first 625 line transmissions and then colour. It would be fun to be involved in what must surely be the final refinement of the television picture – the illusion of three dimensions.

One thing was already clear. Computers could produce very high quality 3D images. A feature film called *Avatar* had just been released which consisted almost entirely of computer-generated images (CGI) and it had a huge commercial success. We ourselves had already learned a little about the technique with our reconstructions of fossils.

Why not make fossils the stars in a 3D documentary? At least they would not grumble or throw tantrums when we stopped to deal with the technical problems that 3D filming would undoubtedly bring. Dinosaurs would obviously be impressive but they were perhaps a bit hackneyed. What about pterosaurs, the great winged reptiles that flew over the dinosaurs' heads? They would really take advantage of all three dimensions of the new medium. That night I wrote a preliminary sketch of a programme about pterosaurs with a story-line based on their evolution and ultimate demise.

But there was a snag. The BBC had no plans to dip its toes into these still somewhat experimental waters. To me, that seemed a pity. So far in the history of broadcasting, the Corporation had led the way in every technical advance, both in sound and in vision. But now it had decided that 3D was one step too far. It would cost a lot of money to start such transmissions and no one could be sure that it would appeal to a significant proportion of viewers. The only network that was actively supporting this new development was SKY and it was they who had asked Anthony for ideas.

SKY, however, had been started by Rupert Murdoch, who had declared himself the sworn enemy of the BBC. Would I be betraying my public service ideals if I were to work on programmes that could only be shown, at least for the time being, on one of his networks?

Fifty years ago, when ITV became the first commercial network to challenge the BBC, I would have considered working for the competitor unthinkable. But now the situation was very different. British television was no longer a contest between a high-minded public-service broadcaster and a single commercial rival financed by advertising. Now there was a wide range of different broadcasters, all operating in the same arena, exchanging facilities, studios, even performers, among one another. The BBC itself was now selling its past programmes to digital channels which were financed by advertisements. Eventually I decided that since the BBC did not have a network that could show my 3D suggestions it would not be improper for me to take the idea to a network that could. I talked with some of my BBC friends. They agreed. I went ahead and completed the script outline in some detail. Anthony sent it to SKY. Approval came by return.

We called our pterosaur programme *Flying Monsters*. Our first location was in a vast quarry in Bavaria where a world famous fossil, half-reptile half-bird and called *Archaeopteryx* had been discovered in the nineteenth century. And there, for the first time, I met an operational 3D camera, complete with crew.

It was gigantic. Two normal cameras had, to put it crudely, been lashed together and hung about with all kinds of wires and attachments to ensure that the two worked in proper synchrony. If they didn't, the two separate pictures would not combine in a viewer's brain to produce the illusion of a picture with depth. But that could only too easily go wrong. So every shot, having been taken, had to be analysed mathematically to check that it matched these criteria. If there was the slightest imperfection, the shot, no matter how complex and demanding, had to be re-taken. I was glad indeed that we hadn't selected real live animals as our stars.

I think we managed to complete one satisfactory shot that first day. It might have been two. But it wasn't more. We all put on the required spectacles and huddled round a monitor to marvel at the three-dimensional effect that had been achieved. But clearly we would have to improve our efficiency if we were to keep within reasonable distance of our budget.

The second day began badly. The weather was dull and the light awful. We started to record a shot in which I sat in a small pit in the quarry floor and began to lift up plates of limestone one at a time to show that fossils in fact were quite rare when – by a great piece of luck – I took away a stony leaf and saw on the plate beneath the perfect coiled shell of an ammonite.

I started to describe with perfectly genuine surprise and delight what I had found – but 'Cut!' said Matt, the director. 'We'll go to a close-up for this.' Had our camera been a normal one, the cameraman would have simply zoomed in. But zooms won't work in 3D. And changing a lens on a 3D camera, at that time, took a minimum of forty minutes. It would be better to shift the whole camera. 'Don't move,' Matt told me. 'Keep continuity.' The crew started hauling the camera into its new position while I waited, sitting in my little pit, with the newly exposed ammonite lying on its limestone plate resting on my knee.

Then it started to rain. Our technical crew had anticipated such a possibility and quickly put up a tent around their delicate and temperamental charge. Anthony, I had no doubt, was counting people and calculating what this shower was doing to his budget. I retreated and joined our coddled camera in its tent. This was not just because I was cold. If my shirt got wet, there would be no continuity between the shot we had taken and the one we were hoping for. The rain lasted no more than about twenty minutes.

But now it was getting late and a new hazard threatened. The sun was sinking behind a huge spoil heap. If we lost the sun, the lack of continuity would make the sequence unusable. 'Shovels, ' shouted Matt, 'Get some shovels. As many as possible. Quickly!' Shovels were produced and those members of the crew who were not immediately needed for the camera's operation scrambled up to the top of the spoil heap and began urgently digging to lower its summit. This could give us an extra few minutes of sunlight while those on the camera completed their technical adjustments. And then I started on a detailed description of my ammonite, which happily I finished just before the sun disappeared behind the spoil heap. 3D images might be the pictures of tomorrow, but in terms of camera mobility, getting them seemed to be taking me back fifty years.

That night, talking to the crew, I discovered just how accurate that feeling was. Fifty years ago there were no zoom lenses for the 16 millimetre camera that Charles Lagus and I had worked with in Sierra Leone. And now Chris Parks, our 3D technical expert, explained to me the fundamental optical reasons why fitting them to a 3D camera was – and always would be – impossible.

We ourselves see objects that are quite close to us in three dimensions because our eyes, spaced as they are approximately three inches apart, produce slightly different images of a nearby object. Our brain then combines the two and gives us a picture with depth. The two synchronised cameras in a 3D rig replicate this. If an object is twenty yards away, however, the pictures from our two eyes are virtually identical and the image sent to our brain is therefore a flat one. If that image is a distant bird singing on a post, or a monkey doing something interesting in the top of a tree, a wildlife cameraman either zooms in to it or puts on a different lens with a longer focal length that acts like a telescope. But two such lenses, three inches apart on a 3D camera will produce two images that are virtually identical.

You might, in theory, solve the problem by increasing the distance between the two cameras of the 3D rig and fitting each with a powerful telephoto lens so that each would produce the slightly different images that our brain requires for 3D amalgamation. However, long lenses on cameras that are far apart but trained on the same distant animal will produce pictures of it with such different backgrounds that the brain cannot amalgamate them. The only subjects that would avoid this difficulty are those where the background is uniform, no matter how apart the two lenses are – birds against a cloudless blue sky, or fish in otherwise empty water. But such images can in any case be treated with a highly expensive process called 'dimensionalisation' to create a 3D effect. It is not as convincing as true 3D and it is highly expensive but nonetheless it works sufficiently well in these particular circumstances for there to be little point in going to the trouble of shooting with a pair of widely spaced 3D cameras. It was a valuable lesson that I would have to be bear in mind if I was ever asked to write another 3D script.

There was one stylistic problem. It was clearly possible to combine CGI images of extinct prehistoric animals with present-day reality. But was it acceptable for them to appear alongside me in, for example, a modern palaeontological laboratory? Clearly, since they were not with me in actuality, I would have to *pretend* that I could see them. This would be acting! And I had always counted on viewers trusting that I never tried to trick or deceive them. So how could I make it obvious, when necessary, that the apparently living creature alongside me did not exist and that I was not trying to convince anyone that it did?

I decided that the way to do so was to use CGI to make the fossilised bones of a specimen detach from their slab, come together to form a joined up skeleton and then to remain like that as it walked and jumped and flew. Viewers then would be well aware of what was happening, and continue to be so even if we ultimately clothed the skeleton with flesh, fur or feathers. Initially there was some resistance to this idea among the production team, but soon everyone accepted that this would make viewers complicit in the illusion we were creating. And when we tried it, we discovered that these animated skeletons had an intriguing, almost endearing charm all of their own and were worth keeping for their own sake as well as to remind viewers every now and then of the true nature of what they were seeing.

Pterosaurs flourished over a period of nearly two hundred million years. During that time they evolved into several hundred species. One of the first to be discovered was one of the smaller ones, a creature about the size of a seagull. Each of its forelimbs had an immensely elongated finger which supported a wing membrane of skin. So it was given the generic name of *Pterodactylus* – 'wing-finger'. Over the following decades, more and more species of flying reptiles were discovered and it became clear that the whole group had to have a name. So they became known as pterosaurs – 'winged reptiles'.

The first pterosaur to be found in Britain was in fact a species of *Pterodactylus*. It was discovered amongst the remains of ammonites and ichthyosaurs in the rocks of the Dorset coast by that redoubtable Victorian fossil hunter, Mary Anning, and we gave it appropriate coverage. But the obvious climax of our programme would be one of the last kinds of pterosaur to appear, a monster with a wingspan of nearly forty feet, the size of a small aeroplane. The first specimen was discovered in Texas close to the Mexican border and given the appropriate if somewhat jaw-breaking name of *Quetzalcoatlus,* after the flying serpent god of the Aztecs.

In order to show how extraordinary this giant was, we would have to find some way of indicating its huge size. Something recognisable would have to appear in the skies alongside it. After some thought, I concluded that the most exciting way for that to happen was for me to be taken up in a two-man hang-glider and then, using CGI, make a monstrous *Quetzalcoatlus* fly into picture and glide alongside me. Atlantic, however, were unable to find any company that would insure me doing so. In the end, I was given a seat in a proper two-man glider.

So it was that our programme about these astounding ancient animals ended with a rhapsodical sequence in which I appeared sitting in a graceful slender-winged glider and was then joined in the sky first by one and finally a small group of these giant creatures. They flapped gently alongside my glider peering in at me in the cockpit while I gazed up at them and we all soared over the green chequer-board of southern England.

SKY had decided to also transmit a 2D version, so we avoided shots that depended solely on depth for their effect. But we were also told that IMAX cinemas, with their giant screens, were keen to show the film. There everyone, with the aid of glasses, would be watching in 3D. So we prepared a special version for them. It ended with a shot of *Quetzalcoatlus* in the distance flying towards camera. As it gets closer, a group of flamingos in the foreground take to flight in terror. But the giant reptile flaps closer and closer until eventually it appears to leave the screen altogether and flies over the heads of the audience – a sensational 3D ending.

The programme was shown by SKY 3D in 2011. Television critics applauded. 'Visually stunning' said the *Sunday Times*. 'Ground breaking' declared the *Telegraph*. But the number of viewers who saw it in the three dimensions we had worked so hard to create was tiny, compared with those who viewed it simultaneously in two dimensions on SKY's main channel. It was going to take time to build a 3D audience on television.

* * *

Anthony was anxious to start work on a successor. CGI had helped us make our first venture into the 3D world without having to deal with the temperaments and fallibilities of living performers. But we had to be a bit braver next time. We had also discovered that the three-dimensional effect was most marked and interesting when the camera was examining relatively small objects, close-up. We should have realised this would be so. Distant objects cannot in fact be *seen* in three dimensions. Our two eyes don't give us sufficiently different views of something fifty yards away for us to appreciate depth visually. We can only deduce that depth from what we know of the size of the objects we are looking at or the way they might move in relationship to one another. So small living things should be the main subjects of our next 3D venture, things that moved in interesting ways and would not run away if a huge camera with its human attendants loomed up quite close to them.

The answer was not, I suppose, immediately obvious – but the more I thought about it, the more pleased I was with its possibilities. Plants! My mind went back to 1995 when we had started work on *The Private Life of Plants*. Much of that series consisted of macro-shots which filled the screen with a single bloom or indeed a grain of pollen as it landed on a flower's stigma. But it also produced some of its most enthralling effects using time-lapse to speed up the motions of plants and show us leaves unfurling, flower buds bursting into bloom, and the tendrils of climbing plants like bindweed throttling their neighbours.

The principle of time-lapse is very simple. Film cameras work by shooting 25 frames a second. If you shoot one frame a second but then transmit the film at normal speed, then you will speed up motion twenty-five times. The cameraman on whom we had relied and who had produced some of the most breath-taking time-lapse sequences for us in the series was Tim Shepherd. Would there be any real problems for him to do the same sort of thing again, but in 3D? On the contrary. He explained to me that, in fact, it was *cheaper* to shoot 3D in time-lapse than in any other way. You do not even need a movie camera. A normal still camera was quite capable of doing the work. All you need do is to mount it on a special motorised base which will, after taking one shot, automatically and immediately shift the three and a half inches laterally, to take the second shot which eventually will be transmitted simultaneously with the first.

Making plants the subjects of our next 3D project had a further advantage for me personally. One of the greatest and most comprehensive collections of living plants in the world, the Royal Botanic Gardens Kew, stands within a few miles of where I live in west London. So I would be able to stay at home and go down to the Gardens whenever necessary and often at very short notice if there was a much-wanted change in the weather or some rare plant suddenly came into flower. We would go into the specialised glass houses to show plants that live in other climates – in deserts or tropical rain forests. The Gardens would also allow us to film something of the work that goes on behind the scenes to persuade difficult and endangered species to flower and set seed.

Martin Williams, who was going to direct the series, suggested that it might be structured around the seasons. Anthony was enthusiastic: *A Year in the Life of the Royal Botanic Gardens* would be a great subtitle, he said. I didn't see how we could do that since it would be February before we could start shooting and we would have to stop in early November in order to meet SKY's Christmas transmission date. Anthony brushed aside the objection as being unreasonably pedantic. But the idea was eventually dropped in favour of *The Kingdom of Plants 3D*. And that, as someone said, was a pity since Anthony was the only independent producer anyone knew who would be able to make a film about the events of twelve months in the space of nine.

32

Spreading Our Wings

We had still not used 3D to film the core subject of natural history programming – scenes of large animals of many different kinds wandering through unspoilt wilderness. Those are the sights that have drawn natural history filmmakers to great open savannahs of eastern and central Africa. But these days, if any producer asked a cameraman to make a top-class, riveting film about African wildlife, but without using any long-focus lenses, I cannot believe anyone would accept. 3D would not rescue a programme that was a dud when viewed in 2D.

There was, however, just one place in the world that I could think of where film-makers have no need of long focus lenses to get close-ups of big dramatic animals – the Galapagos. The islands had remained undiscovered by humanity so long that the animals that evolved there never acquired a fear of human beings. When sailors did reach the islands, they killed the giant tortoises and other animals in such numbers that many species were exterminated. Nonetheless, the survivors still remain fearless to this day. So you can sit surrounded by herds of black marine iguanas, basking in the sun, and crouch beside albatrosses as they dance to one another.

Although I had filmed in the Galapagos several times before, I had done so to illustrate aspects of general natural history – about the way albatross fly, or how lizards could evolve to eat seaweed. I had never ever attempted an overall survey of the islands, looking at the way they had been created geologically, how they had been colonised and the manner in which their communities of animals and plants had changed over millions of years. Now was my chance.

We would need to have two camera crews working simultaneously if we were to follow the script that Martin Williams, the director and I had worked out between us. I had imagined that Anthony would charter one of the larger tourist boats so that we could sail from island to island as we wished. But he eventually decided that the amount of equipment and the number of people in the team were so great that it would be better to take over an entire tourist hotel in Puerto Ayora, the main town in the archipelago and then charter a helicopter which would not only shoot aerial sequences but ferry us among the islands from location to location. But there were no helicopters in the islands and we could not fly one there from mainland South America, because the islands are beyond helicopter range. If we wanted one, it would have to be sent there by sea. What is more, if we were to use it to travel widely among the islands, we would have to establish dumps of aero-fuel at strategic points. And all this would have to be done before filming even started. But getting a helicopter was going to be a major problem.

The islands, of course, are famous for the part they played in helping Charles Darwin formulate his great theory about the origin of species, and by chance we filmed two sequences that were relevant to his subject – one illustrating the last moments of a species before its extinction; and the other the discovery of a new hitherto unknown one.

When Darwin arrived on the island there were a dozen or so species or subspecies of giant tortoise living there. Each had its own characteristic shell shape and each was adapted to the particular set of climatic and ecological conditions of its own island. To sailors crossing the Pacific, these strange creatures which occurred nowhere else in that ocean, were invaluable sources of fresh meat. Thousands upon thousands were taken. Some were killed and eaten immediately. Others were loaded onto the ships and stored in the depths of the holds where they were able to survive without food or water until, after many weeks at sea, they were eventually slaughtered.

Many tortoise species were consequently entirely exterminated, among them one that had evolved on the small island of Pinta, in the far north of the archipelago. But in 1972, a party of scientists undertook an ecological survey of Pinta and to their astonishment, they discovered in a gully on the island an ancient male. An intensive search was mounted but no other individual was found. So the male was brought back to the main island and eventually given his own enclosure in the small zoo maintained by the Scientific Research Station.

He became known as Lonesome George and soon became famous. No one knew, of course, how old he was. The horny plates on a tortoise's back do have ridges on them that form as the years pass, but dating by such means is very problematical and in any case most of Lonesome's plates were worn smooth. Most tortoise experts, however were confident in saying that he was well over a hundred years old.

Two females belonging to the species found on Isabela, the nearest island to Pinta, were introduced into his compound to give him company.

No one saw Lonesome make any attempt to mate with either of them. After fifteen years together, one of them produced a clutch of eggs. Sadly, they proved infertile.

Lonesome's plight brought him such fame that a visit to the Centre to see him in his enclosure became one of the most important events in every visitor's schedule. I myself had called upon Lonesome before and now we asked if I might do again. The authorities, quite rightly, were very protective of him and rather reluctant. We could not in any case film him during the Scientific Centre's normal opening hours as that would interfere with the view visitors might get of him. We would have to come before dawn and only I and Paul Williams, the cameraman, would be allowed into his enclosure. More people than that might upset him.

It was dark when we arrived. Lonesome, his keeper told us, was asleep in his accustomed place beneath a thorn bush. We crept in as quietly as we could. I lay on the ground within a few feet of him and Paul set up his camera a few yards away. Lonesome still slumbered. His head and long neck were not retracted into his shell but were extended and lying unprotected on the ground. His eyes were shut. I had to wonder whether we had arrived too late. But his keeper seemed unworried. Paul was grateful for the delay. There was still barely enough light to give him the exposure he needed.

Then Lonesome opened one eye and blinked. I slowly raised myself into a more decorous position. Paul put his eye to the camera. There was a long pause. Lonesome heaved himself off the ground and – very, very slowly – took a step forward. I cleared my throat and delivered my lines. 'This by any standard is the rarest animal in the world, for he is the very last individual of his species in existence.' Lonesome took a few more steps and settled his shell down on the ground again. We waited. But he had gone back to sleep.

By now the Centre was opening to the public and we left. Ten days later, on 24th June 2012, Lonesome's personal keeper went into the enclosure to see how he was. But he could not rouse him. Lonesome George, the last of his kind, had died in his sleep.

* * *

Soon after filming Lonesome we heard some very exciting scientific news. A new species had just been identified in the Galapagos. This was not a small spider, a tiny coral fish or even a bird. This was a large reptile – a land iguana.

Galapagos' iguanas are descended from species that are common in the forests of South America. They feed on leaves, often along the banks of rivers and it is thought that thousands of years ago, some that had been feeding in such a way were swept out to sea on a raft of floating vegetation as indeed they are occasionally today. Most of these inadvertent travellers doubtless eventually died. But iguanas, like so many reptiles, are tough, hardy creatures that can go without food or water for weeks and some, maybe even only one or two, were cast up alive on the shores of the Galapagos.

Some managed to survive by eating the seaweed growing on the rocky shores of the islands. They were the ancestors of the black marine iguanas that are now found in some numbers on many of the archipelago's islands. But others moved inland and lived on the coarse spiny plants they found there. Two species of these land iguanas are well known. Both are blotchy yellow. But now an Italian scientist, Dr. Gentile, and his team working on the flanks of Wolf Volcano had discovered a third. It was neither black nor yellow. It was pink.

If colour were the only major difference between it and the other two species of land iguana, a pink one might be dismissed as a mere local variant of no particular consequence. But in fact, there were more fundamental differences between the pink and the yellow land iguanas. The pink males had a fatty hump on the back of the neck that gave them a very different profile. And the scales of the male's skin were also different, those on the head being much flatter and those on the neck being small and conical. The Rome team were at this very moment working on the lip of the Wolf Volcano's crater. They had just reported by radio that they had caught another specimen of these rare creatures and would gladly show it to us if we could get there.

Wolf is an immense volcano that forms the northern end of the island of Isabela. It is one of the most remote and difficult-to-reach locations in the whole archipelago. Getting there from Puerto Ayoro would normally involve first several hours in a boat and then a very long and arduous climb up the volcano's flanks. But our helicopter could get us there within an hour or so. And off we went.

Dr. Gentile and his team had camped on the very lip of the crater and excitedly showed us their newly captured iguana. It was not merely a different colour. Its shape, and particularly the odd-shaped head, gave it a very different look from any land iguana I had seen. So I was not surprised when Dr. Gentile explained that their most recent results from genetic tests had made it clear that the pink iguana was not a result of interbreeding with any other known species of Galapagos iguana. So either it was a survivor of a species that had evolved very early in Galapagos history and was probably once more widely spread, or one that had evolved comparatively recently up here on Wolf. Either way, it was a new discovery that had not yet been filmed.

Dr Gentile handed me the precious specimen, warning me that it was quite powerful. I gripped it as firmly as I could manage without actually throttling it and delivered my words. But the poor creature scarcely looked comfortable. If it escaped could the Italian team re-catch it? He thought so. So with his permission I tried a technique that does sometimes work with reptiles. I put him on the ground with his legs drawn to his side in a natural looking position while still gripping him quite firmly. Then, very slowly, I relaxed my grip, withdrew my hands and started talking to camera quite quietly. The iguana sat where he was. Was he free or wasn't he? He didn't seem to know so he sat still on the ground. Then just before I came to the end of what I had to say, he realised that he was in fact at liberty. Off he shot, but fortunately straight into the hands of one of the scientists. We had a scoop, shots of a Galapagos animal that had never before been filmed by any other crew on these much filmed islands.

* * *

I seemed to be busier now than at almost any time since I had become a free-lance. The BBC wanted a sequel to *First Life* which would chart the evolution of backboned creatures and bring the history of all life on this planet up to date. BBC Radio commissioned me to give a series of short ten-minute talks taking a sidelong view of natural history topics – the nature of mermaids, why snakes can't blink, the sad history of the dodo and the extraordinary partnership between a particular species of ant and the Large Blue Butterfly – which were given the collective title of *Life Stories*. And SKY wanted more 3D.

The number of people viewing 3D programmes on television was still comparatively tiny, but now they were beginning to build a new audience. Natural history museums around the world had, over the past few years, been installing cinemas with IMAX and other similar giant screens measuring ten or eleven metres high. How better could such museums convey the wonders of the natural world from the vastness of the Grand Canyon, the immensity of whales and the bizarre shapes of beetles that, enlarged to that size, made dinosaurs look positively cuddlesome.

Until this time, most natural history programmes made for television were shot on a technical standard that made them look disappointingly soft, even blurry on a giant screen. But the extremely high criteria demanded by 3D technology meant that the programmes we had made in this new medium easily matched IMAX's new technical requirements. So suddenly, all the 3D programmes we had made so far had the chance of a new and extended life. *Flying Monsters*, our very first, was already being shown around the world from Australia to China. *Galapagos 3D* already had bookings that would start as soon as it had finished its television showings. And SKY wanted yet another origination.

* * *

Now, at last, I felt we had enough experience of 3D to tackle something really ambitious. If there was one natural history subject that would exploit the third dimension it must, surely, be the way some animals have loosened the bonds of gravity and moved into a third dimension by flying. We would start with the first creatures on this planet to do that – insects. There are fossil dragonflies dating from 320 million years ago with wings nearly a metre across. Since then different insect families have evolved their own specialised aerial techniques, from lumbering beetles to tiny hoverflies. The insects had the skies to themselves for a few million years, until reptiles also produced a group of aeronauts, the pterosaurs. We would not spend a great deal of time on these because they had, after all, been the subject of *Flying Monsters* but at least we already had computer images of them which could form the basis of new sequences. Soon after, the pterosaurs were joined in the air by the birds. And finally, a group of tiny insect-eating mammals took to the skies and evolved into bats. The series almost sold itself. And SKY happily took up the idea.

We would, of course, have to film in the tropics and I immediately suggested we should go to Borneo. Its forests have an unrivalled collection of species with which we could illustrate the problems and the techniques of getting into the air – a frog with feet that, thanks to very long toes connected by skin, catch the air like parachutes when it jumps; lizards that glide by pulling forward their ribs so that their flanks form skinny wings; and mammals like the little known colugo, a creature the size of a small dog that has furry skin stretching from its wrists to its ankles. And like most other places, Borneo has a multitude of birds and bats.

That was quickly agreed and in March 2014, the whole unit of fifty-odd people with several tons of equipment arrived in Borneo. After several weeks filming in the forests we moved to the great caves of Gomantong. This was where, back in 1974, I had climbed up a huge pile of bat droppings covered with cockroaches and nearly choked on the ammoniacal stench that rose from it.

I had returned there in 2012 when making a trio of retrospective programmes to mark my 60 years with the BBC. Gavin Thurston, the cameraman with whom I had travelled over so many years, had devised a very ingenious shot to introduce the sequence. He had erected a cable some twenty feet above the ground that ran from the open clearing in front of the cave, right into its mouth so that the camera mounted on a dolly running along the cable could travel smoothly from the sunlight and into the darkness. It worked beautifully.

But now, in 2014, we had a truly revolutionary way of making that transition – a device called by its builders, an 'octocopter'. It consisted of four long metal rods fixed together to form an eight-armed star. Each arm carried at its end a small motor powering a rotor blade. Beneath, attached to the centre of the star, hung the 3D camera, now a much smaller version of the monsters with which we had first worked. An operator sat on the ground watching the camera's electronic output on a monitor and controlling it and the direction of vehicle's flight using a joy-stick.

The machine was still in development and at this stage could only run for two or three minutes but that would be long enough to give us the shots we needed to take the viewer into the cave. We could then use it to explore its interior and even get shots of the millions of bats that live in the cave that were virtually impossible to get in any other way. I was a little apprehensive. The octocopter was a very noisy beast. How would the bats react to it? Might they not collide with it and cause a very expensive crash? We could but try.

The interior of the cave was just as noisome and unpleasant as I remembered it. A duckboard trackway now kept the visitor's feet above the slimy, inches-thick layer of droppings that covered every part of the floor but the clammy stench of ammonia was still just as unpleasant as I remembered it. The trackway itself was slippery but steadying oneself on the handrail was not a good idea since it too was covered in bat droppings – not as thickly as the rocky floor, it is true, but more recently and therefore more wetly.

The trackway runs for about a quarter of a mile into the cave until – welcome relief – you reach a huge hall illuminated by a broad shaft of sunlight. Centuries, if not millennia ago, a great section of the cave roof had fallen in creating this welcome skylight onto the outside world. Just beyond, lay the dune of droppings that I remembered so vividly. It was still covered by a glistening carpet of cockroaches munching their way through the bat droppings. We would film that too of course, but the great gap in the ceiling above had given the director, David Lee an idea.

It was from here that a significant proportion of the cave's bats left every evening. Shots of them doing so would be hugely impressive. Indeed it was one of the main reasons for coming to the cave and David had already decided how that could best be done.

Someone would climb up the forest-covered hill above the entrance to the cave and fasten a rope to run across the gap in the cave roof. A second rope would then run from the middle of this down to the cave floor. A small seat would be attached at the end of this on which I would sit. Then, when the right time came, I could be hauled up to hang in the path of the departing bats and describe what I saw to the camera on the octocopter hovering in front of me.

To be fair, I must admit that David had sketched out his intentions during preliminary meeting in London before we had left. On paper, it didn't look too daunting and I had given my tentative agreement. It would be virtually the last sequence in our Borneo trip, so I had had plenty of time to think more about it and I had decided on one thing for certain – that I had better have a good look myself at the safety system that should be an essential part of any such set-up.

The rigging was almost completed by the time I arrived at the cave. The man responsible for it was a short, shaven-headed Yorkshireman called Simon Amos. His main business, it seemed, was cleaning the vast glass flanks of the skyscrapers that now stand in the centre of so many cities in Malaysia and Indonesia. But when we came to talk about climbing technicalities we discovered that both of us had spent our early years gripped by the same obsession – rock climbing. He was a good forty years younger than me, but we both knew the same legendary characters in the rock-climbing world and we certainly knew the same routes up precipices in both the Lake District and North Wales. From then on, I had no doubt that I was in safe hands.

We made one or two trial low-level ascents. Simon had three young local men to provide the muscle–power to work the system and I went up and down a few feet as Simon barked orders in Malay with a broad Yorkshire accent. David knew to within a few minutes when the main exodus of bats would begin. I was hoisted up well in time – just in case that evening they decided to start a little early.

Swinging from my rope two hundred feet up, spinning slowly round and round, I had plenty of time to ponder on the thinness of the rope on which I was suspended and on the minute size of the figures on the cave floor far below me who seemed to be concerning them with all kinds of footling problems that had nothing to do with me.

And then bats began to appear. They were not leaving the cave but instead, flying back and forth around a small chamber hollowed out from the side of the main shaft in which I was suspended. It was as if they were all reluctant to be the first to leave, even though the sky above was beginning to darken. If that were so, then they had good cause, for I could myself see in the sky high above the shaft tiny wheeling dots. Bat hawks were assembling for their evening meal.

The number of the bats in the side-chamber increased until suddenly there was a break out. Within seconds, I was enveloped by a blizzard of bats streaming past me some within a few inches of my face, flying at speed and pouring up into the sky above like a great plume of smoke.

Whether or not what I said under these circumstances helped to convey the wonder of this extraordinary place, I did not know. But, as I hung there in the flight path of half a million bats rushing out to start their evening meal, I did reflect that I had all the technological facilities I could possibly wish for or imagine – a radio-microphone pinned to my chest relaying my words down to a recordist on the cave floor, powerful long-running battery lights, a crane thirty feet tall carrying a super-sensitive high-definition slow-motion camera, and a 3D camera hanging in mid-air in front of me, suspended from a hovering octopod. Unless, of course, someone has invented smelly-vision.

* * *

The world of television has changed beyond recognition since I made my first programmes over sixty years ago. And so, of course, has the whole world. Perhaps most significantly and importantly, the human population in that same time-span has more than tripled. And that immense increase hs had far reaching and often devastating effects on the natural world that I had been trying to document. During that time, of course, I made programmes devoted entirely to the changes that were sweeping over the world – *The State of the Planet, The Truth about Climate Change, The Death of the Oceans* and *How Many People Can Live on Planet Earth?* – and every major series ended with an episode that looked at the problems currently facing the creatures it had been examining.

But the fundamental reason why I have spent my life in the way I have, and why I am reluctant to stop making programmes, is that I know of no pleasure deeper than that which comes from contemplating the natural world and trying to understand it.

Index

Index

Index